DISTRIBUTED GENERATION SYSTEMS

DISTRIBUTED GENERATION SYSTEMS

Design, Operation and Grid Integration

G.B. GHAREHPETIAN

S. MOHAMMAD MOUSAVI AGAH

Butterworth-Heinemann
An imprint of Elsevier

Butterworth-Heinemann is an imprint of Elsevier
The Boulevard, Langford Lane, Kidlington, Oxford OX5 1GB, United Kingdom
50 Hampshire Street, 5th Floor, Cambridge, MA 02139, United States

Notices
Knowledge and best practice in this field are constantly changing. As new research and experience broaden our understanding, changes in research methods, professional practices, or medical treatment may become necessary.

Practitioners and researchers must always rely on their own experience and knowledge in evaluating and using any information, methods, compounds, or experiments described herein. In using such information or methods they should be mindful of their own safety and the safety of others, including parties for whom they have a professional responsibility.

To the fullest extent of the law, neither the Publisher nor the authors, contributors, or editors, assume any liability for any injury and/or damage to persons or property as a matter of products liability, negligence or otherwise, or from any use or operation of any methods, products, instructions, or ideas contained in the material herein.

Library of Congress Cataloging-in-Publication Data
A catalog record for this book is available from the Library of Congress

British Library Cataloguing-in-Publication Data
A catalogue record for this book is available from the British Library

ISBN: 978-0-12-804208-3

For information on all Butterworth-Heinemann publications
visit our website at https://www.elsevier.com/books-and-journals

Working together
to grow libraries in
developing countries

www.elsevier.com • www.bookaid.org

Publisher: Joe Hayton
Acquisition Editor: Lisa Reading
Editorial Project Manager: Peter Jardim
Production Project Manager: Mohana Natarajan
Cover Designer: Miles Hitchen

Typeset by SPi Global, India

CONTENTS

LIST OF CONTRIBUTORS

Hamdi Abdi
Razi University, Kermanshah, Iran

Mohammad AlMuhaini
King Fahd University of Petroleum and Minerals, Dhahran, Saudi Arabia

P. Asgharian
University of Zanjan, Zanjan, Iran

Jose Ignacio Candela
Technical University of Catalonia, Barcelona, Spain

Ehsan Dehnavi
Razi University, Kermanshah, Iran

Elham Foruzan
University of Nebraska-Lincoln, Lincoln, NE, United States

G.B. Gharehpetian
Amirkabir University of Technology (Tehran Polytechnic), Tehran, Iran

Barry Hayes
National University of Ireland Galway, Galway, Ireland

Chandrasen Indulkar
Indian Institute of Technology, Delhi, India

Saeid Javadi
Razi University, Kermanshah, Iran

Francisco Jurado
University of Jaén, Jaén, Spain

Salah Kamel
Aswan University, Aswan, Egypt

Amir Reza Khodaei
Razi University, Kermanshah, Iran

Jeremy Lin
PJM Interconnection, Audubon, PA, United States

Fernando H. Magnago
Nexant Inc., Phoenix, AZ, United States; Universidad Nacional de Rio Cuarto, Rio Cuarto, Argentina

Behnam Mohammadi-ivatloo
University of Tabriz, Tabriz, Iran

S. Mohammad Mousavi Agah
University College Dublin, Dublin, Ireland

Mehrdad Mokhtari
Amirkabir University of Technology (Tehran Polytechnic), Tehran, Iran

Mehdi Monadi
Technical University of Catalonia, Barcelona, Spain; Shahid Chamran University of Ahvaz, Ahvaz, Iran

Morteza Nazari-Heris
University of Tabriz, Tabriz, Iran

R. Noroozian
University of Zanjan, Zanjan, Iran

Kakkan Ramalingam
Airports Authority of India, Delhi, India

Ahmed Rashad
Qena Rural Electrification Sector, Qena, Egypt; University of Jaén, Jaén, Spain

Ramtin Rasouli Nezhad
University of Western Ontario, London, ON, Canada

Pedro Rodriguez
Technical University of Catalonia, Barcelona, Spain; Loyola University Andalusia, Seville, Spain

Kumars Rouzbehi
Technical University of Catalonia, Barcelona, Spain

Mohammad Salehimaleh
University of Mohaghegh Ardabili, Ardabil, Iran

Ricardo Albarracín-Sánchez
Universidad Politecnica de Madrid (UPM), Madrid, Spain

CHAPTER 1

Distributed Energy Resources

Mehrdad Mokhtari*, G.B. Gharehpetian*,
S. Mohammad Mousavi Agah†
*Amirkabir University of Technology (Tehran Polytechnic), Tehran, Iran
†University College Dublin, Dublin, Ireland

1.1 INTRODUCTION

In 1882 in New York City Thomas Edison established the first power plant and distribution network called a power company to provide lighting for the residential sector. Several similar power companies were established in later years. At that time, power generation was decentralized or distributed. In other words, the electrical energy was produced and consumed in the same place. Hence, it was not necessary to establish transmission lines for transmitting the produced electrical energy from the power plant to the consumption place.

In the early 20th century, all power companies were integrated for reasons such as the need to reduce production costs and increase reliability for all subscribers as consumers of electrical energy with the development of the power industry.

At the beginning of 1990, integrated power companies were faced with huge growth in electrical energy consumption due to global population growth and prosperity. This provided challenges for the centralized supply of electrical energy because of electrical energy consumption growth combined with the increasing inability of supplying electrical energy by transmission lines. In addition, following technological developments in the electricity industry, electrical energy production from renewable sources were gradually becoming economically feasible as indicated by increased investment in solar and wind power plants established in the late 1990s.

Environmental considerations became a major concern for humanity because more than 70% of all electrical energy produced came from fossil fuels.

Alarming changes in global climate because of environmental pollution and the consumption of nonrenewable resources of fossil fuels caused designers of electrical energy plants to take into account both environmental considerations and safe and highly reliable electricity generation as a universal human obligation.

Distributed Generation Systems
http://dx.doi.org/10.1016/B978-0-12-804208-3.00001-7

1

In a centralized system of energy production, all the abovementioned problems and challenges exist due to the productive resources with fossil fuels and the need for energy transfer from production centers to consumption places through a broad network of interconnected transmission systems at different voltage levels.

From the electrical perspective, there are electrical energy losses from production centers to consumption places. About 5% of the energy produced in power plants with an efficiency of 30%–40% is wasted. In addition, in the transmission system between 4% and 5% of electrical energy is wasted and at the end of the distribution network about 10%–15% is wasted. The production of electrical energy consumption was raised near the consumption place, however, by considering all abovementioned challenges with a lot of casualties caused by the transfer of the energy from production to consumption.

The transmission lines development projects will be postponed by the release of the network capacity with supplying the electrical energy to the consumption place (transferring of enormous investments from development projects to projects with higher priority). For instance, for a transmission line with a capacity of 110 MW with an annual growth of 5 MW, there is a need for another transmission line with the same capacity. If the annual growth of the loads is supplied in the consumption place, then an investment required for the establishment of a new transmission line will be delayed due to the lack of transmitting 5 MW of electrical energy. In the 1990s this issue led to international acceptance of decentralized or distributed power generation as an ecofriendly method and placed it on the agenda. But it should also be noted that the development of distributed power generation will not be a complete negation of centralized power generation. But the proper placement of these two power generation systems together, due to the annual growth of load, will cause the investment in transmission lines projects to be delayed. However, these investments can be greatly reduced by controlling the load growth through demand-side management and demand response method. With the development of distributed generation (DG) in the 1990s and the willingness of consumers to embrace the gradual establishment of these kinds of resources, forecasts in relation to the installed capacity of these resources show increased growth in the future. However, the Electrical Power Research Institute (EPRI), predicted that 25% of the total productive resources in the United States would be supported with distributed resources, but for some reason this has not happened. According to Gas Energy Research (GER), installed capacity of DG resources in the world should have reached 27,000 MW by 2015, which is a substantial amount.

Table 1.1 Increased installation of distributed generation in the United States [1]

Year	2005	2006	2007	2008	2009	2010	2011	2012
DG (MW)	900	1300	2000	3000	4000	5100	5600	6600
Year	2013	2014	2015	2016	2017	2018	2019	2020
DG (MW)	7800	9000	10,300	12,100	13,800	15,300	16,800	19,100

The abovementioned key notes represents a worldwide movement to increase DG systems. Table 1.1 presents the prediction of increased installation of DG in the United States between 2006 and 2020. As seen in this table, the latest estimation of the installation of DG in the United States is 19,100 MW (in 2020), which is more accurate than other estimations.

1.2 DEFINITIONS OF DG

A clear definition is necessary for the DG concept because there is diversity and different combinations of phrases in English in relation to DG.

In general, there are two terms in the scientific literature: *distributed generation* and *decentralized production*. The evolution of the use of these words indicates that DG has gradually eliminated the term decentralized production in almost all relevant scientific literature.

In addition to these terms, *embedded generation* is common in some texts. Also, the term *distributed resource* is sometimes used for the concept of DG.

From the perspective of the electricity market to the DG, the term *distributed utility* is an alternative to all sets of DG, which are used as a distributed company. Finally, we note that the outdated term *power distribution* no longer is used in the scientific literature.

As for a definition of DG, with reference to sources in this field, we are faced with two general definitions. According to the first definition given above, DG refers to the energy source that production capacity is limited. Although this definition is correct in some cases, fails in comprehensiveness. But the second definition of DG sources, with the direct ability of the connections to the distribution network and to the consumer, offers a comprehensive definition. Based on the concepts discussed above, the second definition will be used in this book. So in general, these sources are directly connected to the distribution network or the consumer who must be supplied.

Before expressing the characteristics of these resources, it is necessary to highlight a common mistake in connection with the definition of these resources. It is usual that every source of renewable energy for energy production is known as DG or any DG is necessarily called as a renewable source.

In other words, a DG addition to the use of fossil fuel resources (as the primary nonrenewable energy source) has the ability to use renewable energy sources.

This mistake occurs in the type of technology used in the production of these resources. On the other hand, it is believed that productive resources should necessarily use the latest technology. DG, in addition to using old technologies such as internal combustion engines, and diesel and gas turbines, also uses new technologies including fuel cells and power electronic converters.

1.3 FEATURES OF DG

In this section, the characteristics of DG resources, on the basis of the two criteria of voltage level and production capacity, as discussed in the previous section, are explained.

For the first criterion, the voltage level of the connection as a criterion is very important because the DG can be connected to the distribution network and also directly feed the consumer.

Therefore, the voltage levels of 11, 20, and 33 kV as a medium voltage level to connect the DG sources to the distribution networks and the voltage level of 400 V as a low voltage level for direct connection of these sources to the consumer are considered.

For the second criterion, as production capacity, research centers and researchers have proposed different definitions and ranges. For example, the EPRI has defined a production capacity of 1 kW to 50 MW for the application of a DG as a production source. According to Gas Research Institute (GRI), the production capacity range is from 25 kW to 25 MW. Preston & Wrestler company and Cardel company define the production capacity range of 1 kW to 100 MW and 50 kW to 1 MW, respectively. According to the Institute of Electrical and Electronics Engineers French Community (CIGRE), production capacity of DG should be in the range of 50 kW to 100 MW.

However, apart from the criteria presented above, on the basis of production capacity, DGs are divided into four categories: micro (1 W to 50 kW), small (5 kW to 5 MW), medium (5–50 MW), and large (more than 50 MW).

It is important to always consider that with further increase in the range of production capacity, the injection current of DG to the distribution network does not exceed the limit of the tolerable network's current capacity.

1.4 OPERATION OF DG

The operation of DGs is one of the most important subjects in this field. Detailed knowledge of operational abilities of these resources in a manner of using them for each operation will be efficient. For example, examining the possibility of dispatching capabilities for the resources on how to exploit them is very important.

It is possible to control some DGs remotely and others may not be so controlled. So from this perspective, DGs are divided into two categories as dispatchable and nondispatchable. In addition, it is true in connection with the programmability features of these resources. Depending on the wording of the contract signed, some of these resources can be requested for more energy production at certain hours of the day. These types of DGs are called *programmable sources*. But another group of DGs are not programmable and only receive payments for the production of electrical energy.

In terms of performance, DGs can be classified into two groups. Resources that have the ability for independent performance as a function of the island and resources that are not capable of independent operation. A group of DGs is only capable of supplying their loads once they are connected to their network. But against this category, another group of DG resources are capable to supply downstream loads even if they are disconnected from the network. The independent ability of the DGs are known as island performance.

In addition to the abovementioned abilities, these resources can be used in load management in the network. From the network perspective, DG resources supply a part of the load that causes a part of the network capacity to be freed. Therefore, DGs act to reduce the load of the network and somehow to manage it.

These resources can perform another function as a spinning reserve in the network. In normal operation, although a set of generators are connected to the network, this does not inject any real power into the network.

A major reason for using these types of generators as reserve operational units is to prevent any possible disturbances through quick connection of these units to the network. This role can also be made for DG. Of course, assigning this role to DG needs a contract with the owners. In addition to the main roles of DG mentioned, secondary roles are responsible. As we know, the main goal of DG is the generation of real power, but these resources have the ability to play other roles as well. Some of these roles may be pointed toward the production of reactive power. For example, by connecting a synchronous generator to the end of a gas turbine (as a DG), in addition to active power, reactive power can also be produced.

The production of reactive power from DG due to reduction in active power production, as the main source of income in the electricity market, does not have enough attraction for investors. Therefore, to perform the subtasks of the DG resources, the issues in the ancillary services market should be raised.

Voltage stabilization and power quality enhancement are other subtasks of DG resources. Given that some DGs are connected to the network via power electronic converters cause disturbances in the power waveform in the network because of the nonlinear characteristics of the converters, to maintain the power quality of the network that these resources are connected to is another subtask that should always be considered. It should also be noted that the main task of DG should not be affected by the subtasks.

DG in both public and private ownership is possible. DG resources that are not owned by the public sector are called independent power producers. They are able to sell their production to public companies. But to enter the private sector, convincing answers should be provided to the questions from investors in this field. What motivates the private retail sector to make use of these resources? The formation of this market is the biggest factor encouraging investors to this area. In a very wide distribution network, a combination of active and interested public and private sectors in this area can be operated in a competitive retail market.

1.5 DG TECHNOLOGIES

As mentioned in previous sections, DGs have a large variety in terms of performance and role in the power network. So it is important to have a concise understanding of why these resources are necessary. In the following, we will introduce some of the most important technologies of DG.

1.5.1 Fuel-Based Technologies

There are different types of DG technologies. Fuel-based technology is one of the relatively new types of DG technologies. Combustion engines, microturbines, fuel cells, diesel generators, and biomass are categorized as fuel-based technology.

1.5.1.1 Combustion Engines

In combustion engines, fuel and oxidizer (usually air or oxygen) are reacted and combusted in a closed combustion chamber. In a combustion process, the hot gases at high temperatures and pressures are produced. Because of the expansion of the hot gases, the engine's moving parts are moving and doing mechanical work. Although the purpose of the use of so-called conventional engines is to use the combustion engines in cars, these engines can also be used for DG of electricity.

1.5.1.2 Microturbines

Microturbines are small-scale gas turbines coupled with their own generators at a very high speed that are using a variety of fuels to produce electrical energy. Most microturbines are using high-speed permanent magnet generators to produce alternative electricity. The heat from fuel combustion in turbines is used for energy optimization. Electrical efficiency of the microturbine is 20%–30% and the power range is from 25 to 500 kW [2]. Microturbines are often used in large commercial buildings such as hotels, schools, and offices. Microturbine systems are divided into three categories based on consumption and energy production.

1. Microturbines with a recuperator are more efficient because of using the output heat of the turbines. This type of microturbine is shown in Fig. 1.1.
2. Microturbines without a recuperator (with simple cycle) are less efficient and cheaper as well.
3. Microturbines that work based on combined heat and power (CHP) [3]. In Fig. 1.1, the process of a CHP microturbine is shown.

Microturbines can also be divided into two groups as uniaxial (single-shaft) and biaxial (two-axis). In the single-shaft, both compressor and generator are driven by a turbine, while in the two-axis model, the turbine that drives the compressor is on the same shaft and the turbine that drives the rotor of the generator is on the other shaft. In a two-axis model, the appropriate frequency can be directly produced for consumers by using a conventional synchronous generator (e.g., a rotational speed of 3600 RPM) with a separate turbine for generator and gearbox.

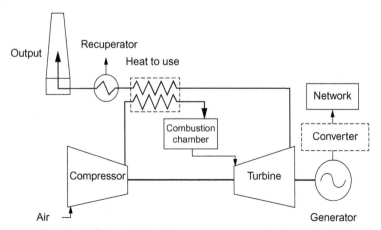

Fig. 1.1 Performance of the microturbine.

Single-axis models usually are designed for operation at very high rotation speed. The frequency of the alternative current produced by these generators is very high. In the high-speed single-shaft generators, it is necessary that its high-frequency output be changed to 50 Hz [4].

Regardless of the high investment costs, microturbines have the advantages of small size, light weight, good efficiency in heat and power cogeneration, low emissions, use of waste fuels, less repair, and good performance at low gas pressures.

Further discussion related to this technology will be given in other chapters.

1.5.1.3 Fuel Cells

In fuel cells, the chemical energy is directly converted into electrical energy. Hydrogen gas, due to its high reactivity, abundance, and lack of environmental pollution, is used as an ideal fuel in fuel cells. Because of the many advantages of fuel cells, this technology is used in many applications, such as transportation, military, portable electronic devices, and especially DG of electricity. The most important of these benefits include:

1. Having high efficiency compared to devices that are using the conventional chemical fuels such as oil and gasoline
2. Environmentally friendly; because water is the by-product generated in the fuel cell
3. No noise pollution; because there are no moving parts in a fuel cell, this device is very quiet
4. Easy setup

Of course, in addition to the abovementioned benefits, fuel cells use expensive equipment that limits the use of this technology. More details about this technology will be presented in subsequent chapters.

1.5.2 Technologies Based on Renewable Energy

The majority of DG is based on renewable energy. Wind turbines, small hydro plants, and solar cells are placed in this category.

1.5.2.1 Wind Turbines

Wind turbines operate on a simple principle. Wind energy rotates the blades that are placed around the rotor. The rotor is connected to a central shaft, which rotates the generator and consequently electricity is produced. Wind turbines are mounted on tall towers to get the most possible energy.

The height of the towers is 30–40 m above the Earth's surface. In addition, the use of small-scale wind turbines is being considered to provide electricity in homes. More details about this technology will be presented in subsequent chapters.

1.5.2.2 Solar Cells

The solid-state solar cell is an electrical component that directly converts sunlight energy into electrical energy by the photovoltaic effect. Small cells are used to provide the required power for smaller devices such as electronic calculators. Solar panels are made from a merger of the cells that have many applications in the electrical industry. It has already been suggested as one of the most important DGs. It will be discussed in more detail in later chapters.

1.5.2.3 Small Hydro Plants

Small hydro plants are used in places where there is the flow of the water or other fluids such as sewage outlet or salty water to the proper height. For example, in place of existing dams, pipelines and canals can be made using this technology. The shafts of these types of power plants (depending on the physical conditions) may be installed vertically, horizontally, or diagonally. These plants have a production capacity up to 5 MW [5].

1.5.3 Energy Storage-Based Technologies

Some DGs use energy storage systems. The most important and best available technology is energy storage batteries that have been made in various types. Following is a short reference for some energy storage technologies.

1.5.3.1 Superconducting Magnetic Energy Storage System

This system is actually a coil that is used in the manufacture of superconducting materials. It is possible to reduce the resistance of the coil to zero by reducing the temperature of the material. So, if the current is injected into the winding and then the winding is short circuited, the energy can be stored in the magnetic field coil.

1.5.3.2 Large Capacitor Banks

Supercapacitors such as batteries store an electric charge. The batteries store the electric charge through a chemical reaction between metal electrodes and a liquid electrolyte. The storage and operations of superconductors are relatively slow because the release of the energy for the chemical reaction needs more time. The high energy stored in batteries is released over a relatively long time. On the other hand, supercapacitors store electrical charges as ions on the surface of the electrodes. In these types of capacitors, the pores on the surface of the electrodes increase the contact surface of the electrolyte. More energy is stored in the capacitors; therefore, capacitors are discharged more quickly than batteries.

1.5.3.3 Storage Systems for Compressed Air
Compressed air storage systems are underground caves that are very large natural reservoirs. In times of low consumption (low load), air is blown into the caves by a pump. This compressed air can be released at the peak hours after colliding with turbines to generate the electricity.

1.5.3.4 Hydraulic Pump Storage Power Plant
The hydraulic pump is a method to store and produce electricity to supply peak times. In times of low electrical demand, excess generation capacity is used to pump water to the water tank at higher altitude. When electricity demand is high, the high-altitude water from the water tank is discharged into the low-altitude water tank through a turbine generator. In this respect, the hydraulic pump storage power plant is similar to compressed air storage systems.

1.6 ADVANTAGES OF DG

In the previous sections the advantages of the DG were briefly explained. In this section, we examine in detail the benefits of the DG.

1.6.1 Functions as a Backup Network Capacity
The first advantage is that DG could be used as capacity support of the network. DGs can reduce power transmission lines by supplying the load where consumers are located because part of the capacity lines will be free. So, from the perspective of the network, the consumer load is reduced.

1.6.2 Emergency Capacity Support
In an emergency, when part of the network is blacked out, using DG in the place of consumption can reduce part of the network's load. So, using these resources can reduce the number of power outages.

1.6.3 Quick Start Up
After a blackout, the network needs to be black started. At this stage, DG can be used to supply a certain part of the network to reduce the total load

of the network. On the other hand, these resources can also be used to start up the power plant.

1.6.4 Combined Electricity and Heat Production

After starting up the power plants and ensuring their stable performance, the grid is connected and the consumers are added to the network step by step. So, the network start up will be easier after blackouts in the entire network by creating small islands supplied by using DGs. Another significant advantage of DG is the use of heat generated by these resources that can be achieved in some of them. In this case, according to Eq. (1.1), the system efficiency increases.

$$\eta = \frac{P_e - P_{\text{thermal}}}{P_{\text{in}}} \tag{1.1}$$

where P_e is the power produced, P_{thermal} is the heat power, and P_{in} is the input power.

1.6.5 Peak Shaving

A typical daily load curve is shown in Fig. 1.2. As can be seen in this figure, the total amount of installed power capacity (dotted line) in the network should be such that the load to be supplied at peak time has a suitable safety margin. In peak condition, electricity generation involves higher costs; therefore, low-cost resources should be used to meet the peak load. Power generation using DGs at peak times, in addition to being economically effective, can reduce the capacity of the transmission lines and transformers. Reducing the current in the main network equipment will reduce their temperature and consequently will increase the life of the equipment.

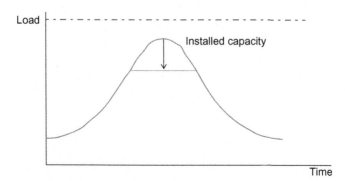

Fig. 1.2 Typical daily load curve.

1.6.6 High Reliability

Whereas high reliability is required for power sensitive loads in a distribution network, using two independent power supplies seems to be essential. With the installation of DG sources in the vicinity of such sensitive loads, they can immediately start up to supply the sensitive loads.

1.6.7 Spinning Reserve

As previously mentioned, DGs have an important role in supplying peak load. The power should be quickly injected into the network to use these resources at peak times. To achieve this purpose, the generator should be synchronized with the network to immediately supply the load. It should be noted that the concept of spinning reserve is not restricted only for rotating generators. For example, fuel cell systems at any moment of the day are capable of producing power to quickly inject the required power into the network.

1.6.8 Nonspinning Reserve

DG sources can be used in emergency conditions to supply the loads due to the quick start up. In this case, unlike the spinning reserve, the generator is turned off and is not connected to the network. The spinning reserve is usually not used to provide peak load.

1.6.9 Load Balance

Because, in distribution networks, the phases are unbalanced, by using single-phase DG sources connected to the distribution network, the phases can be balanced. This is done by adjusting the amount of injected current into the network.

1.6.10 Voltage and Reactive Power Control

The main purpose of the use of DG is active power generation. But in some of these resources, there is also the possibility of reactive power generation. For example, in wind power plants, which are connected to the network via induction generators, the production of reactive power is impossible. But in fuel cells, which are connected to the grid via inverters, simultaneously controlling active and reactive powers is possible. Therefore, the reactive power and voltage profile can be controlled to reduce network losses.

1.6.11 Delaying Investment in the Transmission System

The rated capacity of the transformers and transmission lines is determined according to the load forecasting system and planning studies. With load growth, the construction of the new lines and transformers are needed. If the load is supplied by DG, the construction of the lines and transformers will be delayed.

1.6.12 Fast Installation and Operation

In areas where the growth rate of the load is high, the installation of thermal power plants that need a lot of time is not practical. The use of DG sources that will be constructed in a few months is the perfect solution for this problem.

1.6.13 Less Environmental Pollution Than Centralized Production

In large power plants, all kinds of greenhouse gases are produced. But in DG systems, even those that consume fuel, polluting emissions are very low. This advantage makes it easy to use DG sources in residential areas.

1.7 DISADVANTAGES OF DG

DG sources in addition to all the abovementioned advantages create several problems in the network. Some of the most important ones are mentioned in the following sections.

1.7.1 Difficult to Control

The first issue is the difficulty of controlling DG sources, especially in island-ing mode, because in this case DG sources in addition to active power gen-eration also have the task of controlling the frequency and voltage. To address this issue, the use of quick and advanced controllers will be necessary.

1.7.2 Disturbing Protective Equipment Settings

Because the distribution network protection settings are made before con-necting DG, these resources affect the relay settings. In this case the amount and direction of the current are changed. So all relays, especially the over current relays, need to be adjusted.

1.7.3 Increasing the Short-Circuit Current

In short-circuit condition, after the connection of DG resources to the net-work, in addition to the upstream injection of the short-circuit current, DG also contributes to the error. So, usually a short-circuit current is increased. This is clear in the Fig. 1.3.

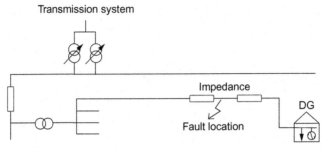

Fig. 1.3 Short-circuit current increase in the presence of DG.

1.7.4 Difficult to Synchronize With the Network

As stated, the DGs can have the ability to supply load independently. For example, diesel generators without a network connection can deliver three-phase power to the consumer. If the generator needs to be connected to the network, automatic synchronization must be done. Otherwise, it is possible that in connection time, the equipment could be damaged.

1.7.5 Choosing a Location for DG Installation

Choosing a suitable location for the installation of the DG sources can increase both the losses and overvoltages in the system. So the location of the DG installation is essential.

1.7.6 Difficult Operation in Urban Areas

The use of some DG technologies in urban areas is not justified because of the amount of space that they occupy. For example, the large blades of wind turbines in the megawatt range need a lot of space, which limits their use in urban areas. Solar cells have no economic justification because they occupy a lot of space in metropolitan areas where land prices are high. However, solar cells can be installed on the roofs of parking garages and buildings.

1.7.7 Increased Cost of Dispatching

The SCADA system should be implemented at the distribution level to provide the remote control of DGs, which ultimately increases the cost of the network.

1.8 COMPARISON AMONG THE DG TECHNOLOGIES

In this section, we compare several of the most common types of DG technologies. Table 1.2 shows a general comparison among DGs. As this table shows, technologies such as gas turbines, microturbines, and fuel cells have two types of efficiency, because these technologies, according to fuel consumption, can be used as CHP. That is, these technologies can use the exhaust heat for heating and also achieve high efficiency.

Table 1.2 also indicates that the cost of installing solar cells per kilowatt is more than the cost of other sources. In this table, the cost of the installation and operation and maintenance (O & M) of these technologies are also mentioned. It should be noted that the information listed in this table is provided on a certain date for the technologies. As time passes and the technology advances we should see falling prices.

Table 1.3 compares the ability of different DG technologies. As can be seen in this table, technologies that consume fuel have a dispatching capability and do not need an energy storage system. In contrast, renewable technologies such as solar cells and wind turbines do not consume fuel, but due to their variable nature, they need an energy storage system. This group of technologies are known as green energy sources due to lack of fuel consumption. The disadvantage of renewable technologies is that they can not be used in peak shaving, reliability improvement, power quality improvement, and CHP applications.

1.9 CONNECTION OF DG TO THE GRID

DG sources can be connected to the network directly, by synchronous generators and induction generators, and indirectly, by using power electronic converters.

1.9.1 Direct Connection to the Network via Synchronous Generator

In direct connection to the network via synchronous generator, the synchronous generator is coupled to the output shaft of the DG and the output of a synchronous generator is connected directly to the electrical network. In this type of connection, synchronization with the electrical network and controlling the excitation of the generator create difficulties. According to the direct connection of the generator to the grid, generators act as a voltage source and when a short circuit occurs in the network, all the power produced will be transferred from the DG to the network.

Table 1.2 General comparison among DGs

Technology	Gas turbine	Micro turbine	Fuel cell	Wind turbine	Solar cell
Power	15 kW–30 WM	25 kW–500 kW	1 kW–20 WM	300 kW–5 WM	300 kW–2 WM
Electrical efficiency (%)	25–30	20–30	30–60	20–40	5–15
Total efficiency (%)	80–90	80–85	80–90		
Installation cost ($/kW)	400–1200	1200–1700	1000–5000	1000–5000	6000–10,000
O & M ($/MWh)	3–8	5–10	5–10	1–4	10
Type of fuel	Natural gas, biogas, propane	Natural gas, biogas, propane, diesel, hydrogen	Natural gas, biogas, propoane	Wind	Sunlight
CO_2 emission (kg/MWh)	580–680	720	430–490	0	0
NO_x emission (kg/MWh)	0.3–0.5	0.1	0.005–0.01	0	0

Table 1.3 A comparison among the capabilities of DGs

	Peak shaving	Reliability improvement	Power quality improvement	Application in CHP	Green energy	Economic fuel	Dispatching	Energy storage
Gas turbine	√	√	Depends on the case	√	×	× Except for biomass	√	×
Micro turbine	√	√	Depends on the case	√	×	× Except for biomass	√	×
Fuel cell	×	√	Depends on the case	√	×	× Except for hydrogen production using sun and wind	√	×
Solar cell	×	×	×	×	√√	√	×	√√
Wind turbine	×	×	×	×	√√	√	×	√√

The abovementioned problem warns us about the need of protecting the generator in short-circuit conditions. In the event of network interruptions, DG has an ability to supply a part of the electrical energy of the network, which is an advantage of this type of connection. Another important advantage of the direct connection of the synchronous generator to the network is the ability to produce the reactive power.

1.9.2 Direct Connection to the Network via Induction Generator

This type of connection is often used for wind turbines and now is outdated with the emergence of new technologies. The control system of the direct connection of the induction generator is simpler than the control system of the synchronous generator because, in direct connection of the induction generator, the exciter of the field does not exist. In addition the induction generator connection is easier compared to the synchronous generator connection to the network, which is another advantage of this type of connection because of the lack of phase rotation and voltage amplitudes. To connect the induction generator to the network, only an increase in the speed of the induction generator's rotor higher than the synchronous speed is needed before connecting the induction generator to the network. In direct connection of the induction generator to the network once the network is cut off, the induction generator will be disconnected; an issue that is not true for the synchronous generator. Because of the inherent ability of the synchronous generator once the network is cut off the synchronous generator can continue to operate independently. Induction generators compared to synchronous generators not only produce reactive power but also consume reactive power. So at every point where the induction generators are connected to the network the voltage will be destroyed. The disadvantages of this type of generator include the possibility of working in motor mode, in the case where the rotor speed is less than synchronous speed, which causes current to be drawn from the network. At the start up of the induction generator, this current causes great shock to the network. It is noted that with the emergence of the doubly fed induction generator, all the problems have been solved and generators can directly connect to the network.

1.9.3 Indirect Connection to the Grid via Power Electronic Converters

In this type of connection, a DG source cannot be directly connected to the network. A DG source is connected to the network by using a converter.

A group of DGs such as fuel cells and photovoltaic cells produce the direct electric energy. Therefore, this type of resource is not allowed to be directly connected to the network. This type of DG source can be connected to the network resources using the converter. An alternative is to increase the level of the voltage by using a DC-DC converter and then to connect it to the network via a DC-AC converter. But the most important thing is the DG sources such as gas microturbines that produce high-frequency electrical energy. To connect these types of DGs to the grid, an AC power with 50 Hz frequency is required. Therefore, the high-frequency power is converted to a 50 Hz AC power via an AC-AC converter (cycloconverter or matrix converter). In the second case, high-frequency AC power (in the range of kilohertz) can be initially converted into DC power and then it is converted into 50 Hz AC power. The costs of this technique are rising, but due to increasing the total control variables the output of the system has higher quality.

REFERENCES

[1] http://www.eia.doe.gov/oiaf/speeches.
[2] Capstone Turbine Corporation. Available from: http://www.microturbine.com/. Accessed 7 November 2016.
[3] The Energy Solutions Center (ESC) Distributed Generation Consortium. Available from: http://www.understandingchp.com/appguide/Chapters/Chap0/0-7_Microturbines.htm. Accessed 7 November 2016.
[4] R.H. Staunton, B. Ozpinec, Microturbine Power Conversion Technology Review, Oak Ridge National Laboratory, 2003.
[5] IEEE guide for control of small hydroelectric power plants, IEEE Std 0171–0922.

FURTHER READING

[1] A.M. Borbely, J.F. Kreider, Distributed Generation, The Power Paradigm for the New Millennium, CRC Press, 2001.
[2] A.F. Zobaa, R.C. Bansal, Handbook of Renewable Energy Technology, World Scientific, 2011.

CHAPTER 2

The Basic Principles of Wind Farms

Ahmed Rashad*,‡, Salah Kamel†, Francisco Jurado‡
*Qena Rural Electrification Sector, Qena, Egypt
†Aswan University, Aswan, Egypt
‡University of Jaén, Jaén, Spain

2.1 HISTORICAL BACKGROUND

Humanity realized the importance of wind and felt its power from the first appearance of civilization. Mankind uses the power of wind in different forms such as propelling ships and grinding grain but the utilization of wind energy did not stop at these limits. With the appearance of electricity in human life and due to an increase of human adoption of electricity, which already depends on traditional fuel, this led to an increase in fuel prices and at the same time an increase in pollution level. From this point, the concept of green energy was born as a mandatory solution for the problem of pollution and the rise in fuel prices.

Wind was the first option for this purpose. The main idea is converting the kinetic energy stored in the wind to a rotating motion and then into electricity. The first try was in the 19th century by developing 12 kW from DC windmill generators. Till 1980, generating electricity from wind was used for home application such as recharging batteries. By 1980 the progress of wind energy technology increased. This progress includes many faces such as improving efficiency, reduction of cost, increase of wind turbine capacity, and even deals with the problems that may face interconnecting wind farms to the grid [1].

Global wind statistics developed by the Global Wind Energy Council (GWEC) can give a good indication to the increase in wind energy generation system (WEGS). Fig. 2.1 shows that the cumulative installed wind capacity has increased from 6.1 GW in 1996 to 423.4 GW in 2015. This rapid increase in wind capacity in the last 20 years indicates the amazing progress in wind energy system technology.

In this chapter, we aim to provide a good background of the WEGS, the basic concepts of WEGS, the components used in it, and the types of induction generators used in WEGS.

Distributed Generation Systems
http://dx.doi.org/10.1016/B978-0-12-804208-3.00002-9

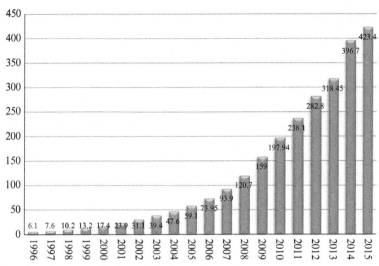

Fig 2.1 Cumulative installed wind capacity.

2.2 SIZE OF WIND TURBINES

Table 2.1 shows the progress in wind turbines' size over the last 30 years. It started with a capacity of 50 kW, 15-m rotor diameter, and 24-m height in 1980; to 10 MW, 145-m rotor diameter, and 162-m height in 2012; and by 2020 it is expected to increase to 20 MW [2].

From Table 2.1 it can be concluded that there is a proportional relation between the capacity of wind turbines and their height and rotor diameter.

2.3 WIND TURBINE APPLICATION

Wind turbines can be installed in wind farms on the land or on the sea. A wind farm installed on land is called an onshore wind farm while a wind farm installed on the sea is called an offshore wind farm. Fig 2.2 shows a photo of onshore and offshore wind farms.

Onshore wind farms are suitable for countries that have a large area associated with desired wind speed. This is because an onshore wind farm is characterized by low cost, easy maintenance, and is easy to interconnect

Table 2.1 Size and capacity of wind turbines

Year	1980	1985	1990	1995	2000	2005	2010	2012	2015
Power	50 Kw	0.1 MW	0.5 MW	0.8 MW	2 MW	3 MW	7.5 MW	10 MW	20 MW
Rotor diameter	15 m	20 m	40 m	50 m	80 m	124 m	126 m	145 m	Expected
Height	24 m	43 m	54 m	80 m	104 m	114 m	138 m	168 m	

(A)   (B)

Fig 2.2 Different types of wind farms based on the type of installation. (A) Offshore wind farm. (B) Onshore wind farm.

to an electrical grid compared with offshore wind farms. But offshore wind farms can be more suitable for countries that have small areas with bodies of water or countries that suffer from high population density.

2.4 WIND TURBINE TECHNOLOGY

2.4.1 Classification According to the Type of Axis Installation

This section presents a brief view of wind turbines classification. Wind turbines can be classified according to the type of their axis installation. If this axis is parallel with the ground, the wind turbine will be known as a horizontal axis wind turbine while if this axis is perpendicular to the ground, the wind turbine will be known as a vertical axis wind turbine. These types of wind farms are shown in Fig 2.3.

(A)   (B)

Fig 2.3 Different types of wind turbines. (A) Horizontal axis wind turbine. (B) Vertical axis wind turbine.

Fig 2.4 Different shapes of vertical wind turbine.

Fig 2.3A shows the horizontal axis wind turbine that can be considered the most famous shape used in wind farms. Fig 2.3B shows the vertical axis wind turbine, but it may be found in other shapes as shown in Fig 2.4.

Naturally, both of these types of wind turbines have advantages and disadvantages. The advantages of a horizontal axis wind turbine can be summarized as better efficiency, maximum utilization of wind speed due to its height, and good power regulation.

The main disadvantage of horizontal axis wind turbines is the high cost compared with vertical axis wind turbines. This cost is represented in the prices of wind turbines and in addition the cost of cables used to transfer the electric power from the turbine to the grid. The main advantage of vertical axis wind turbines is low cost compared with horizontal axis wind turbines, their ease to maintain, and their operation is independent of wind direction. The main disadvantage of vertical axis wind turbines is low efficiency, high torque fluctuation, and lower power regulation compared with horizontal axis wind turbines.

2.4.2 Classification According to the Operating Speed of the Turbines

Wind turbines can be classified according to the operating turbine speed: fixed speed wind turbines and variable speed wind turbines. Fixed speed wind turbines can deliver their rated power only at a very limited range equal to ±1% of rated wind speed. So that any change in wind speed will cause a fluctuation in its output power. Squirrel cage induction generators (SCIGs) can be considered the only suitable generators for this type of turbine. Fig 2.5 shows a simple line diagram of fixed speed wind turbines.

The main advantages of this type of wind turbine are low cost, simple construction, and easy maintenance. The disadvantages are power fluctuation,

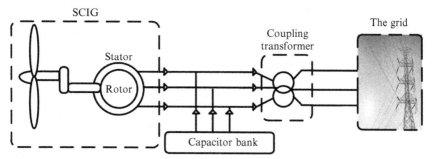

Fig 2.5 Line diagram of fixed speed wind turbines.

high mechanical stress, and lower conversion efficiency. These types of wind turbines are directly connected to the grid through a coupling transformer. The fixed speed wind turbine absorbs the high amount of reactive power in the excitation process so that a capacitor bank will be added as reactive power compensation.

Variable speed wind turbines can operate at a wide range of wind speeds with maximum power conversion efficiency. The advantages of this type are high efficiency, good power quality, and lower mechanical stress while the disadvantages are complicity of construction and high cost. The rotor speed of induction generators (IGs) is adjusted by means of a converter system [1,2].

In variable speed wind turbines, the rotor speed of IGs is always adjusted to keep the power at its rated value even with high changes in wind speed. The process of adjusting the rotor speed can be achieved by controlling the rotor current and hence control the torque-speed characteristic. Controlling the rotor current can be accomplished by means of a power converter system.

From this point of view, variable speed wind turbines can be divided into three divisions according to the degree of the converter system used, as described below.

1. Wound rotor induction generator (WRIG) with variable speed resistance
2. Doubly fed induction generator (DFIG) with rotor converter
3. Variable speed wind turbines with full power converter

2.4.2.1 Wound Rotor Induction Generator With Variable Speed Resistance

WRIGs are the exclusive type for this kind of wind turbine. In this type of generators, the power converter is used to adjust the rotor resistance and hence control the rotor current and the torque-speed characteristic. So that this type is connected to the grid through a two-winding transformer. Fig 2.6 shows a simple line diagram of a WRIG with variable speed resistance.

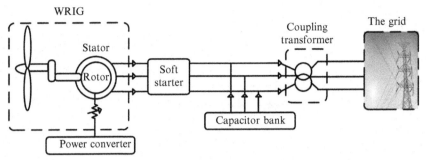

Fig 2.6 WRIG with variable speed resistance.

From Fig 2.6 it can be noted that there is no connection between the rotor and the grid. Just as with SCIGs, WRIGs need to use reactive power compensation. This type of construction gives the ability to adjust rotor speed in a range up to 10% above the synchronous speed. In this type of generator there is no control on output power and it is still proportional to the variation of adjusted speed and reactive power of the system.

2.4.2.2 DFIG With Rotor Converter

WRIGs are the exclusive type for this kind of wind turbine. Fig 2.7 shows a simple diagram of a doubly fed induction generator (DFIG) with rotor converter. The main difference between this type and the previous type is there a connection between the rotor and the interconnected grid. In addition, the

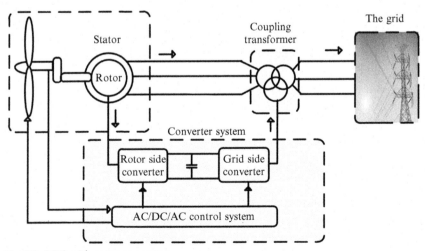

Fig 2.7 DFIG with rotor converter.

wind turbine is connected to the grid through a three-winding transformer, where the stator is connected to the grid through a two-winding transformer and the third winding is used to connect the rotor to the grid through an AC/DC/AC converter. A control system is used to control AC/DC/AC converters. This construction gives the ability not only to control the rotor speed but also control the power independently of reactive power and hence improve the power quality of the wind energy system.

2.4.2.3 Variable Speed Wind Turbines With Full Power Converter

This technology can be used with SCIG, DFIG, and synchronous generator (SG). Fig. 2.8 shows a simple diagram of variable speed wind turbines with a full power converter. Compared to SCIG wind turbines there is no connection between the generator and the grid. This advantage provides better power quality where the output power is transferred to the grid through the full power converter system. The main advantage of using SG for variable speed wind turbines with the full power converter is the freedom to use a gear box or not. If the SG operates at high speed, a gear box is needed and if the SG operates at low speed with multiple poles, there is no need to use a gear box, so that it is known as a direct-driven SG.

2.5 WIND TURBINE COMPONENTS

This section gives a simple explanation of wind turbine construction and the function of each component. Fig. 2.9 shows the components of wind turbines [3].

2.5.1 Rotor of a Wind Turbine

The rotor of a wind turbine consists of two parts, a hub and a blade. The hub is in the front of the turbine connecting the blade to the other parts. The blade is used to convert the kinetic energy stored in the wind into a

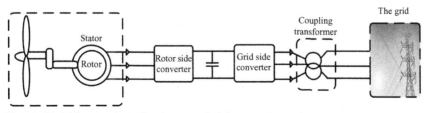

Fig 2.8 Variable speed wind turbines with full power converter.

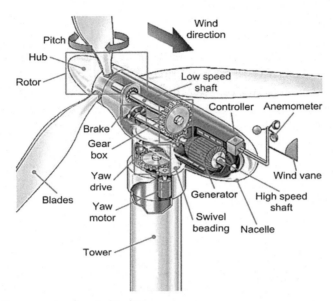

Fig 2.9 Components of a wind turbine.

rotating motion. The rotating motion of the blade results from the difference in the pressure of air above the surface of the blade and under the surface of the blade. This difference in pressure is due to the air velocity air above the surface of the blade being greater than the air velocity under surface of the blade, as shown in Fig. 2.10.

The angle between the blade axis and the wind direction is called the pitch angle (β). If $\beta = 0$ the pressure of air above the surface of the blade is equal to the pressure of air under the surface of the blade, which results

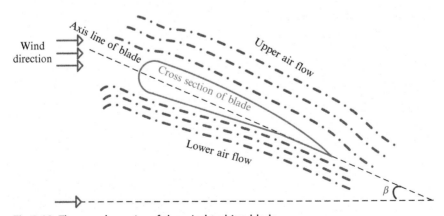

Fig 2.10 The aerodynamics of the wind turbine blade.

in stopping the motion of the blade. So that by controlling this angle, the rotor of a wind turbine can be adjusted.

2.5.2 Pitch and Controller

Pitch is the moving portion of the blade and its function is very similar to the same portion of the plane wings. To understand the importance of this portion, it will be enough to know that the pitch and controller can increase, decrease, or even stop the wind turbine. Pitch can control or change the angle of β to increase or decrease the wind turbine speed according to the wind speed condition. If the wind speed is lower or higher than its rated value, the pitch is controlled to keep angle β at the optimal value that can capture the maximum power from the wind. In addition, it can be used to stop the wind turbine by setting the angle of β to zero with the aid of a brake.

2.5.3 Gear Box

The basic function of the gear box is to convert the low speed of the wind turbine rotor (in low-speed shaft) into high speed (in high-speed shaft), which suits the generator operation. The ratio between speed of turbine rotor and speed of generator rotor is called the gear box ratio (G_r) and is given by

$$G_r = \frac{\omega_g}{\omega_T} = \frac{(1-s)120f}{P\omega_T} \tag{2.1}$$

where ω_T is the speed of turbine rotor, ω_g is the generator speed, S is the slip of the generators and it is usually 1% for IGs and zero for SGs, f is the frequency (50 or 60 Hz), and P is the number of pair poles, which is usually 4 or 6. Usually the gear box has several G_r to increase the power conversion.

Example 2.1 A 4-pole, 50 Hz, DFIG wind turbine produces 2.5 MW at a wind speed of 15 m/s and turbine rotor speed 14.6 rpm. If the wind is decreased to 9.6 rpm, which causes the turbine rotor speed to drop, then determine the number of poles needed to keep the gear box ratio fixed at the same frequency. Assume that the slip is zero.

$$G_r = \frac{\omega_g}{\omega_T} = \frac{(1-s)120f}{P\omega_T} = \frac{120 \times 50}{4 \times 14.6} = 102.73$$

$$G_r = 102.73 = \frac{\omega_g}{\omega_T} = \frac{(1-s)120f}{P\omega_T} = \frac{120 \times 50}{P \times 9.6}$$

$$P = \frac{120 \times 50}{102.73 \times 9.6} = 6.08 \cong 6\,\text{pole}$$

2.5.4 The Brake System

The brake system is used to support the brake condition when it is needed to slow or stop the wind turbine in high wind speed or in a maintenance period. The brake system is usually installed on the low-speed shaft to avoid the high braking torque.

2.5.5 Wind Turbine Generator

Wind turbine generator is used to convert the rotating motion into electrical energy. SCIG was the first type of IG used to convert the rotating motion into electrical energy. With progress in power electronics, WRIGs took their place in the wind energy market. WRIGs can be used as DFIGs or wound rotor synchronous generators (WRSGs). Nowadays the permanent magnet synchronous generator (PMSG) has entered the wind energy market with aid of the power converter. DFIG and PMSG are more favored types in the wind energy market than SCIG, but SCIG is very cheap compared with DFIG and PMSG.

2.5.6 Anemometers and Wind Vanes

Anemometers and wind vanes are used to measure the speed and direction of the wind, respectively. The output data is transferred to the aerodynamic control system and the yaw system that are controlling the speed and direction of the wind turbine according to wind condition.

2.5.7 Yaw System

The function of the yaw system is providing flexibility to wind turbines to keep the wind perpendicular to the turbine blades as much as possible. The yaw system includes an electrical motor, yaw gear, gear rim, bearing, and yaw breakers. The main function of yaw breakers is to fix the wind turbine after changing direction, during operation or maintenance.

2.5.8 Tower

The tower carries the nacelle that represents the house of previously mentioned components and installing it in the ground. Table 2.2 gives some examples to the wind turbines found in the market. It can be observed from Table 2.2 that the output power of the wind turbine is proportional to the tower height (hub height). This is due to the fact that the wind speed

Table 2.2 Examples of wind turbines found in the market [4–6]

Wind turbine	Wind turbine height	Wind turbine capacity (MW)
Nordex N43	40 m/50 m	0.6
Nordex N43	46 m/70 m	1.3
Vestaa V80	60 m/100 m	2
Gamesa G114	80 m/125 m	2.5

increases with increase in height from the Earth's surface. The expression that described the relation between the height and wind speed is given in Eq. (2.2) [7].

$$v_h = v_0 \left(\frac{h}{h_0} \right)^{\alpha} \tag{2.2}$$

where v_h is the wind speed at the height h, v_0 is the average wind speed at h_0, which is usually equal to 10 m, and α is the friction coefficient or Hellman exponent. This coefficient is dependent on the topography at a specific site and it can be taken as 1/7 (0.143) for open land [1]. It should be mentioned that α varies from one place to another and from 1/7 (0.143) during the day to 1/2 (0.5) at night.

2.6 WIND POWER CALCULATION

It is known that the air has a mass (m) and the movement of this mass according to the wind speed (v) produces a kinetic energy (ke) that can be given by

$$ke = \frac{1}{2} m v^2 \tag{2.3}$$

The mass of air can be expressed in terms of air density (ρ) and the volume of the mass of air (V_{ol}) that crosses the turbine blade, as shown in

$$m = \rho V_{ol} = \rho A x \tag{2.4}$$

where x is the distance crossed by the air from the first point of the hub to the point where the air leaves the turbine blade, ρ is the air density at temperature equal to 25 degrees at sea level, and A is the swept area and that is equal to $A = \pi r^2 = \pi L^2$ where r and L are the rotor radius and blade length, respectively, as shown in Fig. 2.11. Now Eq. (2.3) can be written as follows:

$$ke = \frac{1}{2} \rho \pi r^2 v^2 x \tag{2.5}$$

The power stored (P_w) in the air is equal to the change of kinetic energy with the time.

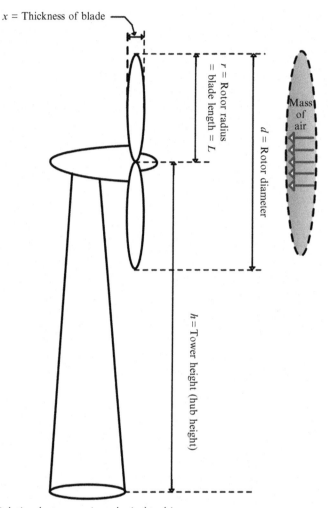

Fig. 2.11 Relation between air and wind turbine.

$$P_w = \frac{1}{2}\rho A v^2 \frac{dx}{dt} = \frac{1}{2}\rho \pi r^2 v^3 \qquad (2.6)$$

Eq. (2.6) represents the total power stored in the wind but does the wind turbine have the ability to extract all of this power?

This can occur only if the wind turbine is ideal where the air flow velocity is assumed to be uniform across the rotor blades and the air is incompressible. Actually the air velocity before impacting with the blade is greater than the air velocity after leaving it. Also the air pressure before impacting the blade is lower than the air pressure after leaving it so that a wind turbine can capture only a fraction of this power. The fraction of power depends on a factor called power coefficient (Cp). According to the Betz limit, Cp, is equal to 0.59 theoretically but in reality Cp varies from 0.2 to 0.5. Therefore, the power that can be captured by the wind turbine can be calculated by modifying Eq. (2.6) in terms of Cp as shown in

$$P = \frac{1}{2}\rho\pi r^2 v^3 Cp \tag{2.7}$$

where P is the power extracted from the wind.

Example 2.2 If we want to install a 200 MW wind farm in a site with air density of 1.225 Kg/m³, average wind speed 9 m/s at height 10 m and 1/7 and 1/2 friction coefficient for day and night, respectively, how many wind turbines are needed for this aim if we use a wind turbine that has a hub height of 80 m with 44 m blade length and 0.32 power coefficient?

First, find the wind speed at the hub height.

$$v_h = v_0 \left(\frac{h}{h_0}\right)^\alpha = 9\left(\frac{80}{10}\right)^{1/7} = 12.11 \simeq 12\,\text{m/s}$$

12 m/s is the rated wind speed of the wind turbine.

Second, find the generated power at the rated wind speed.

$$P = \frac{1}{2}\rho\pi r^2 v^3 Cp = \frac{1}{2} \times 1.225 \times \pi \times (44)^2 \times (12)^3 \times 0.32$$

$$P = 2.05 \simeq 2\,\text{MW}$$

Third, find the number of wind turbines

$$\text{Number of wind turbines} = \frac{200}{2} = 100\,\text{unit}$$

2.7 WIND TURBINE POWER CHARACTERISTIC CURVE

As can be seen from Eq. (2.7), the output power of the wind turbine is proportional to wind speed, rotor diameter, and power coefficient (Cp). Already the wind speed varies during the day and that means that the output power will be varied. So, there are upper and lower limits for the output power to maintain wind turbine safety. The power characteristic curve of the wind turbine describes the relation between the mechanical power developed by the wind turbine and the wind speed variation. The recommendations of power characteristic curves of wind turbines are developed by the International Energy Association (IEA). These recommendations are continuously updated by the International Electrotechnical Commission (IEC). IEC 61400-12 is considered as the basic source that defines and describes the power characteristic curve of wind turbines.

Fig. 2.12 shows the wind turbine power characteristic curve according to IEC 61400-12 [8]. As shown in the power characteristic curve, two operating modes of wind turbines are defined as parking mode and operating mode. These modes depend on determining or defining of three values of wind speed: cut-in, rated, and cut-out wind speed.

Cut-in speed is the lowest value of wind speed that enables the wind turbine to operate. Under this value the wind turbine is stopped and enters the parking mode.

Cut-out speed is the highest value of wind speed that the wind turbine can operate at without damage. Over this value the wind turbine has to be stopped and enters the parking mode.

Rated speed is the value of wind speed that the wind turbine operates at and produces its nominal power. After this value the output power of a wind turbine is fixed with any increase in wind speed till the cut-out speed.

According to the abovementioned definitions the operation of a wind turbine is divided into two modes: parking and operating mode. Our

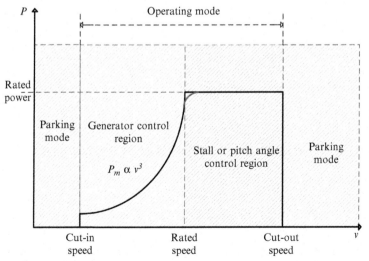

Fig. 2.12 The wind turbine power characteristic curve according to IEC 61400-12.

concern here is with operating mode. In Fig. 2.12 the operating mode is also divided into two regions: generator control region and stall or pitch angle control region.

Generator control region starts from the cut-in wind speed to the rated wind speed. The main advantage of this region is that the mechanical power is proportional to the cubic of wind speed v^3.

Stall or pitch angle control region starts from the rated wind speed to the cut-out wind speed. In this region, the mechanical power is no longer proportional to the cubic of wind speed v^3. It has to be fixed with the increase of wind speed to prevent wind turbines from any damage. There are two types of control that can be issued or applied to the wind turbine to achieve the mission stall or pitch angle control. Stall or pitch angle control is known as an aerodynamics control system.

In Fig. 2.12 the red line on the curve represents the actual power characteristic curve of the wind turbines found in the market but the sharp black portion represents the theoretical power characteristic curve of the wind turbines. An example of the actual power characteristic curve is shown in Fig. 2.13. As shown in Fig. 2.13 the brochure of any actual wind turbine contains all data that may be needed, even the weight of each component.

Example 2.3 In Example 2.2 is the used wind turbine able to operate during the night if its cut-out speed is 26 m/s and what is the output power if it is able to operate?

To answer this question we need to compute wind speed at night according to the average wind speed

$$v_h = v_0 \left(\frac{h}{h_0} \right)^\alpha = 9 \left(\frac{80}{10} \right)^{1/2} = 25.45\,\text{m/s}$$

The wind speed is lower than the cut-out speed and greater than the rated wind speed so that the wind turbines of this wind farm will operate in the stall or pitch angle control region.

2.8 STALL OR PITCH ANGLE CONTROL

As said before, the pitch angle (β) plays an important role in power extracting from the wind so this power can be controlled by controlling this angle. Stall or pitch angle control is used to control this angle.

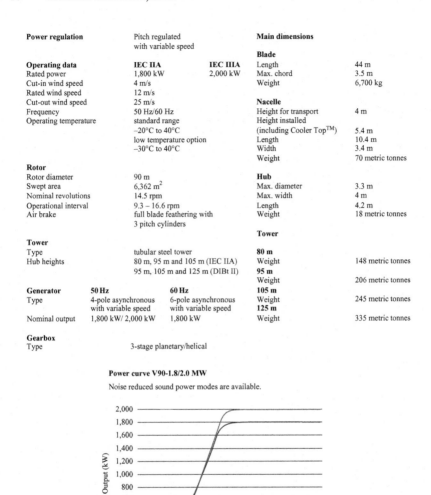

Power regulation	Pitch regulated with variable speed		Main dimensions	
			Blade	
Operating data	**IEC IIA**	**IEC IIIA**	Length	44 m
Rated power	1,800 kW	2,000 kW	Max. chord	3.5 m
Cut-in wind speed	4 m/s		Weight	6,700 kg
Rated wind speed	12 m/s			
Cut-out wind speed	25 m/s		**Nacelle**	
Frequency	50 Hz/60 Hz		Height for transport	4 m
Operating temperature	standard range		Height installed	
	−20°C to 40°C		(including Cooler Top™)	5.4 m
	low temperature option		Length	10.4 m
	−30°C to 40°C		Width	3.4 m
			Weight	70 metric tonnes
Rotor				
Rotor diameter	90 m		**Hub**	
Swept area	6,362 m²		Max. diameter	3.3 m
Nominal revolutions	14.5 rpm		Max. width	4 m
Operational interval	9.3 – 16.6 rpm		Length	4.2 m
Air brake	full blade feathering with 3 pitch cylinders		Weight	18 metric tonnes
			Tower	
Tower				
Type	tubular steel tower		**80 m**	
Hub heights	80 m, 95 m and 105 m (IEC IIA)		Weight	148 metric tonnes
	95 m, 105 m and 125 m (DIBt II)		**95 m**	
			Weight	206 metric tonnes
Generator	**50 Hz**	**60 Hz**	**105 m**	
Type	4-pole asynchronous with variable speed	6-pole asynchronous with variable speed	Weight	245 metric tonnes
			125 m	
Nominal output	1,800 kW/ 2,000 kW	1,800 kW	Weight	335 metric tonnes
Gearbox				
Type	3-stage planetary/helical			

Power curve V90-1.8/2.0 MW

Noise reduced sound power modes are available.

Fig. 2.13 Data brochure and power curve Vestas V90-1.8/2.0 MW [6].

2.8.1 Stall Control

It should be understood that stall is a phenomena; this is a phenomenon used to save the wind turbine from damage. To explain the stall phenomenon, it will be useful to go back to Fig. 2.10, where the uniform distribution of the wind above and under the blade surface occurs at ordinary or rated wind speed. With high wind speed (above the rated speed) the uniform distribution will disappear, causing turbulence in the wind distribution on the above surface of the blade. This turbulence decreases the torquing force that causes the rotational motion of the blade and hence decreases the speed of the turbine rotor. By good design of the wind turbine blade, stall phenomenon is used to keep the safety of the wind turbine during extreme wind speed. There are two types of stall control: passive and active stall controls.

Passive stall control. In this type of control, the blades are fixed to the hub and hence β is fixed at its rated value. In passive stall control, the blade is designed in such a way that guarantees two functions:

1. The blade can extract the maximum power at rated wind speed as much as possible.
2. The stall can occur only at high wind speed to prevent the turbine from damage.

In passive stall control there are no mechanical parts in the blade, which can be considered to be an advantage. During a wind speed variation above the rated wind speed, passive stall control is affected by high fluctuation in power. Where a little increase in wind speed above the rated value will cause an increase in output power during extreme wind speed, the output power will decrease with further increase in wind speed while the wind turbine does not reach the cut-out wind speed. Fig. 2.14 shows the power characteristic curve of the wind turbine using passive stall control.

Active stall control. The deference between passive and active stall controls is that in active stall control the blades are not fixed so that β can be changed or controlled. This will provide the ability to generate the stall phenomenon by increasing angle β in the way that the blades will be turned into the wind. So that with the increase in wind speed, angle will be increased β to generate

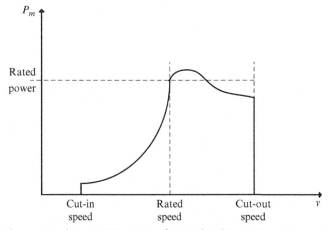

Fig. 2.14 The power characteristic curve of a wind turbine using passive stall control.

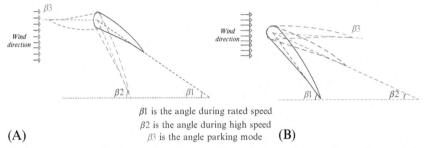

$\beta 1$ is the angle during rated speed
$\beta 2$ is the angle during high speed
(A) $\beta 3$ is the angle parking mode   (B)

Fig. 2.15 Impact on blade position due to pitch angle control and active stall control. *B1* is the angle during rated speed, *B2* is the angle during high speed, and *B3* is the angle parking mode. (A) Blade position with active stall control. (B) Blade position with pitch angle control.

the stall phenomenon and decrease the moving force and hence keep the power at its rated value as shown in Fig. 2.15.

2.8.2 Pitch Angle Control

Pitch angle control has the same idea as active stall control except that the blades' movement is not into the wind as in active stall control but out of the wind. It can be accomplished by decreasing angle β.

It can explained as follows: When the wind speed is higher than the rated value, the angle β will be decreased to decrease the pressure on the lower surface of the blade and hence decrease the torque force. The decrease in torque force will decrease the rotor speed and hence output power so that it will be kept at its rated value. The opposite procedure will happen when wind speed is lower than its rated value where angle β will be increased to increase the output power and keep it at its rated value. Fig. 2.15 shows the difference between impact on the blade position due to pitch angle control and active stall control.

From Fig. 2.15 it can be observed that in active stall control angle β is increased with the increase of wind speed but in pitch angle control angle β is decreased with the increase of wind speed. This is the main difference between the principle operations of active stall control and pitch angle control angle.

2.8.3 Tip Speed Ratio

From Eq. (2.7) it can be concluded that the power extracted from the wind is proportional to three factors, namely, the rotor radius r, wind speed v, and power coefficient Cp. Cp is also a function of two factors of angle β and an important parameter called tip speed ratio (TSR), λ. λ describes the relation

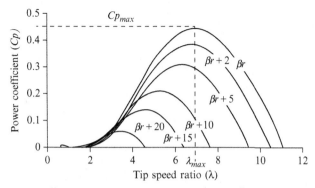

Fig. 2.16 Power coefficient Cp via tip speed ratio λ with different angle β.

between the wind speed, v, and wind turbine rotor speed ω_T (blade speed). The mathematical equation between v and ω_T can be represented as

$$\lambda = \frac{\text{Speed of rotor tip}}{\text{Wind speed}} = \frac{r \times 2\pi\omega_T}{v} = \frac{\text{m/s}}{\text{m/s}} \quad (2.8)$$

Two-bladed wind turbines have λ around 6, three-bladed wind turbines have λ around 5, and four-bladed wind turbines have λ around 3.

Fig. 2.16 explains the relation between the three parameters Cp, λ, and β.

From Fig. 2.16 it can be observed that the maximum value of Cp can be obtained when the wind turbine operates at its rated wind speed that produces the maximum value of λ with rated pitch angle β_r.

Example 2.4 In Example 2.2 find the optimal TSR if the wind turbines use gear box ratio of 103.45 and 4-pole, 50 Hz asynchronous.

First, find the rotor speed of the wind turbine using Eq. (2.1). By considering the slip of 1% for the IG:

$$\omega_g = \omega_s \times (1-s) = \frac{120f}{P} \times (1-s) = \frac{120 \times 50}{4} \times (1-0.01) = 1485\,\text{rpm}$$

$$G_r = \frac{\omega_g}{\omega_T} = 103.45$$

$$\omega_T = \frac{1485}{103.455} = 14.35\,\text{rpm}$$

From Example 2.2 we know that the wind speed is 12 m/s and the rotor radius is 44 m. The optimal TSR occurred at rated wind speed with rated angle β_r.

$$\lambda = \frac{r\omega_T}{v} = \frac{44 \times (2\pi \times 14.35)}{12 \times 60} = 5.51$$

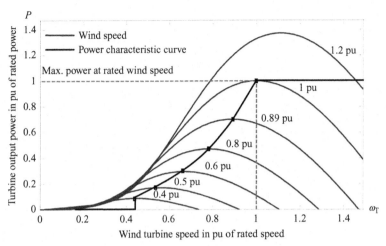

Fig. 2.17 Maximum power point tracking (MPPT).

2.9 MAXIMUM POWER POINT TRACKING

To understand the meaning of maximum power point tracking (MPPT), let us assume that a wind turbine produces its rated power (1 pu) at rated wind speed (base wind speed 1 pu) and rated pitch angle β_r. During the generator mode the wind speed is varied from cut-in speed to the rated speed as shown in Fig. 2.17 (0.4–1 pu) and hence the produced power. The maximum power that can be extracted from each wind speed can be obtained at certain values of wind turbine speed ω_T. This point is called the maximum power point for each wind speed. The path that this point takes from one speed to another is known as MPPT. MPPT represents the generator mode in the power characteristic curve. The MPPT must be defined by the manufacturers for each wind turbine.

Example 2.5 Use the MATLAB Simulink program and the SCIG model to implement the wind turbine in the previous examples. Set the simulation time to 30 s and wind speed varies from 8 to 25 m/s. Draw the power, wind speed, and pitch angle curve with the time. Fig. 2.18 shows the MATLAB Simulink model SCIG,

Open the wind farm; it consists of three wind farms as shown in Fig. 2.19.

First, change the parameters of the wind farm to fit the examples.

Second, create a subsystem that represents the wind speed variation from 8 to 25 m/s.

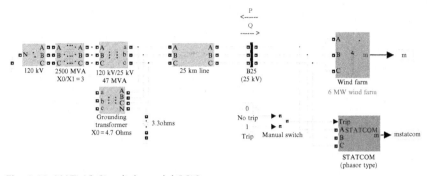

Fig. 2.18 MATLAB Simulink model SCIG.

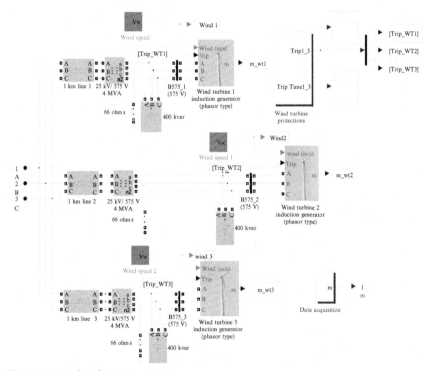

Fig. 2.19 Wind turbines.

Third, run the model and draw the power, wind speed, and pitch angle curve with the time. The output power is shown in Fig. 2.20.

It is important to note that the change in pitch angle does not occur until the wind speed exceeds the rated wind speed of the wind turbine.

This simple example helps to understand the relation between wind speed and the other parameters of wind turbines.

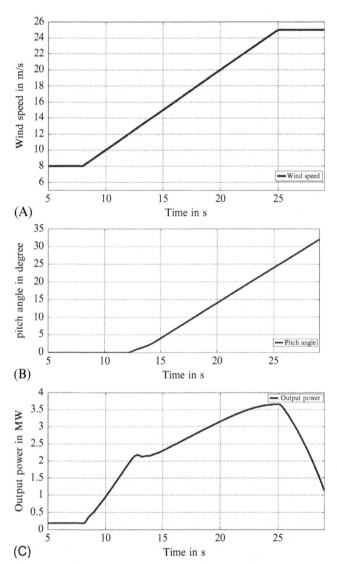

Fig. 2.20 Impact of wind speed variation on wind turbine operation. (A) Wind speed variation, (B) pitch angle variation, and (C) output power variation.

2.10 MATHEMATICAL MODELING OF WIND TURBINE GENERATORS

To understand the operation of a wind energy system, mathematical modeling of wind turbine generators should be known. Electric generators used in wind turbines can be divided into sections as IGs and SGs. SCIGs and DFIGs are the main members of IGs (see Fig. 2.21).

(A)   (B)   (C)

Fig. 2.21 Wind turbine electrical generators. (A) SCIG, (B) DFIG, and (C) PMSG.

It is very important to know that the direct and the quadratic transformation (d–q transformation) of generator voltage and current play an important role in a wind turbines control system; therefore it is helpful to have a good background in d–q transformation. There are two methods for modeling generators, the space vector model and the d–q transformation model. First, the electrical generators will be modeled using the space vector model, then the transfer space vector model into d–q transformation.

2.10.1 Mathematical Modeling of IGs

To understand the space vector model of IG, let us consider a vector x. Vector x is rotating in the hypothetical form of abc axis. This vector is rotating in space with speed equal to ω, which is the same speed of stator. Voltage v, flux linkage φ, and electrical torque T_e are variables that define the operation of IG. Both stator and rotor have their variables, but the stator rotates with the same speed of vector x while the rotor rotates in the same farm but with speed equal to ω_r. Therefore these variables can be expressed by their vectors that rotate in the hypothetical farm of abc axis. Fig. 2.22 shows a schematic diagram of IG according to the hypothetical farm of abc axis.

It is known that the rotation movement results from the interaction between the stator flux and rotor flux, so that there will be a flux linkage between φ_s and φ_r. φ_s and φ_r consist of two main components, namely, self-inductance of the stator and rotor coils (L_s and L_r, respectively) and the magnetizing inductance (L_m). The self-inductance stator and rotor consists of two components of the magnetizing inductance and the leakage inductance between stator and rotor coils (L_{ls} and L_{lr}, respectively). The vector space equations of stator and rotor flux linkage (φ_s and φ_r, respectively) can be written as

$$\vec{\varphi}_s = L_s \vec{i}_s + L_m \vec{i}_r \quad \text{where } L_s = (L_{ls} + L_m)$$
$$\vec{\varphi}_s = L_{ls} \vec{i}_s + L_m \vec{i}_m$$

(2.9)

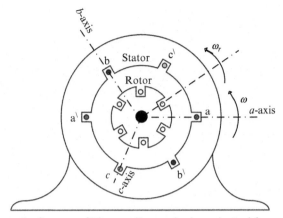

Fig. 2.22 Schematic diagram of IG according to the hypothetical farm of *abc* axis.

and

$$\vec{\varphi_r} = L_r\,\vec{i_r} + L_m\,\vec{i_s} \quad \text{where } L_s = (L_{ls} + L_m)$$
$$\vec{\varphi_r} = L_{lr}\,\vec{i_r} + L_m\,\vec{i_m}$$

(2.10)

where $\vec{i_s}$, $\vec{i_r}$, and $\vec{i_m}$ are the current vectors of the stator, rotor, and magnetization current.

The equivalent circuit of the IG space vector in the hypothetical reference frame is shown in Fig. 2.23. From the equivalent circuit of IG, the space vector of voltage can be concluded as shown below.

$$\vec{v_s} = R_s\,\vec{i_s} + p\vec{\varphi_s} + j\omega\,\vec{\varphi_s}$$

(2.11)

$$\vec{v_r} = R_r\,\vec{i_r} + p\vec{\varphi_r} + j(\omega - \omega_r)\,\vec{\varphi_r}$$

(2.12)

where $\vec{v_s}$ and $\vec{v_r}$ are the voltage vectors of stator and rotor and $\vec{\varphi_s}$ and $\vec{\varphi_r}$ are the flux linkage of stator and rotor, and where R_s and R_r are the winding resistant of stator and rotor and $p = d/dt$.

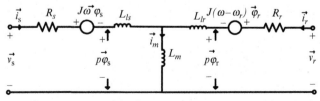

Fig. 2.23 Equivalent circuit model of IG according to space vector in hypothetical reference frame.

The generated electromagnetic torque T_e can be written as follows:

$$T_e = \frac{3P}{2} Re\left(j\,\vec{\varphi_s}\,\vec{i_s}\right) = -\frac{3P}{2} Re\left(j\,\vec{\varphi_r}\,\vec{i_r}\right) \tag{2.13}$$

The next equations describe the relation between the electromagnetic torque T_e and the mechanical torque T_m in induction machines.

$$J\frac{d\omega_T}{dt} = T_e - T_m \tag{2.14}$$

where J is the moment of inertia of the wind turbine rotor and ω_T is the speed of the wind turbine rotor. Now space vector of the three variable flux, voltage, and torque where Eqs. (2.9) and (2.10) represent the space vector of flux, Eqs. (2.11), (2.12) represent the space vector of voltage, and Eqs. (2.13), (2.14) represent the space vector of torque. Take into account that all rotor parameters are referred to stator side. Also, these equations are general equations the give a representation of induction machines.

Remember that these equations are established for a hypothetical farm rotating in space with speed ω; now let's convert this speed into synchronous speed ω_s. This can be done by replacing ω with ω_s and using the relation between ω_s and ω_r in the slip equation.

$$s = \frac{\omega_s - \omega_r}{\omega_s} \tag{2.15}$$

where $\omega_s = 2\pi f$ and f is the stator frequency.

This synchronous farm is very helpful in simulating wind turbines. So that we can rewrite the voltage space vector equations according to a synchronous farm as follows:

$$\vec{v_s} = R_s\,\vec{i_s} + p\,\vec{\varphi_s} + j\omega_s\,\vec{\varphi_s} \tag{2.16}$$

$$\vec{v_r} = R_r\,\vec{i_r} + p\,\vec{\varphi_r} + js\omega_s\,\vec{\varphi_r} \tag{2.17}$$

Not that the space vector equations of flux and torque remain unchanged. Fig. 2.24 shows the equivalent circuit of IG according to space vector in synchronous reference frame.

Equivalent circuit of IG according to space vector in a synchronous reference frame (synchronous mode) is very helpful in conclusion of the steady state equivalent circuit of IG and d-q transformation of IG.

To obtain the steady state equivalent circuit of IG, the space vectors of voltages in synchronous mode have to be converted to its respective value in steady state mode.

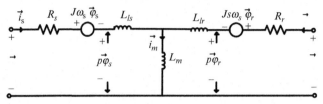

Fig. 2.24 Equivalent circuit of IG according to space vector in a synchronous reference frame.

This can be accomplished in three steps as follows:

1. It is already known in steady state mode that $p = d/dt = 0$; this means that any part in Eqs (2.16), (2.17) will be zero.

2. The relation between vectors and steady state quantities can be obtained by dividing the vector quantities by $\sqrt{2}$. As the vector has magnitude and angle, its equivalent steady state value has real (Re) and imaginary (Im) values ($V = V + jV$ and $I = I + jI$).

3. Transfer the flux linkage φ in Eqs. (2.9), (2.10) into their respective steady state leakage reactance value X where $X = \omega_s \varphi$

$$X_s = \omega_s \varphi_s = \omega_s (L_{ls} I_s + L_m I_m) = I_s X_{ls} + I_m X_m \tag{2.18}$$

$$X_r = \omega_s \varphi_r = \omega_s (L_{lr} I_r + L_m I_m) = I_r X_{lr} + I_m X_m \tag{2.19}$$

4. In a wind energy system, the rotor current is drowned out from the rotor; then the direction of the rotor current in Fig. 2.24 will be reversed. So that the stator current will be equal to $I_r + I_m$.

5. Now rewrite voltage Eqs. (2.16), (2.17) according to the above steps

$$V_s = R_s I_s + j I_s X_{ls} + j I_m X_m \tag{2.20}$$

$$V_r = -R_r I_r - j I_r s X_{ls} + j I_m s X_m \tag{2.21}$$

where X_s, X_r, and X_m are the stator, rotor, and magnetization leakage reactance, respectively. Now let us draw the steady state equivalent circuit according to the concepts of the above equations. This equivalent circuit is shown in Fig. 2.25. By dividing Eq. (2.21) by slip, s, the equivalent circuit can be transferred into an equivalent circuit as shown in Fig. 2.25B.

2.10.2 Power Flow of Wind Turbines Based on IG

Fig. 2.25B can be used to understand the power flow of IG wind turbines. The power captured from the wind (P_T) by the wind turbine blade is

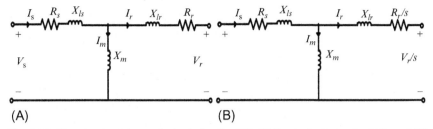

Fig. 2.25 The steady state equivalent circuit of IG. (A) Equivalent circuit of IG, and (B) Equivalent circuit of IG with respect to slip speed

decreased due to the gearbox and moment of inertia of the wind turbine rotor to produce the mechanical power (P_m). P_m is also reduced due to the copper losses in the coil generator rotor (P_{rcul}). The rest of the power represents the power transferred to the stator and it is known as air gap power (P_{ag}). Due to the copper losses in the stator (P_{scul}), P_{ag} is reduced to the output power (P_s) transferred to the system. Fig. 2.26 shows the power flow from the wind to the electrical grid.

The equations that represent Fig. 2.26 can be derived as

$$P_s = P_{ag} - P_{scul} \tag{2.22}$$

$$P_{ag} = P_m - P_{rcul} \tag{2.23}$$

To find P_m and P_{rcul}, the rotor resistance in Fig 2.25B can be divided into two components as shown below.

$$\frac{R_r}{s} = R + \frac{1-s}{s} R_r \tag{2.24}$$

where the first part from the equation is used to drive the P_{rcul} and the second part is used to drive the P_m as shown below.

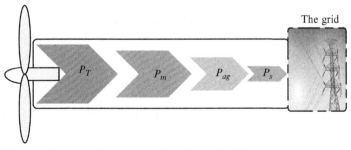

Fig. 2.26 Power flow from the wind to the electrical grid.

$$P_m = 3I_r^2 \frac{1-s}{s} R_r \tag{2.25}$$

$$P_{rcul} = 3I_r^2 R_r \tag{2.26}$$

The P_s is equal to $P_{ag}-P_{scul}$. The P_{scul} is given by the next equation:

$$P_{scul} = 3I_s^2 R_s \tag{2.27}$$

By substituting Eqs. (2.23), (2.25), (2.26) in Eq. (2.22), P_s will be equal to

$$P_s = 3I_r^2 \frac{1-s}{s} R_r - 3I_r^2 R_r - 3I_s^2 R_s = 3V_s I_s \cos\phi_s \tag{2.28}$$

where ϕ_s is the angle between V_s and I_s and $\cos\phi_s$ is the power factor. Note that coefficient 3 in Eq. (2.28) represents the three phases, which means that all these equations are derived for phase values and V_s is taken as a reference with an angle equal to zero degrees. ϕ_s will be equal to the next equation:

$$\emptyset_s = \angle V_s - \angle I_s \tag{2.29}$$

The wind turbine efficiency is the ratio between the electrical power delivered to the grid (P_s) to the power capture from the wind (P_T).

$$\eta = \frac{P_s}{P_T} \tag{2.30}$$

The mechanical torque can be computed as follows:

$$T_m = \frac{P_m}{\omega_m} = \frac{1}{\omega_m}\left(3I_r^2 \frac{1-s}{s} R_r\right) = \frac{1}{\omega_m}\left(3I_r^2 \frac{R_r}{s} 1 - s\right) = \frac{P}{\omega_r}\left(3I_r^2 \frac{R_r \omega_r}{s \, \omega_s}\right)$$

$$= \frac{3PR_r}{s\omega_s} I_r^2$$

where P is the number of poles and $\omega_m = \dfrac{P}{\omega_r}$

$$T_m = \frac{3PR_r}{s\omega_s} I_r^2 \tag{2.31}$$

Eq. (2.30) gives the torque in terms of the rotor current; also, the torque can be obtained in terms of stator voltage. For simplicity, the magnetizing current can be neglected (it is very small compared with rotor current). This will result in

$$I_s = I_r \quad \text{and} \quad z_{eqv} = \left(R_s + \frac{R_r}{s}\right) + j(X_s + X_r) \tag{2.32}$$

$$I_r = \frac{V_s}{\sqrt{\left(R_s + \dfrac{R_r}{s}\right)^2 + (X_s + X_r)^2}} \tag{2.33}$$

By substituting Eq. (2.33) in Eq. (2.31), the mechanical torque will be equal to

$$T_m = \frac{3PR_r}{s\omega_s} \frac{V_s}{\sqrt{\left(R_s + \dfrac{R_r}{s}\right)^2 + (X_s + X_r)^2}} \tag{2.34}$$

Example 2.6 A wind farm sits with air density of 1.225 Kg/m³, average wind speed of 11.14 m/s at height 10 m and 1/7 friction coefficient. The wind turbine used in this wind farm has a hub height of 80 m and rotor diameter of 90 m. Power coefficient, gearbox ratio, and tip speed are 0.2295, 109.193, and 5.05, respectively. The wind turbine generator has a 4-pole, 60 Hz and 1000 V asynchronous generator. The generator has a 0.99 mΩ, 0.81 mΩ, and 0.059 mH stator and rotor resistances and leakage inductance, respectively. The slip and synchronous speed are − 0.003 and 1700 rpm, respectively.

1. Find the efficiency of this wind turbine if the total mechanical losses are 20 kW if the magnetizing current is neglected.
2. Compute the mechanical torque.

First, compute the P_{scul} and P_{scul}:

$$V_{sphsae} = \frac{v_{rated}}{\sqrt{3}} = \frac{1000}{\sqrt{3}} = 577.35\,\text{V}$$

I_m is neglected and the negative slip means that the generator operates in generation mode

$$Z_{eqv} = \left(R_s + \frac{R_r}{s}\right) + j(X_s + X_r)$$
$$= \left(0.00099 + \frac{0.00081}{-0.003}\right) + j(0.000059 + 0.000059)$$

$$Z_{eqv} = 0.335\angle143.71 \text{ degrees} = 0.269 + j0.2\,\Omega$$

$$I_s = I_r = \frac{577.35\angle0 \text{ degrees}}{0.335\angle143.71 \text{ degrees}} = 1723.433\angle - 143.71 \text{ degrees}$$

$$P_{scul} = 3I_s^2 R_s = 3 \times (1723.433)^2 \times 0.00099 = 8.82155\,\text{kW}$$

$$P_{rcul} = 3I_r^2 R_r = 3 \times (1723.433)^2 \times 0.00081 = 7.21763\,\text{kW}$$

Now compute the P_s

$$P_m = P_s + P_{scul} + P_{rcul}$$

So, P_m needs to be computed from the site data and turbine data

$$v_h = v_0 \left(\frac{h}{h_0}\right)^\alpha = 11.14 \left(\frac{80}{10}\right)^{\frac{1}{7}} = 14.99 \simeq 15 \, \text{m/s}$$

$$P_m = \frac{1}{2}\rho\pi r^2 v^3 Cp = \frac{1}{2} \times 1.225 \times \pi \times (45)^2 \times (15)^3 \times 0.2295$$

$$P_m = 3018.126 \, \text{kW}$$

$$P_s = P_m - P_{scul} - P_{rcul} = 3018.126 - 8.82155 - 7.21763$$

$$P_s = 3002.08 \, \text{kW} \simeq 3 \, \text{MW}$$

$$P_{total} = P_m + P_{mecl} = 3018.126 + 20 = 3036.126 \, \text{kW}$$

$$\eta = \frac{P_s}{P_T} = \frac{3002.08}{3036.126} \times 100 = 98.87\%$$

Example 2.7 In Example 2.6 if the wind turbine has tip speed of (λ) 5.05, compute the mechanical torque.

$$\omega_T = \frac{r\lambda}{v} = \frac{15 \times 5.05 \times 60}{45 \times 2 \times \pi} = 16.1 \, \text{rpm}$$

$$G_r = \frac{\omega_g}{\omega_T} = \frac{\omega_g}{16.1} = 109.193$$

$$\omega_g = 16.1 \times 109.193 = 1758 \, \text{rpm}$$

$$\omega_m = \frac{2\pi}{60}\omega_g = 184.1 \, \text{rad/s}$$

$$T_m = \frac{P_m}{\omega_m} = \frac{3018.126}{184.1} = 16.4 \, \text{kNm}$$

2.10.3 *d-q* Transformation of IG

As said before, the control system of wind turbines depends on the direct and quadratic components of IGs variables (voltage, current, and flux). IG equations according to space vector in a synchronous reference frame can be easily transferred to a *d-q* reference frame. The *d-q* reference frame divides the equivalent circuit of IG in two circuits, one for each reference. This means that stator voltage and current will be represented by two equations and the same word applied for rotor. Also, the torque can be represented using *d-q* transformation.

As said before, any vector in the space has magnitude and angle so that it can be written in the form of real (*Re*) and imaginary (*Im*) values. So that all space vectors of voltage, current, and flux leakage of IG will be decomposed into their equivalent components on the d-q reference frame. The d-q components of stator and rotor variables (voltage, current, and flux leakage) can be described as follows:

$$\left.\begin{array}{l} v_{ds} = R_s i_{ds} + p\varphi_{ds} - \omega_s \varphi_{qs} \\ v_{qs} = R_s i_{qs} + p\varphi_{qs} - \omega_s \varphi_{ds} \end{array}\right\} \text{ Stator voltage} \tag{2.35}$$

$$\left.\begin{array}{l} \varphi_{ds} = L_s i_{ds} + L_m i_{dr} \\ \varphi_{qs} = L_s i_{qs} + L_m i_{qr} \end{array}\right\} \text{ Stator flux leakage} \tag{2.36}$$

$$\left.\begin{array}{l} v_{dr} = R_r i_{dr} + p\varphi_{dr} - \omega_{sl} \varphi_{qr} \\ v_{qr} = R_r i_{qr} + p\varphi_{qr} - \omega_{sl} \varphi_{dr} \end{array}\right\} \text{ Rotor voltage} \tag{2.37}$$

$$\left.\begin{array}{l} \varphi_{dr} = L_r i_{dr} + L_m i_{ds} \\ \varphi_{qr} = L_r i_{qr} + L_m i_{qs} \end{array}\right\} \text{ Rotor flux leakage} \tag{2.38}$$

Eqs. (2.36), (2.38) can be written in matrix form and then used to obtain the d-q components of stator and rotor currents as follows:

$$\begin{bmatrix} \varphi_{ds} \\ \varphi_{qs} \\ \varphi_{dr} \\ \varphi_{qr} \end{bmatrix} = \begin{bmatrix} L_s & 0 & L_m & 0 \\ 0 & L_s & 0 & L_m \\ L_m & 0 & L_r & 0 \\ 0 & L_m & 0 & L_r \end{bmatrix} \cdot \begin{bmatrix} i_{ds} \\ i_{qs} \\ i_{dr} \\ i_{qr} \end{bmatrix} \tag{2.39}$$

From Eq. (2.39) the current matrix can be obtained as

$$\begin{bmatrix} i_{ds} \\ i_{qs} \\ i_{dr} \\ i_{qr} \end{bmatrix} = \begin{bmatrix} L_s & 0 & L_m & 0 \\ 0 & L_s & 0 & L_m \\ L_m & 0 & L_r & 0 \\ 0 & L_m & 0 & L_r \end{bmatrix}^{-1} \cdot \begin{bmatrix} \varphi_{ds} \\ \varphi_{qs} \\ \varphi_{dr} \\ \varphi_{qr} \end{bmatrix}$$

$$= \frac{1}{L_s L_r - L_m^2} \begin{bmatrix} L_r & 0 & -L_m & 0 \\ 0 & L_r & 0 & -L_m \\ -L_m & 0 & L_s & 0 \\ 0 & -L_m & 0 & L_s \end{bmatrix} \cdot \begin{bmatrix} \varphi_{ds} \\ \varphi_{qs} \\ \varphi_{dr} \\ \varphi_{dr} \end{bmatrix} \tag{2.40}$$

The electrical torque of SCIG and DFIG can be obtained as follows:

$$T_e = \begin{cases} \dfrac{3P}{2}\left(i_{qs}\varphi_{ds} - i_{ds}\varphi_{qs}\right) \\[2mm] \dfrac{3PL_m}{2L_r}\left(i_{qs}\varphi_{dr} - i_{ds}\varphi_{qr}\right) \\[2mm] \dfrac{3PL_m}{2}\left(i_{qs}i_{dr} - i_{ds}i_{qr}\right) \end{cases} \tag{2.41}$$

Eq. (2.41) shows that there is more than one expression for electrical torque using d-q transformation. All of abovementioned equations can be used according to the objective. The most effective one in designing the control system of wind turbines is the third one. This is because it depends on the d-q components of stator and rotor currents that are easy to compute. All of the previous equations and transformations are general equations for IG but how can we use them with SCIG and DFIG and even in a control system of SCIG and DFIG? This is our aim in the next section.

2.11 SQUIRREL CAGE IGs IN A WEGS

Wind turbines based on SCIGs are classified as self-exited fixed speed wind turbines. Only the stator of a SCIG is connected to the grid through a two-winding transformer; a capacitor bank is used for the self-excitation process. A static synchronous compensator (STATCOM) is connected at the pion of common connection (PCC) between the wind turbine and the interconnected grid, as shown in Fig. 2.27. It is already known that the rotor of a SCIG is a short circuit so that the rotor voltage is zero and hence the d-q component of rotor voltage. The stator and rotor voltages can be described by Eqs. (2.35), (2.37) with small changes in Eq. (2.37) by substituting V_{dr} and V_{qr} with zero. The d-q equivalent circuit of SCIG according to a vector space model is shown in Fig. 2.28 while Fig. 2.29 shows the d-q equivalent circuit of SCIG according to a synchronous farm model.

SCIG absorbs a large amount of reactive power from the interconnected grid. This will result in an undesirable impact on voltage at the PCC so that it

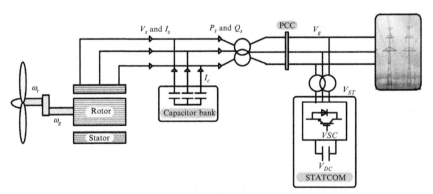

Fig. 2.27 SCIG wind turbine interconnected grid.

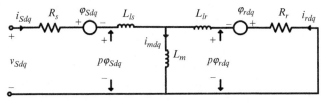

Fig. 2.28 The *d-q* equivalent circuit of SCIG according to vector space mode.

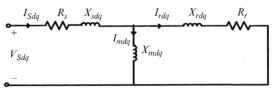

Fig. 2.29 The *d-q* equivalent circuit of SCIG according to a synchronous farm model.

needs reactive power compensation. Flexible AC transmission system (FACTS) devices are used for this function.

STATCOM is famous in this area while the modern members of FACTS are understudied. The impact of STATCOM on wind farm operation will be discussed in the next section but now let us understand the principle operation of STATCOM.

2.11.1 Principle Operation of STATCOM

This section deals with principle operation of STATCOM and its control system. STATCOM is connected in parallel with a transmission line through a coupling transformer that is called a shunt device. STATCOM consists of a DC bus and voltage source converter (VSC). VSC uses power electronic devices to synchronize the DC voltage with the grid voltage.

STATCOM injects or absorbs reactive power to regulate the voltage at PCC (V_{PCC}). If the V_{PCC} is lower than its rated value, SATCOM will inject reactive power to the grid to regulate the voltage, but if the V_{PCC} is higher than its rated value, SATCOM will absorb reactive power from the grid to keep the voltage at its rated value. This function is accomplished by injecting voltage that is in phase with live voltage in steady state and quadratic to the line current in operation mode. The sign of this quadratic voltage is determined if the reactive power is injected or absorbed and this sign depends on the grid voltage. For simplicity, a fixed SCIG wind farm MATLAB Simulink model will be our case study. The simple equations that represent the operation of STATCOM can be written as flows:

$$P_{ST} = \frac{V_g V_{ST} \sin \delta}{X_t}$$

$$Q_{ST} = \frac{V_g (V_g - V_{ST} \cos \delta)}{X_t} \Bigg\} \tag{2.42}$$

In Eq. (2.42) V_g and V_{ST} are the grid and STATCOM voltage, respectively, X_t is the reactance of the coupling transformer, and δ is the angle between V_g and V_{ST}. In steady state, δ can be considered equal to zero so that there is no active power transferred between the grid and STATCOM ($P_{ST} = 0$) and only the reactive power is transferred (Q_{ST}). Q_{ST} will be equal to the next equation.

$$Q_{ST} = \frac{V_g (V_g - V_{ST})}{X_t} \tag{2.43}$$

Example 2.8 In Example 2.6 the wind farm is connected to 3 VAr STATCOM through coupling transformer has reactance 0.024 Ω. In the next cases calculate the reactive power of STATCOM needed to keep the voltage at PCC at its steady state value.
1. The voltage at PCC was decreased to 85% of its steady state value due to wind speed variation.
2. The voltage at PCC was increased to 115% of its steady state value due to wind speed variation.

First, compute the reactive power of STATCOM due to decrease in voltage.

$$V_g = 577.35 \times 0.85 = 490.75\, v$$

$$Q_{ST} = \frac{V_g (V_g - V_{ST})}{X_t} = \frac{490.75\,(490.75 - 577.35)}{0.042} = -1770831.68$$
$$\simeq -1.77\,\text{MVAr}$$

Second, compute the reactive power of STATCOM due to increase in voltage.

$$V_g = 577.35 \times 1.15 = 663.95\, v$$

$$Q_{ST} = \frac{V_g (V_g - V_{ST})}{X_t} = \frac{663.95\,(663.95 - 577.35)}{0.042} = 2394670.41$$
$$\simeq 2.39\,\text{MVAr}$$

From Example 2.7 it can be noted that when the voltage at PCC was decreased, the reactive power of STATCOM was negative. This means that STATCOM injects reactive power to regulate the voltage at PCC. When the voltage at PCC was increased, the reactive power of STATCOM was positive. This means that STATCOM absorbs reactive power to regulate the voltage at PCC.

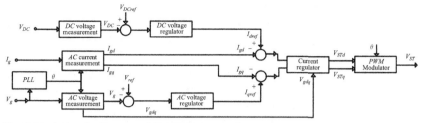

Fig. 2.30 A simple diagram of STATCOM control system.

2.11.2 The Control System of STATCOM

As we said before the d–q transformation plays an important role in the wind energy control system. It is illustrated in the control system of STATCOM.

As shown in Fig. 2.30 a phase-locked loop (PLL) is used to generate the angle θ that is a phase angle between voltage and current of the grid V_g and I_g, respectively. θ is used to produce the d–q components of V_g and I_g (V_{gdq} and I_{gdq}). Both DC voltage (V_{DC}) and V_g are compared with the reference of V_{DCref} and V_{ref}. These reference voltages are designed to keep the V_g at 1 pu.

The result of this compression is injected to AC and DC voltage regulators to produce the direct quadratic components of the reference current I_{dref} and I_{qref}. Both I_{dref} and I_{qref} are compared with direct quadratic components of the reference current I_{gd} and I_{gq}, respectively. The outputs of this compression are injected to a current regulator, which is assisted by V_{gdq} as a feed forward from the AC voltage regulator. The output power of a current regulator is injected to pulse width modulation (PWM) with θ to generate and synchronize the VST with V_g. It must be noted that all voltage and current regulators are based on a proportional–integral–derivative (PID) controller.

Example 2.9 A small wind farm is connected to a 120 kV grid via 25 kV, 30 km transmission line. The wind farm composed of 3×2 MW wind turbines based on SCIG to produce 6 MW and 575 V at base wind speed equal to 12 m/s. The wind farm is connected to 3 MVAr STATCOM. Use a fixed SCIG wind farm MATLAB Simulink model to obtain the following:

1. Active and reactive power and voltage at bus 575 V PCC.
2. Active and reactive power and pitch angle.
3. Reactive power of STATCOM.
    Consider that the wind speed variation is from 8 to 20 m/s.

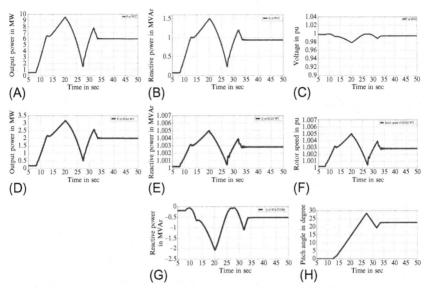

Fig. 2.31 Characteristics of Example 2.9. (A) Active power at PCC, (B) reactive power at PCC, (C) voltage at PCC, (D) active power of SCIG WT, (E) reactive power of SCIG W, (F) rotor speed of SCIG WT, (G) reactive power of STATCOM, and (H) pitch angle of SCIG WT.

The solution is shown in Fig. 2.31. From this example, the operation of these wind turbines based on SCIG can be understood by explaining the impact of wind speed variation on the example requirements. This can be explained as follows:

1. With the increase in wind speed from 8 m/s until the rated speed 12 m/s, active power, reactive power at PCC, active power, reactive power, and rotor speed are increased but they are still in the allowable limits so that there is no change in voltage at bus 575 V PCC, pitch angle, and reactive power of STATCOM.

2. The operation of a pitch angle control system and STATCOM starts with the increase of wind speed over rated speed 12 m/s.

3. With the increase of wind speed over rated speed 12 m/s, the pitch angle control system will increase the pitch angle to keep the rotor speed in its allowable limit.

4. Also, with the increase of wind speed over rated speed 12 m/s, SCIG will absorb a large amount of reactive power from the grid; this will cause a decrease in the voltage at PCC so that the STATCOM will inject reactive power to regulate the voltage and keep it at its allowable limits, as shown in Fig. 2.31.

2.12 DFIGs IN A WEGS

Wind turbines based on DFIG can be classified as variable speed wound rotor wind turbines. Because the rotor is wound rotor type, voltage and current of the rotor of DFIG not only must be taken into consideration but they play an important role in the control system and hence in the stability of the wind farm. Eqs. (2.35)–(2.41) are used in modeling DFIG. In wind turbines based on DFIG, the power is delivered to the interconnected grid from the stator and rotor. The DFIG is connected to the grid through a three-winding transformer. Two-windings of the coupling transformer connect the stator of DFIG to the grid. The third-winding is used to connect the rotor of DFIG to the grid through AC/DC/AC converters.

Fig. 2.32 shows the DFIG wind turbine interconnected grid, while Fig. 2.33 shows the equivalent circuit of DFIG based on d-q transformation.

This construction of wind turbines based on DFIG allows controlling the active power independent of reactive power, as shown in the next section.

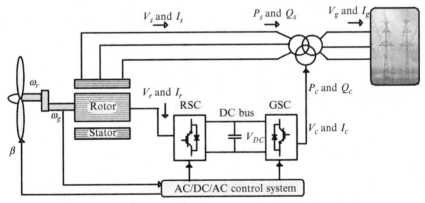

Fig. 2.32 DFIG wind turbine interconnected grid.

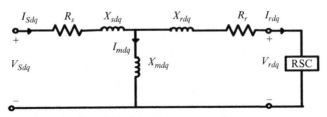

Fig. 2.33 Equivalent circuit of DFIG based on d-q transformation.

2.12.1 AC/DC/AC System

An AC/DC/AC system consists of two VSCs connected to each other through a DC bus. VSCs are forced by power electronic devices. One VSC is connected to the rotor side, usually known as the rotor side converter (RSC). The other one is connected to the grid side, usually known the grid side converter (GSC). The relation among the rotor, AC/DC/AC, and the grid depend on the rotor speed. There are three operating conditions of DFIG that can explain this relation. These three operating conditions can be explained as follows:

1. When the rotor speed is greater than the synchronous speed, the rotor power (P_r) is transferred from the rotor to the DC bus through RSC so that the voltage of the DC bus (V_{DC}) will be increased.
2. When the rotor speed is lower than synchronous speed, P_r is transferred from the DC bus to the rotor through RSC so that the V_{DC} will be decreased.
3. When the rotor speed is equal to the synchronous speed, there is no transfer.
4. The function of GSC is to control the voltage at PCC when the grid voltage (V_g) is lower than the reference value the GSC transmits the power from the DC bus to the grid and vice versa.

2.12.2 DFIG Control System

Fig. 2.34 shows the control system of DFIG wind turbines.

The control system of DFIG can be divided into two sections, namely, section rotor side control system and grid side control system.

2.12.2.1 Rotor Side Control System

The rotor side control system is responsible for controlling the output power and keeps it at its allowable limit of tracking characteristic. As shown in Fig. 2.34, this function can be established by controlling the direct and quadratic components of the rotor voltage (V_{dr} and V_{qr}). Controlling the quadratic components of the rotor current is responsible for generating the electromagnetic torque, T_e, according to Eq. (2.41). The process of controlling rotor voltage (V_r) can be explained as follows:

1. Compare the grid voltage (V_g) with reference voltage (V_{ref}) and inject the result to an AC voltage regulator. The output is the direct component of reference current of RSC (I_{dr-ref}). Also, I_{dr-ref} can be obtained from

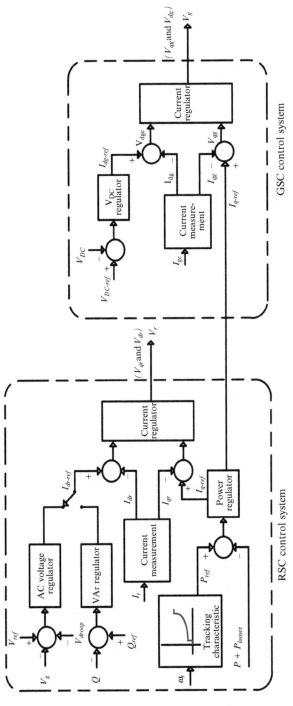

Fig. 2.34 Control system of DFIG wind turbines.

comparing the reactive power (Q) with reference reactive power (Q_{ref}) and injecting the result to a VAr voltage regulator.

2. The power tracking characteristic is used to obtain the reference output power (P_{ref}) according to the rotor speed (ω_r).
3. The total power (actual power + the power losses) is compared with P_{ref} and the result is injected to a power regulator. The output is the quadratic component of reference current (I_{q-ref}).
4. Both I_{dr-ref} and I_{qr-ref} are compared to the direct and quadratic components of rotor current (I_{dr} and I_{qr}), respectively. Then the output of these comparisons is injected to the current regulator to generate the controlled rotor voltage (V_r).

2.12.2.2 Grid Side Control System
The grid side control system is used to regulate the DC bus voltage and regulate the grid voltage at PCC. The process of controlling grid voltage (V_g) can be explained as follows:
1. Compareg the DC bus voltage (V_{DC}) with a reference voltage (V_{DC-ref}) and inject the result to a DC voltage regulator. The output is the direct component of reference current of GSC (I_{dg-ref}). Then I_{dg-ref} is compared with the direct component of grid current (I_{dg}).
2. I_{q-ref} produced by the power regulator is compared with quadratic component of grid current (I_{qg}).
3. Then the output of these comparisons is injected to the current regulator to generate the controlled V_g.

Example 2.10 A small wind farm is connected to a 120 kV grid via a 25 kV, 30 km transmission line. The wind farm composed of 3×2 MW wind turbines based on DFIG to produce 6 MW and 575 V at base wind speed equal to 12 m/s. Use variable the DFIG wind farm MATLAB Simulink model to obtain the following:
1. Active and reactive power and voltage at bus 575 V PCC.
2. Active and reactive power and pitch angle.
3. Reactive power of STATCOM.
Consider that the wind speed varies from 8 to 20 m/s.

The solution is shown in Fig. 2.35, From this example, the main advantage of wind turbines based on DFIG over the wind turbines based on SCIG is as follows:
1. There are no reactive power compensation devices used with wind farms based on DFIG as there is in wind farms based on SCIG.

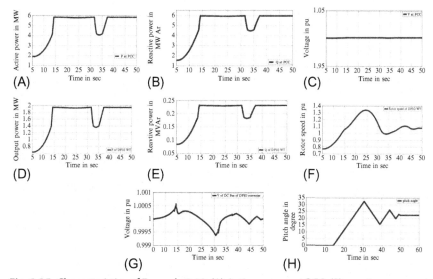

Fig. 2.35 Characteristics of Example 2.10. (A) Active power at PCC, (B) reactive power at PCC, (C) voltage at PCC, (D) active power of DFIG WT, (E) reactive power of DFIG WT, (F) rotor speed of SCIG WT, (G) voltage of DC bus of DFIG converter, and (H) pitch angle of SCIG WT.

2. With the increase of wind speed over rated speed 12 m/s, the output power of the DFIG wind turbine and the active power at PCC are fixed at its rated value. In the SCIG wind turbine of Example 2.7 power is increased, although they operate at the same condition.

3. Although the wind speed increased to 20 m/s, there is a change in voltage at PCC of wind farms based on DFIG as opposed to wind farms based on SCIG.

4. The reactive power absorbed by wind farms based on DFIG is very low compared with wind farms based on SCIG.

5. All of these differences are due to the operation of an AC/DC/AC converter system and the pitch angle control system. The impact of wind speed variation on the voltage of a DC bus is shown in Fig. 2.35G.

2.13 SYNCHRONOUS GENERATOR WIND TURBINE

The most application of synchronous generator wind turbine (SGWT) is standalone wind energy system and home applications. This is because it has a narrow range of power capacity. The types of SGWT are divided into two types according to the type of the excitation field permanent magnet excitation and external field excitation. In the external field excitation type, the rotor of SG is a wound rotor, which is connected to the DC supply through brushes so that this type is known as a WRSG. In permanent magnet excitation, the magnetic field is produced by a permanent magnet so that this type is known as PMSG. The rotor of both types could be nonsalient or salient pole.

SG can operate at low speed with multiple poles (high numbers of poles) so that there is no need to use the gearbox; this type is known as direct driven wind turbines. Also, SG can operate at high speed, but the wind turbine rotor is connected to the generator through a gear box; this type is known as a converter driven wind turbine. There are two types of PMSGs, namely, surface mounted PMSG and inset PMSG. The difference between these two types is the manner of the permanent magnet mounting on the rotor. In surface mounted PMSG, as its name indicates the permanent magnet is mounted of the surface of the rotor. The disadvantage of surface mounted PMSG is that the surface mount could be separated due to the centrifugal force so that this type is used in low-speed operation. As opposed to surface mounted PMSG, in inset PMSG the permanent magnet is mounted inside the rotor so that it can be used in high-speed operation.

2.13.1 *d-q* Transformation of a SG

d-q transformation of a SG is based on synchronous farm reference as the same way of IG. The main difference is in rotor current where there is no quadratic component of rotor current. This is because the excitation process depends on one of two things: permanent magnet for PMSG or DC external filed excitation for WRSG. Both of them are represented by a constant current source (I_f).

$$\left.\begin{array}{l} v_{ds} = -R_s i_{ds} + p\varphi_{ds} - \omega_r \varphi_{qs} \\ v_{qs} = -R_s i_{qs} + p\varphi_{qs} + \omega_r \varphi_{ds} \end{array}\right\} \tag{2.44}$$

$$\left.\begin{array}{l} \varphi_{ds} = L_{dm} I_f - (L_{ls} + L_{dm}) i_{ds} \\ \varphi_{qs} = -(L_{ls} + L_{qm}) i_{qs} \end{array}\right\} \tag{2.45}$$

Eqs. (2.44), (2.45) can be used to represent the equivalent circuit of SG as shown in Fig. 2.36. For simplicity, $L_d = L_{ls} + L_{dm}$ and $L_q = L_{ls} + L_{qm}$ so that Eq. (2.46) can be rewritten as follows [9]:

$$\left.\begin{array}{l} \varphi_{ds} = L_{dm} I_f - L_d i_{ds} \\ \varphi_{qs} = -L_q i_{qs} \end{array}\right\} \tag{2.46}$$

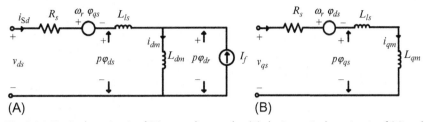

Fig. 2.36 Equivalent circuit of SG according to d-q. (A) d axis equivalent circuit of SG and (B) q axis equivalent circuit of SG.

Because $L_{dm}\,I_f$ is constant so that $p(L_{dm}\,I_f) = d/dt\,(L_{dm}\,I_f) = 0$, by substituting Eq. (2.46) in Eq. (2.44) we get

$$\left.\begin{aligned}
v_{ds} &= -R_s i_{ds} - L_d p i_{ds} + \omega_r L_q i_{qs} \\
v_{qs} &= \omega_r L_{dm} I_f - R_s i_{qs} - L_q p i_{qs} - \omega_r L_d i_{ds}
\end{aligned}\right\} \tag{2.47}$$

The electrical torque of SG can be obtained by Eq. (2.41) that is used with IG but the most used expression is the first one of Eq. (2.41).

$$T_e = \frac{3P}{2}\left(i_{qs}\varphi_{ds} - i_{ds}\varphi_{qs}\right) \tag{2.48}$$

The steady state equivalent circuit of SG can be obtained in the same way that is used with IG. Where $p = d/dt = 0$ and the relation between vectors and steady state quantities can be obtained by dividing the vector quantities by $\sqrt{2}$. So that Eq. (2.46) will transfer to the next equation.

$$\left.\begin{aligned}
V_{ds} &= -R_s I_{ds} + X_q I_{qs} \\
v_{qs} &= X_{dm} I_f - R_s I_{qs} - X_d I_{ds}
\end{aligned}\right\} \tag{2.49}$$

where $X_q = \omega_r L_q$, $X_{dm} = \omega_r L_{dm}$ and $X_d = \omega_r L_d$.

2.14 GRID CODE REQUIREMENTS

The increase in wind generation technology encourages many countries to increase their utilization of wind energy by expanding their number of wind farm installations. This increase in wind farms interconnected to the grid causes power system operators to make roles for connecting wind farms to the grid. These roles are known as grid code requirements (GCR). Low voltage ride through (LVRT) or fault ride through (FRT) are the most important elements in GCR that everyone working in the wind energy field

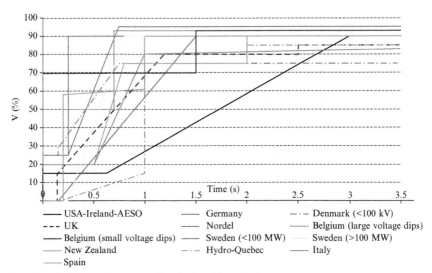

Fig. 2.37 LVRT requirements of various grid codes.

should have a good background about. LVRT requirements of various grid codes defers from one country to another according to the need of their power system operators, as shown in Fig. 2.37 [10].

All GRDs agree with basic concepts, which are:

1. The continuity of connection: The continuity of connection means that the ability of wind farms to remain in operation during a grid fault, disturbance. or voltage dips.
2. The contribution to voltage stability during LVRT: This means the ability of wind farms to inject or absorb a reactive power at PCC according to the grid disturbance condition to keep the stability of the system.

According to some recent research (such as [10–12]) and publications of a large energy company in Europe [13], the behavior of a wind farm during a grid disturbance must adhere to the voltage patterns shown in Fig. 2.38. Lines 1 and 2 as represented in Fig. 2.38 show two voltage patterns. In the area above limit line 1, the following requirement should be fulfilled:

− The voltage dips must not lead to voltage instability at PCC or disconnection of the wind farm from the grid.

In the area above limit line 2, the following requirements should be fulfilled:

a. The wind farm should remain connected to the grid.
b. For all wind farms that are still connected to the grid during the voltage dip, the active output power of a wind farm must be continued after the fault clearance.

Highest value of the three line-to-line grid voltage
U/U_N

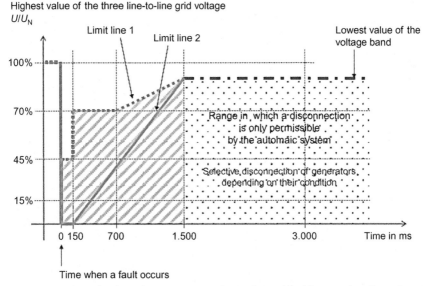

Fig. 2.38 The limit for the voltage pattern at the grid. *(Modified from Grid Code: High and extra high voltage. E. ON Netz GmbH. April, 2006.)*

c. The active output power of a wind farm is allowed to increase beyond the original value with a gradient of at least 20% of the rated power per second.

d. The wind farm must support the grid voltage by injecting reactive power during a voltage dip (providing a reactive current at CCP at least 2% of rated current for each percent of voltage dips.

After voltage dip clearance, the voltage support must be maintained for at least 500 ms. The ability of wind farms to fulfill GCR has become the focus of researcher's attention.

2.15 SUMMARY

A historical background, development of wind power, types of wind turbines, power in the wind, types and molding of generators used in WEGS, and maximum wind turbine efficiency were introduced in this chapter. A control system of a WEGS was derived based on *d-q* transformation. MATLAB Simulink models were used to illustrate the operation of wind turbines. At the end of this chapter, a brief view on GCR was presented.

Abbreviations

$\rightarrow$	space vector
A	swept area

Cp	power coefficient
DFIG	doubly fed induction generator
d-q	direct and quadratic components
FRT	fault ride through
GCR	grid code requirements
G_r	gear box ratio
GSC	grid side converter
GWEC	global wind energy council
IEA	International Energy Association
IEC	International Electrotechnical Commission
i_m	magnetization current
i_r	rotor current
i_s	stator current
J	moment of inertia
ke	kinetic energy
L	blade length
L_m	magnetizing inductance
L_r	self-inductance rotor
L_s	self-inductance of stator coils
LVRT	low voltage ride through
m	air mass
MPPT	maximum power point tracking of wind energy
P	number of pair pole
p	d/dt
P_{ag}	gap power
PCC	pion of common connection
P_m	mechanical power
PMSG	permanent magnet synchronous generator
P_{rcul}	copper losses of generator rotor
P_{scul}	copper losses in stator
r	rotor radius
R_r	winding resistant of rotor
R_s	winding resistant of stator
RSC	rotor side converter
S	slip speed of induction generator
SCIG	squirrel cage induction generator
SGWT	synchronous generator wind turbine
β	the angle between the blade axis and the wind direction is called pitch angle
STATCOM	static synchronous compensator
T_e	electrical torque
T_m	mechanical torque
TSR	tip speed ratio
v	wind speed
v_0	average wind speed at h_0, where v_0 is usually measured at $h0 = 10$ m
v_h	wind speed at the height h
V_{ol}	volume of the mass of air
WEGS	wind energy generation system

WRIG	wound rotor induction generator
WRSG	wound rotor synchronous generator
x	distance crossed by the air from the first point of the hub to the point where the air leaves the turbine blade
X_m	magnetization leakage reactance
X_r	rotor leakage reactance
X_s	stator leakage reactance
α	friction coefficient or Hellman exponent
η	wind turbine efficiency
ρ	air density
φ	flux linkage
φ_r	rotor flux linkage
φ_s	stator flux linkage
ω_g	generator speed
ω_r	rotor speed of induction generator
ω_T	speed of turbine rotor

REFERENCES

[1] T. Ackermann, Wind Power in Power System, John Wiley & Sons Publishers Ltd., The Atrium, Southern Gate, Chichester, UK, 2005.
[2] Available from: http://www.iea.org. Accessed April 2016.
[3] G.O. Suvire, Wind Farm—Technical Regulations, Potential Estimation and Siting Assessment, 2011.
[4] Available from: http://www.nordex-online.com/. Accessed April 2016.
[5] Available from: http://www.gamesacorp.com. Accessed April 2016.
[6] Available from: https://www.vestas.com/. Accessed April 2016.
[7] T. Burton, N. Jenkins, D. Sharpe, E. Bossanyi, Wind Energy Handbook, second ed., John Wiley & Sons Publishers Ltd., Chichester, UK, 2011.
[8] IEC 61400-2, part 2, Second edition, Geneva, Switzerland, 2006.
[9] B. Wu, Y. Lang, N. Zargari, S. Kouro, Power conversion and control of wind energy systems, Wiley-IEEE Press, Hoboken, New Jersey, 2011. August.
[10] F. Iov, A.D. Hansen, P. Sørensen, N.A. Cutululis, Mapping of grid faults and grid codes, Tech. Rep. Risø-R-1617(EN), Risø Nat. Lab, Tech. Univ. Denmark, Roskilde, Denmark, July 2007.
[11] Grid Code: High and extra high voltage. E. ON Netz GmbH. April, 2006. www.eon-netz.com. Germany.
[12] J. Hossain, A. Mahmud, Renewable energy integration challenges and solutions, Green Energy and Technology, Springer, 2014. http://www.springer.com/series/8059.
[13] E.M.G. Rodrigues, G.J. Osório, R. Godina, A.W. Bizuayehu, J.M. Lujano-Rojas, J.P. S. Catalão, Grid code reinforcements for deeper renewable generation in insular energy systems, Renew. Sustainable Energy Rev. 53 (2016) 163–177.

FURTHER READING

[1] F. Blaabjerg, Z. Chen, Power Electronics for Modern Wind Turbines, Morgan & Claypool, 2006.

CHAPTER 3

Solar Energy and Photovoltaic Technology

Kakkan Ramalingam*, Chandrasen Indulkar[†]
*Airports Authority of India, Delhi, India
[†]Indian Institute of Technology, Delhi, India

3.1 INTRODUCTION

Environmental concerns and climate change have put pressure on utility power system managers to look for alternative sources of energy. Recent research advances and developments in exploiting renewable energy sources for improving power system operations have seen encouraging results. Distributed generation (DG) is a method of generating electricity from multiple renewable energy sources that are very near to load demands. DGs interconnected to utility power systems have multiple advantages such as increased system reliability, reduced peak power requirement, improved power quality, requisite supply of reactive power, and environmentally clean energy. The renewable energy resources used for generation of electricity are solar, thermal, photovoltaic (PV), wind farms, hydro, biofuels, wave, tidal, ocean, and geothermal sources. However, PV systems have been considered a better renewable energy source for electricity generation, because of the abundant long-time availability of free solar energy at the earth's crust. PV generation is based on the PV effect, which is a process with PV cells that uses solar light photons to strike on the doped semiconductor silicon to produce electricity.

The PV effect, discovered in 1839, was developed to produce power using doped semiconductors in 1954 [1]. PV power has been the fastest growing renewable energy technology that grew from 50 MW in 1990 to 177GW (IEA) in 2014 [2]. PV research at NREL (United States) states that solar energy systems will be cost competitive with other energy sources by 2020 [2]. Recent research on PV systems at various laboratories in the world are actively focusing research on

- Material sciences for doping semiconductor materials for higher efficiency.
- Inverter technology for efficient conversion of DC to AC.

Distributed Generation Systems
http://dx.doi.org/10.1016/B978-0-12-804208-3.00003-0

- Integration technologies on potential issues with interconnection to utility grid systems.
- Battery technology for energy storage systems for large-scale PV systems.
- Promoting policies on PV systems by governments.

These factors are encouraging the development of high-potential PV markets. PVs have proved a better choice in terms of research advances in cell technology, modular characters, standalone and grid connected opportunities, reliability, ease of use, lack of noise and emissions, long life cycle, and reduced cost per unit of energy generated. Grid connected PV systems have their own challenges and opportunities in the development of modern utility power systems and smart grids. The future trends in grid connected PV systems promise a wide and large-scale business market as forecasted by research institutions. A lot of research is going on to bring down the cost of PV energy generation by way of research in materials science that improves the cell efficiency on the conversion of solar energy to electrical energy, reduced costing of efficient energy storage devices such as batteries and associated energy conversion devices of inverters and PV components with standards for controllers to convert solar DC power to AC power, and integrating methods of solar energy systems to utility power systems.

This chapter, considering the above factors, presents a study of solar radiation in Section 3.2, a brief discussion on semiconductor physics in Section 3.3, a review of PV materials in Section 3.4, the electrical characteristics of PV cells in Section 3.5, a note on PV components and standards in Section 3.6, a report on PV systems and technologies in Section 3.7, materials for future PV systems in Section 3.8, and a summary in Section 3.9.

The contents of this chapter on recent advances in PV technologies will be very useful to graduate students, practicing engineers, researchers, planners, manufacturers, energy managers, policy makers, and developers of future power systems planning.

3.2 SOLAR RADIATION [1,3–5]

PV cells made out of semiconductor materials convert light energy to electrical energy by a process called the photoelectric effect. A detailed review for understanding the Solar System and solar radiation concepts is presented in this section.

The Solar System consists of an average star: the Sun, and the planets: Mercury, Venus, Earth, Mars, Jupiter, Saturn, Uranus, Neptune, and Pluto. It includes the satellites of the planets, numerous comets, asteroids, meteoroids, and the interplanetary medium. The Sun is the richest source of electromagnetic energy, mostly in the form of heat and light in the Solar System. The Sun contains 99.85% of all the matter in the Solar System. The planets, which condensed out of the same disk of material that formed the Sun, contain only 0.135% of the mass of the Solar System. Jupiter contains more than twice the matter of all the other planets combined. Satellites of the planets, comets, asteroids, meteoroids, and the interplanetary medium constitute the remaining 0.015%.

The Sun is the center of gravity of the Solar System. Thermonuclear fusion between hydrogen atoms and helium atoms takes place at the core of the Sun unceasingly at millions of degrees of temperature. The loss of matter during the fusion releases a huge amount of energy in the form of electromagnetic radiation. A part of the energy reaches the crust of the earth's atmosphere with an average irradiance of 1367 W/m^2, which varies as a function of the Earth-to-Sun distance. The solar irradiance, insolation (incident of solar radiation) is the intensity of the solar electromagnetic radiation incident on a surface of 1 square meter (kW/m^2). The intensity of the solar radiation reduces before reaching Earth, as it passes through the atmosphere by absorption, diffusion, and reflection of particles in the air and atmosphere.

Solar radiation is the integral of the solar irradiance over a period of time ($kW\ h/m^2$). The energy reaches from the core to the surface of the Sun and is released into space primarily as light. The Sun's outermost and relatively thin 400 km layer is called the photosphere and has a temperature of ~5770 °K. This is the layer that emits the spectrum of radiation, which is visible to the human eye and is termed as "light." The solar radiation from the surface of the Sun is transmitted to the earth as electromagnetic waves of light energy at the speed of light. The speed of light is about 11 million miles/min, and light takes 8.45 min to travel from the Sun to Earth. The energy emitted by the Sun is 3.72×10^{20} MW, which equates to a radioactive power of 63 MW/m^2 of its surface. At the mean distance of 150 million kilometers between Earth and the Sun, this radiation reaches outside the earth's atmosphere with an intensity of 1.367 kW falling onto a 1 m^2 surface oriented normally to the Sun's beams. This is called the solar constant. Figs. 3.1 and 3.2 show the Solar System and Earth parameters with reference to the Sun, respectively.

The radioactive energy from the Sun's outer surface travels as electromagnetic waves and reaches the earth through the atmosphere. The radiation energy varies as a function of temperature; the higher the temperature the higher the energy. A solar luminosity in "L" is the constant energy flux

Fig. 3.1 The Solar system. *(Taken from https://commons.wikimedia.org/wiki/File:Ceres_solarsystem.jpg.)*

Fig. 3.2 Sun and Earth relationship (http://www.itacanet.org/the-sun-as-a-source-of-energy/part-2).

from the Sun, and is "$L = 3.9 \times 10^{26}$" in watts. The solar flux density in "S_d" is the amount of solar energy per unit area on a sphere centered at the Sun with a distance in "d." Thus,

$$S_d = L / \left(4\pi d^2\right) \text{W/m}^2 \tag{3.1}$$

The solar constant is the solar flux density reaching the earth. The mean distance of Earth from the Sun is 1.50×10^{11} m. Therefore, the solar constant in "S" is the energy flux density at the earth's surface, and is obtained from Eq. (3.1).

$$S = \left(3.9 \times 10^{26}\right) / \left[4\pi \times \left(1.50 \times 10^{11}\right)^2\right]$$
$$= 1367 \text{W/m}^2$$

The solar energy incident on the earth is therefore determined by Eq. (3.2).

$$S_E = S \times \pi R_E^2 \qquad (3.2)$$

where R_E is the radius of the earth disk, in which the solar energy flux is incident.

The solar energy incident on the earth is partially absorbed and partially reflected back by albedo to space. The albedo is the percentage of sunlight reflected back to space by the planet. The albedo ratio for the planet Earth is 30%. The solar energy absorbed by Earth is determined by Eq. (3.3).

$$Sa = S_E(1 - A) \qquad (3.3)$$

where "A" is the planetary albedo of the earth, which is about 0.30.

The radiation emitted by Earth is called terrestrial radiation, which is assumed to be like a black body radiation. A black body is something that emits or absorbs electromagnetic radiation with 100% efficiency at all wavelengths. The amount of radiation energy flux emitted by a black body depends on the absolute temperature of the black body. The energy emitted by the black body is related to the fourth power of the body's absolute temperature as per Stephan–Boltzmann's law in Eq. (3.4).

Stephan–Boltzmann law states that

$$F = \sigma T^4 \qquad (3.4)$$

where σ is Stephan–Boltzmann's constant and equals to 5.67×10^{-8} in "W/ m^2/K," and T is the temperature in Kelvin.

The energy emitted from the earth is given by Eq. (3.5).

$$\begin{aligned} F_E &= (\text{Black body emission}) \times (\text{Total area of the earth}) \\ &= (\sigma T^4 e) \times (4\pi R_E^2) \end{aligned} \qquad (3.5)$$

As planetary energy balances, the energy emitted by Earth is equal to the energy absorbed by Earth.

$$\begin{aligned} (\sigma T^4 e) \times (4\pi R_E^2) &= S \times \pi R_E^2 (1 - A) \\ (\sigma T^4 e) &= S(1 - A)/4 \end{aligned} \qquad (3.6)$$

The equivalent black body temperature is related by Stephan–Boltzmann's law. The solar energy at the earth's surface is related to the distance of the Sun from the earth and the radius of the earth as shown in Eq. (3.7).

$$F = S(d/R_s)^2 \qquad (3.7)$$

Example 3.1

The distance from Sun to Earth is $d = 1.50 \times 10^{11}$ km, and the solar radius is $R_s = 6.98 \times 10^8$ km. Calculate the solar energy at the outer surface of the Sun and the temperature. Solar constant $S = 1367$ W/m^2.

The solar energy of the Sun is given in Eq. (3.7).

$$F = 1367 \times [(1.50 \times 10^{11})/(6.98 \times 10^8)]^2$$
$$= 63.13 \, \text{MW/m}^2$$

As per Stephan-Boltzmann's law in Eq. (3.4), the temperature at the surface of the Sun is

$$T = (F/\sigma)^{1/4}$$
$$= (63.13 \times 10^6/5.67 \times 10^{-8})^{1/4}$$
$$= 5776.48°\text{K}$$

Example 3.2

The equatorial radius of the earth is $R_E = 6378.173$ km. Calculate the solar energy incident on earth, the energy absorbed by earth and the energy reflected by albedo from the earth. Calculate the temperature at the earth's surface.

The solar energy incident on the earth is given by Eq. (3.2).

$$S_E = S \times \pi R_E^2$$
$$= 1367 \times 3.14 \times (6,378,173)^2$$
$$= 1.75 \times 10^{11} \, \text{MW}$$

The energy absorbed by earth is given by Eq. (3.3)

$$Sa = S(1 - 0.30) \times \pi R_E^2$$
$$= 1367 \times 0.70 \times 3.14 \times (6,378,173)^2$$
$$= 3.06 \times 10^{10} \, \text{MW}$$

$$\text{Energy reflected by albedo} = 1367 \times 0.30/4$$
$$= 102.53 \, \text{W/m}^2$$

From Eq. (3.6), the earth's surface temperature is as follows:

$$\sigma T^4 e = S(1 - A)/4$$
$$= 1367 \times 0.70/4 \quad \text{since } (A = 0.30)$$
$$= 239.23\,\text{W}/\text{m}^2$$
$$Te = \left(239.23 \times 10^8/5.67\right)^{1/4}$$
$$= 254.86°\text{K}$$

Earth's atmosphere has a considerable influence on the intensity of solar radiation reaching the ground. Its height is ~70–80 km, and it mainly consists of nitrogen (~78%) and oxygen (~21%). Other gasses and water vapor together make up only 1%, but can have a large effect on climate and the environment as "greenhouse gasses" such as methane and carbon dioxide.

Fig. 3.3 shows the layers of the atmosphere, which influence the reach of solar radiation to the earth surface. The bottom layer is called the troposphere. It is the cloudiest layer and reaches up to the tropopause. The tropopause is the boundary between the troposphere and the stratosphere. It is located between the heights of 11 and 16 km depending on the location above the earth and the atmospheric conditions. The stratosphere is nearly

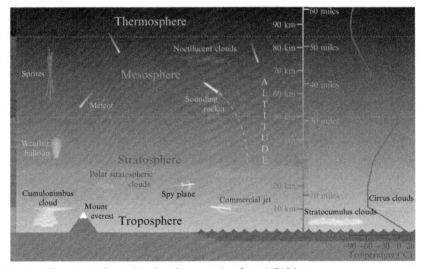

Fig. 3.3 The atmosphere. *(Used with permission from UCAR.)*

cloudless with very low air humidity. The ozone layer is located in the stratosphere at altitudes of 15–85 km. The mesosphere is above the stratosphere, beginning at a height of ∼50 km above the ground surface. The ionosphere, also known as the thermosphere is above the mesosphere, and reaches up to the height of around 640 km. The exosphere above the ionosphere is the outer boundary and reaches up to 9600 km. Beyond this is the interplanetary space.

Earth's atmosphere is situated at 5–6 km above the ground surface. The extraterrestrial solar radiation is reduced by scattering and absorption caused by air molecules, aerosol particles, water droplets, and ice crystals in clouds. When passing through the atmosphere, scattering of solar radiation takes place within the whole spectral range. The scattering occurs in the following ways:

- Scattering by water droplets and/or ice crystals in clouds relatively evenly across the whole spectral range.
- Scattering by molecules (Rayleigh scattering), predominantly of radiation at shorter wavelengths.
- Scattering by aerosol particles (Mie scattering) at wavelengths dependent upon the particle size and distribution.

Gaseous molecules and aerosols cause a relatively high absorption of solar radiation. Scattering and absorption in the atmosphere affect the spectral balance of the solar radiation reaching the ground.

Fig. 3.4 shows the solar spectrum. The solar spectrum is the distribution of solar radiation as a function of the wavelength. It consists of a continuous emission from the Sun and the Sun's total radiation output is approximately equivalent to that of a black body at 5776°K. The solar radiation fits closely with the black body emission in the visible and infrared region at this temperature, but deviates in the UV region. The solar constant is the amount of solar radiation received outside the earth's atmosphere on a surface normal to the incident radiation per unit time and per unit area at Earth's mean distance from the Sun. The solar constant as measured is 1367 W/m^2 with ±3 W/m^2. The radiant energy from the Sun lies ∼47% in the infrared region (>0.70 μm), 43% in the visible region (0.40–0.70 μm), and about 10% in the UV region (<0.40 μm).

The solar insolation (incident solar radiation) is the actual amount of solar radiation incident upon a unit horizontal surface over a specified period of time for a given locality. It depends on the solar constant, ratio of the actual distance from the Sun to the mean distance of Earth from the Sun, and solar zenith angle of the location. The zenith angle depends on the latitude, the

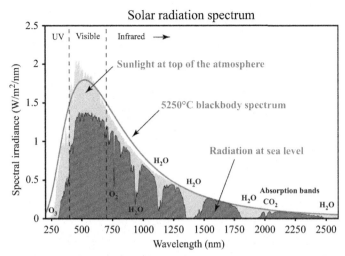

Fig. 3.4 Solar spectrum distribution [1]. *(Taken from https://commons.wikimedia.org/wiki/File:Solar_Spectrum.png.)*

angle of solar declination, day of the year, and time of the day. These factors vary at each location due to the eccentricity of Earth's elliptical orbit around the Sun.

The solar radiation is absorbed and scattered by atmospheric gasses, aerosols, clouds, and the earth's surface. Fig. 3.5 depicts the pattern of scattering of the radiation in the atmosphere, before reaching the earth's surface. The fraction of the incident solar radiation is about 0.31 that is reflected back and scattered to space is called the albedo. The scattering of solar radiation by air molecules is inversely proportional to the fourth power of the wavelength as described by the Rayleigh theory of scattering. The sky is blue because atmospheric molecules scatters solar radiation much more in the blue than the red part of the spectrum. The sky is visible in the scattering process. The blue light is removed by scattering during the long path through the atmosphere at sunset or sunrise, leaving the reddish colors of the spectrum and thus the sky is reddish during sunset or sunrise. The atmospheric gasses absorb solar UV radiation. The depletion of solar radiation is dominated by ozone absorption in the UVs, Rayleigh scattering both in the UV and visible region, and water vapor absorption in the near infrared region. Aerosols are a suspension of liquid particles such as sulfates, black carbon, organic carbon, dust, and sea salt in the atmosphere. Aerosols scatter and absorb solar radiation and cause climate changes. Clouds reflect incoming solar radiation back to

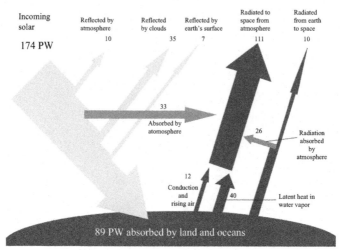

Fig. 3.5 Solar scattering in the atmosphere [1]. *(Taken from https://en.wikipedia.org/wiki/ Solar_energy#/media/File:Breakdown_of_the_incoming_solar_energy.svg.)*

space and thus cool the earth's atmosphere system, the so-called cloud albedo effect. The cooling due to cloud albedo effect occurs primarily at the surface.

The direct solar radiation is received directly from the Sun by the earth. The diffused solar radiation is the scattered radiation coming from all other directions. The global solar radiation is the sum of both the components received on a horizontal surface. Variations of air pressure and temperature within the atmosphere influence absorption, and therefore affect the spectrum at sea level and at different heights above sea level.

The amount and type of radiation falling on the ground surface depend on the changing characteristics of the atmosphere. Other important factors are the size of the planet and its location within space, but the crucial factor for the amount of radiation being absorbed by the earth, or reflected from it, is the composition of the ground surface.

The Sun provides over 99.98% of all energy to the earth's surface, the rest is internal geothermal energy. This results in an average surface temperature of 14°C, although extreme variations may occur locally and temporally. The angle of incidence of the solar radiation is changing continually as the earth is circling around the Sun and also spinning around its own axis. The ratio of radiation intensity and angle of incidence may be described by Lambert's law as a cosine function. The 23.5-degree inclination of the earth's axis also has an influence. The all-important factor is the change of the angle of incidence during different times of the day.

Earth is almost spherical in shape, and gravitational force binds the atmosphere like a shell. The intensity of the solar radiation at a point on the surface is therefore influenced by the curvature of the surface and the effective thickness of the atmosphere.

The solar radiation reaches its highest intensity when the Sun is at its zenith and the angle of incidence is 90 degrees and the thickness of the atmosphere is at its minimum. When the Sun's position in the sky is close to the atmosphere more radiation must pass through, and so more radiation is scattered and absorbed by the atmosphere, and thus less radiation reaches the ground surface.

The solar radiation energy to Earth is transmitted by electromagnetic waves over a wide range of wavelengths at varying intensities. The electromagnetic solar radiation impinging on the upper edge of the atmosphere is called extraterrestrial radiation. The mean integral for the complete spectrum is 1367 W/m^2, which is called the solar constant.

The complete spectrum comprises the ultraviolet (UV), visible (Vis), and infrared (IR) wavelengths. However, these wavelength ranges need to be subdivided depending on the individual application fields. Best known is the prismatic colors of visible light, the colors of the rainbow. IR is split into near infrared (NIR) and far infrared (FIR). UV is normally subdivided into UV-A, UV-B, and UV-C radiation. Approximately 6% of the total solar radiation falling on the earth is UV. Shorter wavelengths at higher frequency have higher energy, thus increasing the effect on biological and chemical systems. The normal measurement of the wavelength of solar and atmospheric radiation is the nanometer (nm, 10^{-9} m) and for infrared radiation is the micrometer (μm, 10^{-6} m). Table 3.1 shows the spectrum range.

Table 3.1 Wavelength of solar radiation [3] (http://www.kippzonen.com/Knowledge-Center/Theoretical-info/Solar-Radiation)

Short waves		
UV-C	100–280 nm	Emitted from Sun, totally absorbed by atmosphere
UV–B	280–315 nm	90% absorbed by atmosphere
UV–A	315–400 nm	Mostly reaches the ground
Visible	400–780 nm	Visible light. Colors of rainbow
Long waves		
NIR	780 nm–3 μm	Heat radiation from Sun
FIR	3 μm–50 μm	Heat radiation from atmosphere, clouds, and earth

The meteorologically significant spectral range extends from 300 to 3000 nm (short-wave radiation). Approximately 96% of the complete extraterrestrial radiation is situated within this spectral range. The maximum radiation intensity of the solar spectrum occurs at 500 nm, toward the blue end of the visible range. A major effect of solar radiation reaching the earth's surface is vital for the existence of life on Earth. About 30% of the extraterrestrial radiation solar radiation is reflected back into space, but ~51% is absorbed by land and water, and another 19% is absorbed by the clouds and atmosphere. Long-wave radiation is FIR, and is mostly the transformed short-wave energy that is re-radiated from the land, water, clouds, and atmosphere. Only a small amount of the total energy remains on the earth, but this is enough to maintain all the biological processes on the earth and to drive the weather systems. Radiation from the Sun sustains life on Earth and determines the climate. The energy flow within the sun results in a surface temperature of around 5800°K, similar to that of a 5776°K black body due to absorption in the cool peripheral solar gas.

The black body radiation is a function of wavelength of the solar spectrum. The radiation curve shifts its peak toward short wavelength, as the temperature increases. Wein's displacement law states that the product of the temperature and the wave length at which the radiation curve peaks is a constant.

$$\lambda_{max} = b/T \tag{3.8}$$

where, b is a displacement constant equal to 2898, and T is in °K, and λ is in (μm) micrometer.

Example 3.3

The earth as a black body with average surface temperature of 15°C radiates energy from an average area equal to 5.1×10^{14} m². Find the rate at which energy is radiated by the earth and the wavelength at which the maximum power is radiated. Compare this with the wavelength of the power radiated by the Sun as a black body at a temperature of 5776°K.

The energy emitted by Earth by radiation is given by Eq. (3.5).

$$\text{Energy emitted by earth} = \sigma T^4 \times \text{Area of the surface}$$
$$= 5.67 \times 10^{-8} \times (15 + 273)^4 \times 5.1 \times 10^{14}$$
$$= 2.0 \times 10^{11}\,\text{MW}$$

The wavelength at which maximum power is emitted is given by Wein's law in Eq. (3.8).

$$\lambda_{max}(\text{earth}) = 2898/(273 + 15)°\text{K} = 10.1\,\text{μm}$$

Similarly,

$$\lambda_{max}(sun) = 2898/5776°K = 0.50\,\mu m$$

The wavelength at which maximum power radiates shifts towards short-wave range in the solar spectrum as the temperature of the body increases.

Absorption and scattering levels change as the constituents of the atmosphere change. Clouds block most of the direct radiation. Seasonal variations and trends in ozone layer thickness have an important effect on the terrestrial UV level. The ground level spectrum also depends on the depth of travel through the atmosphere. The spectrum of the radiation changes through each day because of the changing absorption and scattering path length. With the Sun overhead, direct radiation that reaches the ground passes straight through the entire atmosphere, all of the air mass, overhead. This radiation is called "Air Mass 1 Direct" (AM 1D) radiation at a sea level reference site. The global radiation with the Sun overhead is similarly called "Air Mass 1 Global" (AM 1G) radiation. Because it passes through no air mass, the extraterrestrial spectrum is called the "Air Mass 0" spectrum. The entry into the atmosphere is called AM-0.

Solar power is measured as the intensity at a point of time and the energy delivered over a period of time. Irradiance "in (W/m^2)" is the intensity of solar radiation contributed by all wavelengths within the spectrum, and is expressed in watts per square meter of a surface. Power in watts "W" is the momentary total irradiance incident on a particular area. Energy per unit area, "in $(kW\,h/m^2)$" is the energy unit per square meter area as a measure of irradiance incident on a surface over a period of time.

3.3 MATERIALS AND SEMICONDUCTORS [1,4–9]

Solar cells are made out of semiconductor materials that have special material characteristics for converting the solar light energy into electrical energy. Certain types of materials have special properties with their atomic and crystalline structure, and electron configurations that exhibit photoelectric effects to absorb light energy and convert it into electricity. The knowledge on materials and its properties is presented in this section to help understand the principles of manufacturing PV cells and designing of the PV system technology with identification of PV materials.

3.3.1 Characteristics of Materials [8]

Materials are either naturally available such as stone, wood, clay, and skin or materials such as glass, plastics, and fibers that are processed or produced with superior properties or performance characteristics than the natural ones. There are strong relationships between the structural elements of the materials and their properties. The properties of the materials can be altered by heat treatment and addition of other substances. Different new materials such as composites and advanced materials are developed with specialized characteristics by adopting developed technologies for processing to meet the needs of a modern and complex society.

The structure of a material usually relates to the internal arrangement of components. The subatomic structure involves electrons within individual atoms and interactions with their nuclei. The structure of material encompasses the organization of atoms or molecules relative to one another. The property of a material is a type of response when subjected to external stimuli such as deformation of the material when a load is applied to it or a polished metal surface that reflects light when exposed to light. A property is the characteristics of the material independent of its shape and size in terms of the kind and magnitude of the response to specifically imposed stimuli. The important properties of solid materials are grouped into six different categories. They are mechanical properties of deformation on stimuli of imposing a force, electrical properties of electric conductivity and dielectric constant for stimuli of an electric field, thermal behavior in terms of heat capacity and thermal conductivity, magnetic properties of stimuli by magnetic field, optical property on stimuli by electromagnetic or light radiation, and deteriorative properties on stimuli by chemical reactivity of materials.

In addition to the structure and properties of the material, there are two more important components of processing and performance of materials. The structure of the materials depends on the processing and the performance is the function of properties. The relationships among the above four components do matter in the design, production, and utilization of materials.

Solid materials are classified primarily based on their chemical makeup and atomic structure, into three groups: metals, ceramics, and polymers. In addition, there are the composites, which are a combination of two or more of the above three basic classes. Most of the materials with some intermediaries fall in any one of the above classes. Another classification is advanced materials that are used in high-technology applications. Metals are composed of one or more metallic elements such as copper, iron, or gold and some nonmetallic elements such as carbon, nitrogen, or oxygen in relatively small amounts.

Atoms in metals and alloys are arranged in a very orderly manner and are relatively denser and stronger than the other two groups. Ceramics are compounds between metallic and nonmetallic elements such as oxides, nitrides, and carbides. Ceramic materials are alumina, silica and silicon nitride and traditional ceramics, such as clay minerals, porcelain, cement, and glass. Polymers are the organic compounds chemically based on carbon, hydrogen, and other nonmetallic elements—oxygen, nitrogen, and silicon—such as plastics, polyvinyl chloride (PVC), and rubber materials. Composites are composed of two or more individual materials that come from, namely, metals, ceramics, and polymers to achieve a combination of superior properties and with the best characteristics of the group, which cannot be otherwise achievable by a single material. Wood and bone are the naturally occurring composites and the man-made composites are fiberglass and carbon fiber reinforced polymer (CFRP). These materials are used in aircraft and aerospace applications as well as in high-tech sporting equipment.

It is important to note that the properties and characteristics of the four groups vary widely in terms of the following parameters: density, stiffness, strength (toughness), and electrical conductivity.

Advanced materials are used in high-technology applications such as electronic equipment, computers, aircraft, spacecraft, fiber-optic systems, and military rocketry. These are traditional materials that are newly developed, expensive high-performance materials with enhanced properties. Advanced materials include semiconductors, biomaterials, smart materials, and nanoengineered materials. Semiconductors have electrical properties intermediate between good metal electrical conductors and insulators. The use of semiconductors in electronics and the computer industry have revolutionized the latest emerging solar cell technology for power generation. Biomaterials are made out of a combination of materials from the four groups. They are used for implanting into the human body for replacement of diseased or damaged body parts. Smart materials are a group of new and state-of-the art materials of the future that are developed to significantly influence many technologies. These materials sense the changes in their environments and respond to the changes in a predetermined manner as a trait that is found in living organisms. Smart systems with piezoelectric sensors and computers used in helicopters reduce the aerodynamic cockpit noise that is created by the rotating rotor blades. The advent of the scan probe microscope has demonstrated the ability to observe, manipulate, and move the individual atoms or molecules to form new structures and design new materials. The ability to carefully arrange atoms provide opportunities to develop materials with mechanical, electrical, optical, and other properties that are not otherwise possible. The study of the properties of such materials is called nanotechnology. Technological advances will use these nanoengineered materials increasingly in the future.

The technological challenges in materials development still remain demanding for sophisticated and specialized materials in consideration of the environmental impact of the production of materials. Some challenges are the development of materials for disposal of radioactive wastes in nuclear plants, materials to work at high temperatures to increase efficiency and reduce weight of transport vehicles, materials that are highly efficient and less costly in conversion of light energy into electrical energy by solar cells, new materials that are highly efficient and less costly for hydrogen fuel cells and batteries for electric vehicles, and new manufacturing processes and methods to produce nonpolluting materials and developing environmentally friendly uses of materials.

3.3.1.1 Atomic Structure

Each atom of an element consists of a nucleus composed of protons and neutrons, which is encircled by moving electrons. Both electrons and protons are electrically charged. Neutrons are electrically neutral. The magnitude of charge of the positively charged protons and negatively charged electrons is 1.60×10^{-19} C. Protons and neutrons have approximately the same mass of 1.67×10^{-27} kg, where the mass of an electron is 9.11×10^{-31} kg. The atomic number of an atom is equal to the number of protons in the nucleus and is equal to the number of electrons. The atomic mass of a specific atom is the sum of the masses of the neutrons and protons in the nucleus. The number of protons is the same in all atoms of the element, but the number of neutrons may be variable. Thus, atoms of some elements have two or more different atomic masses, which are called isotopes. The atomic weight of an element is the weighted average of the atomic masses of naturally occurring isotopes. The atomic weight of an element or the molecular weight of a compound is measured in the atomic mass unit (AMU); that is, AMU per atom or mass per mole of material. One mole of a substance consists 6.023×10^{23} atoms or molecules. This is Avogadro's number.

Electrons revolve around the atomic nucleus in discrete orbital, and the position of any particular electron is defined by its orbitals. Electrons are permitted to have only specific values of energy. An electron can change energy by quantum jumping to allowable higher energy level by absorption of energy or to a lower energy level by the emission of energy. These allowed electron energy levels are called energy levels or states. These states are

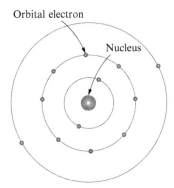

Fig. 3.6 Shells and electron configuration of an atom. *(Reproduced with permission from W.D. Callister Jr., Materials Science and Engineering: An Introduction, John Wiley & Sons Inc., 2007.)*

separated by finite energies. Every electron in an atom is characterized by quantum numbers. Shells are quantized by principal quantum number n, an integer starting with 1, 2, 3, 4 and so on, also designated as K, L, M, N, O, and so forth, which corresponds to 1, 2, 3, 4, and so on. This quantum number relates to the distance of the shell from the nucleus. The second quantum number, l, signifies the subshells, denoted by s, p, d, f, g, h. Each subshell has specified states such as $s = 1, p = 3, d = 5, f = 7, g = 9$, and $h = 11$. Each state has two electrons. The states within each subshell are identical but split to slightly different energy levels when a magnetic field is applied. Fig. 3.6 shows the shells and electron configuration of an atom. Table 3.2 shows the details of shells of an atom.

3.3.1.2 Electron Configuration

Each electron state can hold only two electrons with opposite spins. The electron configuration or structure of an atom represents the manner in which the states are occupied. The valence electrons are those that occupy the outermost shell. These valence electrons are extremely important to participate in the bonding between the atoms to form atomic or molecular aggregates. The physical and chemical properties of the solid materials are based on these valence electrons. The states in the valence electron shell are completely filled for atoms such as inert gasses with stable electron configurations. Some atoms of elements with unfilled valence shells attain stable electron configuration by gaining or losing electrons to form charged ions or by sharing electrons with other atoms. This is the basis for chemical reactions, and also for atomic bonding in solids.

Table 3.2 Shells and subshells of an atom (https://en.wikipedia.org/wiki/Electron_shell)

Quantum no. (n)	Shell	Subshells	No. of states	No. of electrons Per subshell	Per shell
1	K	s	1	2	2
2	L	s	1	2	8
		p	3	6	
3	M	s	1	2	18
		p	3	6	
		d	5	10	
4	N	s	1	2	32
		p	3	6	
		d	5	10	
		f	7	14	
5	O	s	1	2	50
		p	3	6	
		d	5	10	
		f	7	14	
		g	9	18	

The Periodic Table

All the elements have been classified according to their electron configuration in the periodic table. Fig. 3.7 shows the periodic table.

The elements are arranged in seven horizontal rows called periods with increasing atomic numbers. The arrangement in vertical columns called groups is situated in accordance with arrays that have similar valence electron structure as well as chemical and physical properties. The elements in group 0, the right most corner of the table, have stable electron structure and are the inert gasses. The elements in Group VIIA and VIA are deficient of one and two electrons, respectively, for a stable structure. The elements in Groups IA and IIA are one and two electrons, respectively, in excess of a stable structure. The elements in the three long periods Groups IIIB through IIB are transition metals, which have partially filled d electron states and in some cases one or two electrons are in the next higher energy shell. Elements in Groups IIIA, IVA, and VA display intermediate characteristics between metals and nonmetals by virtue of their valence electron structures.

Most of the elements in the periodic table belong to the metal classification. They are called electropositive elements that are capable of giving up their few valence electrons to become positively charged ions. The elements on the right-hand side of the table are called electronegative elements that

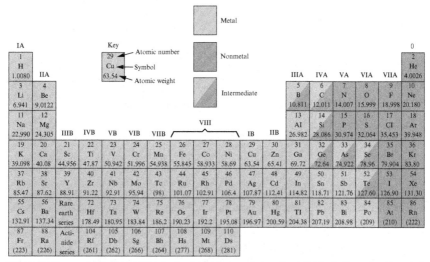

Fig. 3.7 The periodic table. *(Reproduced with permission from W.D. Callister Jr., Materials Science and Engineering: An Introduction, John Wiley & Sons Inc., 2007.)*

readily accept electrons to form negatively charged ions or sometimes share electrons with other atoms. Electronegativity of elements increases in moving left to right and from bottom to top in the periodic table. Atoms are more likely to accept electrons, if their outer shells are almost full, and they are less shielded from the nucleus. More shells are less shielded from the nucleus, as the distance of the electron from the nucleus increases.

3.3.1.3 Electron Configuration by the Periodic Table

An atom's electron configuration is a numeric representation of its electron orbitals. Electron orbitals are differently shaped regions around an atom's nucleus, where electrons are mathematically likely to be located. There are basic principles behind electron configuration. Fig. 3.8 shows the electron configuration by periodic table.

As an atom gains electrons, they fill different orbitals sets according to a specific order. Each set of orbitals, when full, contains an even number of electrons. The orbital sets are

- The s orbital set contains a single orbital, and by Pauli's exclusion principle, a single orbital can hold a maximum of 2 electrons.
- The p orbital set contains 3 orbitals, and can hold a total of 6 electrons.
- The d orbital set contains 5 orbitals, so it can hold 10 electrons.
- The f orbital set contains 7 orbitals, so it can hold 14 electrons.

Electron configuration in the perodic table

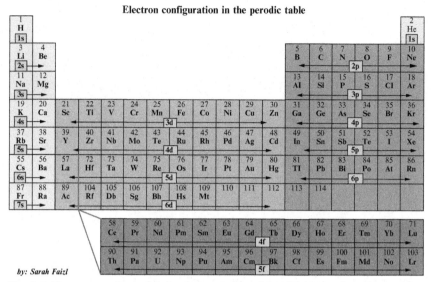

by: Sarah Faizl

Fig. 3.8 Electron configuration in the periodic table [8]. *(Taken from http://chem. libretexts.org/@api/deki/files/1281/PeriodicTable2.jpg?revision=1.)*

Electron configurations are written to clearly display the number of electrons in the atom as well as the number of electrons in each orbital. Each orbital is written in sequence, with the number of electrons in each orbital written in superscript to the right of the orbital name. The final electron configuration is a single string of orbital names and superscripts.

3.3.1.4 Energy Levels in Many-Electron Atoms

The shielding effect and electron–electron interactions cause the energy levels of subshells such as 2s and 2p to be different from those of hydrogen-like atoms. This is done by treating the electron shield cores as a proton but the core has an effective nuclear charge Z. For the hydrogen-like atoms, energy levels for 2s, 2p stay the same, but the separation between 2s and 2p energy levels increases as the atomic number (Z) increases. Similar situations happen for 3s, 3p, and 3d energy levels. The energy level in the Aufbau process varies, because electrons tend to occupy the lowest energy level available. The Pauli exclusion principle and Hund's rule guide in the Aufbau process, which show the electron configurations for all elements.

The Pauli exclusion principle suggests that only two electrons with opposite spin can occupy an atomic orbital. That is, no two electrons have the same four quantum numbers n, l, m, s. A state accepts two electrons of different spins.

Hund's rule suggests that electrons prefer parallel spins in separate orbitals of subshells. This rule guides in assigning electrons to different states in each subshell of the atomic orbitals.

In an ordinary periodic table, the s, p, and d block elements are in the main body of the periodic table, whereas the f block elements are placed below the main body. An electron configuration table is a type of code that describes how many electrons are in each energy level of an atom and how the electrons are arranged within each energy level. It packs a lot of information into a little space and it takes a little practice to read. Table 3.3 gives the electron arrangement and energy level for gold.

The first number is the energy level. An atom of gold contains six energy levels. The lowercase letter is the subshell. The subshells are named s, p, d, and f. The number of available subshells increases as the energy level increases. For example, the first energy level only contains an s subshell, while the second energy level contains both an s subshell and a p subshell.

The number in superscript is the number of electrons in a subshell. Each subshell can hold only a certain number of electrons. The s subshell can hold no more than 2 electrons, the p subshell can hold 6, the d subshell can hold 10, and the f subshell can hold as many as 14.

Table 3.3 Electron arrangement and energy level for gold

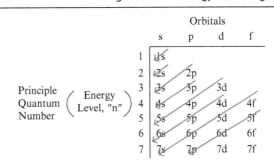

Order: 1s 2s 2p 3s 3p 4s 3d 4p 5s 4d 5p 6s 4f 5d 6p 7s 5f 6d 7p

Example 3.4

Write down the electron configuration of silicon atom (Si), whose atomic number is 14 and gold (Au), whose atomic number is 79. Refer to the periodic table.

Solution

a. Silicon has 14 electrons equal to the atomic number. The electron configuration assumes the number of shells as per energy level and the electrons occupy subshells as per the order till all electrons are allocated to the shells and subshells.

Energy level	Subshell				Total electrons
Shell	s	p	d	f	14
1	s^2				2
2	s^2	$2p^6$			8
3	s^2	$3p^2$			4

Configuration of Si (14) is: $1s^2 2s^2 2p^6 3s^2 3p^2$.

b. Gold: The atomic number of gold is 79 equal to 79 electrons. The electrons occupy as per the energy level and get arranged in subshells as per the order in Table 3.3 to all electrons equal to 79.

Energy level	Subshell						Total electrons
Shell no.	s	p	d	f	g	h	79
1	$1s^2$						2
2	$2s^2$	$2p^6$					8
3	$3s^2$	$3p^6$	$3d^{10}$				18
4	$4s^2$	$4p^6$	$4d^{10}$	$4f^{14}$			32
5	$5s^2$	$5p^6$	$5d^{10}$				18
6	$6s^1$						1

The electron configuration of gold is, therefore, as given below.

$$1s^2 2s^2 2p^6 3s^2 3p^6 4s^2 3d^{10} 4p^6 5s^2 4d^{10} 5p^6 6s^1 4f^{14} 5d^{10}$$

Note: The numbers indicate energy levels and the powers add up to total electrons or atomic number.

3.3.1.5 Atomic Bonding

There are interatomic forces that bind the atoms in the elements together. When two isolated atoms are brought together from infinite separation to close proximity, each exerts forces on each other, as their distance between the nuclei reduces. These are two types of forces, attractive and repulsive, and the magnitude of each force is a function of the separation or interatomic distance. The net force between the atoms is the sum of both the attractive and repulsive forces. Fig. 3.9 explains that a state of equilibrium occurs when the two forces are equal, and the net force becomes zero.

The atomic forces are related to potential energy as the integral of the force with reference to its distance, and therefore, the force is interpreted in terms of energies as shown in Fig. 3.9B. The minimum distance r_o is the equilibrium, where the net force becomes zero. The minimum potential energy E_o is the bonding energy that is the minimum required to separate these two atoms to an infinite separation. In solid materials, there will be a

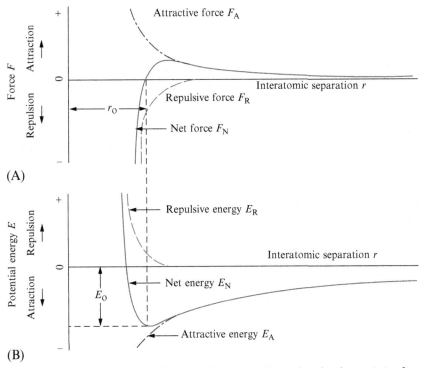

Fig. 3.9 Interatomic distance: (A) force and (B) energy. *(Reproduced with permission from W.D. Callister Jr., Materials Science and Engineering: An Introduction, John Wiley & Sons Inc., 2007.)*

complex situation, where the interatomic forces or energy interactions among many atoms will be complex, as it involves more number of atoms. The analogous minimum potential energy E_o is the bonding energy that associates with each atom. The magnitude of the bonding energy and the shape of the energy versus interatomic separation curve vary with the type of materials and the type of atomic bonding. The properties of materials such as high melting temperature point for larger bonding energy or states of solid for large bonding energy, liquid or gas depend on the smaller value of E_o, the shape of the energy–interatomic separation curve, and bonding type.

Three different types of primary or chemical bonds are found in solids. They are ionic, covalent, and metallic. Each type involves valence electrons, and the nature of the bond depends on the electron structure. Ionic bonding is found in compounds composed of elements of metals and nonmetals. Atoms of metallic elements easily give up their valence electrons to the non-metallic atoms and become stable. Ionic bonding is a transfer of electrons from the electropositive, excess valence electron, atoms to electronegative, short of valence electron atoms. They are nondirectional, and the magnitude is equal in all directions around the ion. The bonding energy is high around 3–8 eV per atom, and high melting temperatures. The ionic materials are hard, brittle, and electrically and thermally insulative.

Covalent bonding is sharing of electrons between two atoms in a stable electron configuration. Each atom contributes one electron to the bond and the shared electrons belong to both the atoms. The covalent bond is directional between the two atoms that share the electrons. This type of bonding is found in elements such as carbon, silicon, and germanium. The possibilities of a covalent bond are determined by $8 - N'$, where N' is the number of valence electrons of the atom. For example, $N' = 4$ for carbon; that is, there are four possibilities of covalent bonds as found in methane CH_4. Covalent bonds are strong as in diamond, very hard and very high melting temperature around 6400°F. It is possible to have interatomic bonds that are partially ionic and partially covalent bonds.

Metallic bonding is found mostly in metals and alloys. Metals have one, two, or a maximum of three valence electrons. These electrons are not bound to belong to any atoms but are free to drift as a sea of electrons throughout the entire material. The valence electrons and the nucleus form the ion core with a net positive charge equal to the magnitude of total valence electrons charge per atom. The free electrons hold the ion core together. Bonding may be weak or strong. Bonding energies vary in a range from 0.70 eV/atom for mercury to 8.60 eV/atom for tungsten with melting temperature of –38°F to 6170°F, respectively. Metallic bonding is found in all elemental metals in Group IA and IIA elements in the periodic table. Metals with metallic bonding are good conductors of electricity, whereas ionically or covalently bonding materials are electrical and thermal insulators in the absence of free electrons.

There is secondary bonding, namely, van der Waals bonding and hydrogen bonding. Such bondings are found in inert gasses.

3.3.1.6 The Structure of Crystalline Solids

The properties of some materials are related directly to their crystal structure. The single crystal structure magnesium is more brittle than gold or silver that have yet another crystal structure. Significant property differences exist between crystalline and noncrystalline materials of the same composition. A crystalline material is one where the atoms are situated in a periodic array over large atomic distances. The atoms position themselves in a repetitive pattern on a normal solidification process and maintain crystalline structure, and each atom is bonded to its nearest neighbor atoms. Those that do not maintain the crystalline structure under the solidification process are called noncrystalline or amorphous materials. There is a large number of different crystalline structures and, accordingly, the properties of the materials vary.

In crystalline structures, atoms ions are solid spheres having well-defined diameters. The term lattice, a three-dimensional array of points coinciding with atom position, is used in explaining the crystalline structure.

3.3.2 Semiconductor Materials for Solar Cells [1,9]

The semiconductor layer is important in a solar cell, where the light photon is absorbed to convert light energy into electrical energy. Different semiconductor materials are suitable for this purpose. The important semiconductor properties determine the solar cell performance. The details are discussed here.

The semiconductor properties and the parameters that determine the design and performance of a solar cell are

(i) Concentrations of doping atoms that are of two types: N_D, donars donate free electrons or acceptors accept electrons, N_A.

(ii) Mobility and diffusion coefficient D of charge carriers due to drift and diffusion.

(iii) Lifetime and diffusion length that characterizes the recombination-generation processes.

(iv) Band gap energy E_G, absorption coefficient and refractive index that characterizes the ability to absorb visible and other radiation.

3.3.2.1 Semiconductor Properties of Crystalline Silicon (c-Si) [9]

i. Atomic structure

Silicon has four valence electrons that covalently bonds with other four atoms by sharing two electrons by each atom in a crystalline form. The single crystalline silicon has crystal lattice structure with long-range order. In practice, semiconductors always contain some impurities. Semiconductors with insignificant impurities are called intrinsic semi-conductors. Semiconductors with doped atoms with impurities are called extrinsic semiconductors. At room temperature, there are always some broken covalent bonds, thus liberating valence electrons from the bonds and making them mobile in the crystal lattice structure. The position of a missing electron from the bond is positively charged and is referred to as a hole. The atoms have four covalent bonds with eight valence electrons duly intact in maintaining the lattice at $0°K$. At a temperature higher than $0°K$, the covalent bond breaks due to thermal energy and thus liberates free electrons, leaving a hole in the bond. The electron from another atom may jump into this empty position and restore the bond, leaving a hole in its original bond. Thus, the holes are moving opposite to the motion of the valence electron that tries to jump into the neighbor atom to restore the bond. Because the breaking of bonds leads to the formation of an electron hole pair in an intrinsic semiconductor, there is an equal number of concentration of holes positively charged p and an equal number of concentrations of electrons negatively charged n. There are 1.5×10^{-10} broken bonds per cm^3 at a temperature of $300°K$ in intrinsic c-silicon. It means that the concentration of charge carriers $p = n = 1.5 \times 10^{10}$ cm^{-3} in intrinsic c-Si at 300 K. This is called intrinsic carrier concentration n_i

ii. Doping

The concentration of holes and electrons in c-Si can be manipulated by doping. Doping means adding impurities of atoms of the appropriate element as a substitute in the crystal lattice c-Si atom. Because silicon has four valence electrons, the substitute atom may be either phosphorous of five valence electrons or boron of three valence electrons. When a phosphorous atom is substituted, four of the five valence electrons easily bond with an Si atom, but the fifth one bound weekly with the phosphorous atom, which gets easily liberated as free electron due to the thermal energy sufficient in the Si atom at room temperature, and moves through the crystal lattice. This enhances the electron concentration and such electron donors are denoted as N_D. Similarly, when born atoms with three electrons are substituted in the Si crystal lattice, it readily accepts one electron from the neighboring Si atom and completes the bond, but leaves a hole in the Si atom and it can move around the lattice. The impurities that enhance the concentration of holes are called acceptors and are denoted as N_A. Because only one type of impurity is added, only one type of charge carrier is increased and the charge neutrality is nevertheless maintained. The donor atoms become positively ionized and the acceptor atoms become negatively ionized. Fig. 3.10 explains the concept.

The control of electrical conductivity in a semiconductor is possible by doping. It is an important semiconductor feature. The electrical conductivity in a semiconductor depends on the concentration of electrons and the holes, and their mobility. The concentration of electrons and holes influences a number of impurity atoms added in the atomic structure of the semiconductor. The p-type semiconductor has a concentration of majority carriers of holes and minority carriers of electrons, and the n-type semiconductor has a concentration of majority carriers of electrons and minority carriers of holes. The range of doping in c-Si is shown in Fig. 3.11.

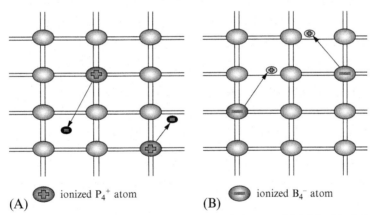

(A) ⊕ ionized P_4^+ atom   (B) ⊖ ionized B_4^- atom

Fig. 3.10 Doping with donar and acceptor atoms with Si atoms [9]. (A) Positively ionized P atom and free electron and (B) negatively ionized B atom and a hole.

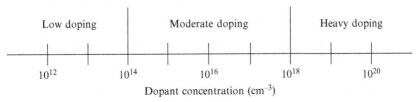

Fig. 3.11 The range of doping levels in c-Si [9].

3.3.2.2 Carrier Concentration [1,9]

The semiconductor is said to be in a state of equilibrium, where no external perturbing forces such as voltage, magnetic field, mechanical stress, or illumination is applied. Under equilibrium state, the parameters do not change with time.

The carrier concentration is determined based on the density of energy states function, which describes the number of allowed states per unit of volume and energy. The occupation function, known as the Fermi-Dirac distribution function, describes the ratio of states filled with an electron to the total allowed states in a given energy. The electrons in Si atoms are allowed to have only discrete energy values. The periodic atomic structure of a single crystal silicon atom results in the ranges of allowed energy states for electrons that are called energy bands and the excluded energy ranges as forbidden gaps or band gaps. The liberated electrons from the bond determine the charge transport in a semiconductor. The valence electrons involved in the covalent bond have their allowed energies in the valence band (VB). The allowed energies of electrons liberated from the covalent bonds form the conduction band (CB). The VB is separated from the CB by a band of forbidden energy levels. Fig. 3.12 gives the energy band diagram for c-Si.

E_V is the maximum attainable energy level in the VB and E_C is the minimum attainable energy level in the CB. The difference between the two is the band gap energy, E_G. The band gap energy is an important parameter in a material. The band gap energy for a single crystal silicon is 1.12 eV (1 eV $= 1.602 \times 10^{-19}$ J). The Fermi energy is the electrochemical potential that is the averaged energy of electrons in a material. Fermi energy in an intrinsic semiconductor lies in the middle of the band gap. The Fermi energy is a function of absolute temperature and is given by kT, where k is the Boltzmann's constant, ($k = 1.38 \times 10^{-23}$ J/K). Fermi energy equals to 0.0258 eV at 300°K.

However, the Fermi energy level can be manipulated by doping of the semiconductor. The electrons in the CB and the holes in the VB are the

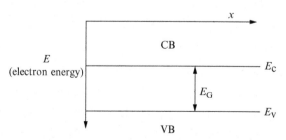

Fig. 3.12 The basic energy band diagram [9].

charge carriers. The concentration of charge carriers is obtained by the appropriate density states function and its Fermi distribution function of appropriate charge carriers, respectively. Inserting donor or acceptor atoms into the lattice structure of crystalline silicon introduces allowed energy levels into the forbidden band gap. The donor atoms liberate electrons and the energy level E_D moves close to the CB in the n-type material. Similarly adding acceptor atoms liberate holes and the energy level E_A moves close to the VB. Doping also influences the position of Fermi energy. The concentration of electrons increases by increasing the donor atoms in n-type material and the concentration of holes increases with the increase of acceptor atoms in p-type material. The Fermi energy level with n-type doping moves close to the CB, and with p-type doping moves close to the VB in the band diagram. Fig. 3.13 explains the concept.

The concentration of electrons and holes, respectively, are given by the following equations:

$$n = N_C \exp\left[(E_F - E_C)/kT\right] \quad \text{for } E_C - E_F \geq 3kT \tag{3.9}$$

$$p = N_V \exp\left[(E_V - E_F)/kT\right] \quad \text{for } E_F - E_V \geq 3kT \tag{3.10}$$

where N_C and N_V are the effective density state in the CB and VB, respectively. When the Fermi energy level lies in the band gap more than $3kT$ from either band edge is satisfied, the semiconductor is referred to as nondegenerate. $N_C = 3.22 \times 10^{19}$ cm^{-3}, $N_V = 1.83 \times 10^{19}$ cm^{-3} for c-Si at 300°K. $kT = 0.0258$ eV at 300°K.

The Fermi energy level is given by

$$E_F = E_C - kT \ln(N_C/N_d) \quad \text{for } n\text{-type materials} \tag{3.11}$$

$$E_F = E_C + kT \ln(N_V/N_a) \quad \text{for } p\text{-type materials} \tag{3.12}$$

where N_d is the concentration of donor atoms and N_a is the concentration of acceptor atoms. Energy level for intrinsic semiconductor is E_i.

$$E_i = E_C - E_G/2 + 0.0258/2 \ln(N_V/N_C) \quad \text{for intrinsic c-Si} \tag{3.13}$$

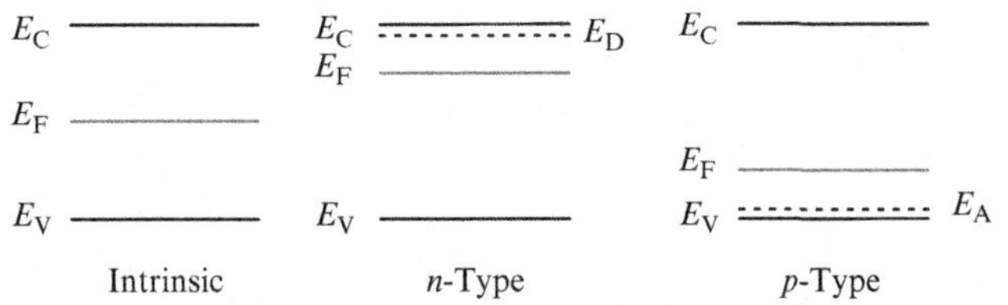

Fig. 3.13 The position of Fermi energy and the energy level in the band gap diagram by doping [1,9].

If electrons are added to the CB by donors, then the Fermi energy level shifts to higher energy level close to the CB. Similarly, if electrons are removed by acceptor atoms in the VB, Fermi energy level shifts to lower energy level close to the VB. The intrinsic carrier concentration n_i is the concentration of electrons at the CB edge and is also equal to the concentration of holes at the VB edge in the pure semiconductor. It depends on absolute temperature and the materials constant.

3.3.2.3 p-n Junction Semiconductor

p-Type material is a semiconductor doped by acceptor impurity materials that increases holes in the VB, and the n-type material is a semiconductor doped with donor impurity materials that increases electrons in the CB. When these types of materials are joined, a p-n junction is formed. Fig. 3.14 shows the p-n junction. The large electrons in n-type material flow into the holes of the p-type material. The n-type material becomes positively charged and the p-type material becomes negatively charged in the vicinity of the p-n junction. The process of diffusion of carriers continues till the junction potential reaches equilibrium, at the time when the flow of electrons and holes are equal on both sides in the junction.

p-n junction

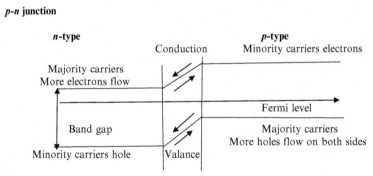

Fig. 3.14 p-n junction unbiased condition.

Example 3.5

Determine the concentration of electrons and holes in a *p-n* junction device with the doping levels as $N_a = 10^{16}$ cm^{-3}, $N_d = 0.5 \times 10^{16}$ cm^{-3}. Intrinsic semiconductor carrier concentration $n_i = 1.08 \times 10^{10}$ cm^{-3} Calculate E_F for donor doped and E_i for c-Si assuming $E_C = 0.0$ eV.

The charge concentrations in a *p-n* junction is illustrated below.

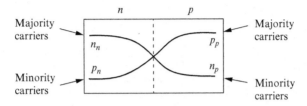

n_n : Concentration of electrons on *n* side
P_n : Concentration of holes on *n* side
P_p : Concentration of holes on *p* side
n_p : Concentration of electrons on *p* side

p-Type material:

$$\text{Majority carriers } p_p \sim N_a = 10^{16}\,\text{cm}^{-3}$$
$$\text{Minority carriers } n_p = (n_i)^2/N_a$$
$$= (1.08 \times 10^{10})^2/1.0 \times 10^{16}$$
$$= 1.16 \times 10^4\,\text{cm}^{-3}$$

n-Type material:

$$\text{Majority carriers } n_n \sim N_d = 0.50 \times 10^{16}\,\text{cm}^{-3}$$
$$\text{Minority carriers } p_n = (n_i)^2/N_d$$
$$= (1.08 \times 10^{10})^2/0.5 \times 10^{16}$$
$$= 2.33 \times 10^4\,\text{cm}^{-3}$$

$$E_i = E_C - E_G/2 + 0.0258/2 \, \ln(N_V/N_C)$$
$$= 0.0 - 1.12/2 + 0.0258/2 \, \ln(1.83 \times 10^{19}/3.22 \times 10^{19})$$
$$= -0.57\,\text{eV}$$

$$E_F = E_C - kT \, \ln(N_C/N_d) \text{ for } n\text{-type materials}$$
$$= 0.0 - 0.0258 \, \ln(N_C/N_d)$$

$$E_F = E_C + kT \, \ln(N_V/N_a) \text{ for } p\text{-type materials}$$
$$= 0.0 + 0.0258 \, \ln(N_V/N_a)$$

The electrical currents are generated in a semiconductor due to the transport mechanism of charge from place to place by electrons and holes by drifts and diffusion. Drift is charged particle motion in response to an electric field. The positively charged particles move along the direction of the electric field and the negatively charged electrons move opposite to the direction of the electric field. The average motion of electrons and holes is due to drift velocity and is proportional to the strength of the electric field either along the direction of the electric field or opposite. Diffusion is a process whereby the high-concentration region particles move toward the low-concentration particle region as a result of random thermal motion. The diffusion currents are proportional to the gradient in particle concentration.

3.3.2.4 The Photoelectric Effect [10]

The photoelectric effect describes the conversion of light into an electric current. Solar light has a stream of photons, where each photon carries one quantum of energy. Each photon is associated with just one wavelength or frequency. High-frequency photons have more energy than photons with low frequency. Fig. 3.15 explains the photoelectric effect in a semiconductor.

Intrinsic Semiconductor

In a pure semiconductor, the outermost electron of the underlying molecule is not heavily bound. An incoming photon with enough energy can promote the electron from the VB to become a free electron in the CB. This, in turn, leaves a positive hole in the VB. The minimum energy that is necessary for this to happen is called the band gap. The band gap varies from material to material and also varies with temperature, and therefore the

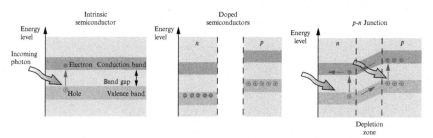

Fig. 3.15 Photoelectric effect in a semiconductor [10] (http://www.greenrhinoenergy. com/solar/technologies/pv_cells.php).

performance of solar modules deteriorates with higher temperatures. However, there is no electric current in an intrinsic semiconductor, because the promoted electrons recombine again with the holes. However, a doped semiconductor generates current due to the photoelectric effect.

3.3.2.5 Spectral Sensitivity of PV Materials [10]

Fig. 3.16 shows spectral sensitivity and band gaps of certain PV materials.

In Fig. 3.16, photons in longer wavelengths cause electrons to flow due to energy from ambient temperature. On the other hand, shortwave photons may not be able to be absorbed, as they have too much energy. The response rate measures a material's ability to convert light into an electric current: The rate is given by Eq. (3.14). Table 3.4 gives comparative spectral sensitivities of PV materials.

$$\zeta = \frac{\text{Electric current}}{\text{Irradiance}} \tag{3.14}$$

Due to the spectral sensitivities, the response rate is highly dependent on the wavelength of the incoming light. Most of the energy of sunlight is in the visible spectrum. Although silicon captures a wider spectrum than the human eye, it is more sensitive to infrared than to visible light, thus less efficient in converting the all-important visible spectrum.

Other III–V compound materials such as GaAs, GaInP, or GaAsAl cover wider ranges where the spectral sensitivity is better matched to the incident

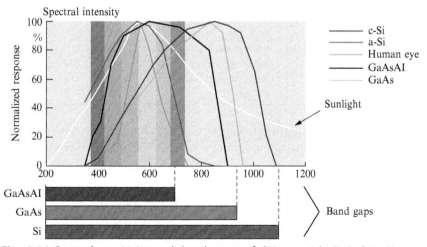

Fig. 3.16 Spectral sensitivity and band gaps of PV materials [10] (http://www.greenrhinoenergy.com/solar/technologies/pv_cells.php).

Table 3.4 Comparison of spectral sensitivities of PV materials [10]

Crystalline silicon	Thin film	Multijunction
Monocrystalline: The atoms form a regular lattice. Due to the regular structure, monocrystalline has a better response rate *Polycrystalline*: This is a series of crystals rather than one crystal	Thin film modules use compounds with very strong light absorption characteristics. The absorption materials are deposited on glass or foil. They include *Amorphous silicon (a-Si)*: Atoms do not form diamonds, but are randomly put together. The response rate is much lower than in monocrystalline structure. It captures more of the highly intensive light than crystalline silicon and can be changed by alloying it with germanium or carbon *CdTe*: Cadmium-telluroid: Inexpensive technology with medium efficiency *CIS/CIGS (cadmium-indium-gallium-selenide and copper indium gallium diselenide)*: Have efficiencies as high as crystalline silicon	Differences in spectral sensitivities between materials can be exploited by adding multiple absorption layers on top of each other, resulting in multiple *p-n* junctions The material with the largest band gap is positioned closest to the incoming light, absorbing all the high-energy photons. For low-energy (long wavelength) photons, this first layer will be transparent until they hit the second layer with a lower band gap. This way, overall efficiency can be increased to 40%, compared to 20% for monocrystalline silicon

sunlight, thus increasing the efficiency of the cell. By applying two materials in a tandem or multijunction cell, an even wider spectrum can be captured.

3.4 PHOTOVOLTAIC MATERIALS [1,10–12]

PV materials are used in PV cells for conversion of light energy into electricity by a process called the photoelectric effect. A review on these materials is presented in this section.

3.4.1 Semiconductor Materials [1,10]

All materials may be classified into three major categories: conductors, semi-conductors, and insulators, depending upon their ability to conduct an electric current. The materials germanium and silicon are often used in semiconductor devices. Germanium has higher electrical conductivity than silicon. Silicon is more suitable for high-power devices than germanium. A relatively new material with the desirable features of both germanium and silicon is gallium arsenide.

The ability of a material to conduct current is directly proportional to the number of free electrons in the material. Good conductors, such as silver, copper, and aluminum, have large numbers of free electrons; their resistivities are of the order of a few millionths of an ohm-centimeter. Insulators such as glass, rubber, and mica, which have very few loosely held electrons, have resistivities as high as several million ohm-centimeters.

Semiconductor materials lie in the range between these two extremes, as shown in Fig. 3.17. Pure germanium has a resistivity of 60 Ω cm. Pure silicon has a considerably higher resistivity, in the order of 60,000 Ω cm. As used in semiconductor devices, however, these materials contain carefully controlled amounts of certain impurities, which reduce their resistivity to about 2 Ω cm at room temperature and this resistivity decreases rapidly as the temperature rises.

Carefully prepared semiconductor materials have a lattice crystal structure. In this type of structure, which is called a lattice, the outer or valence electrons of individual atoms are tightly bound to the electrons of adjacent atoms in electron pair bonds. Such a structure has no loosely held electrons; semiconductor materials are poor conductors under normal conditions. High temperature or strong electric fields are applied to semiconductor materials to separate the electron pair bonds and provide free electrons for electrical conduction.

The lattice structure is altered by adding small quantities of impurities of different atomic structure and thereby obtains free electrons and thus the basic electrical properties of pure semiconductor materials can be modified

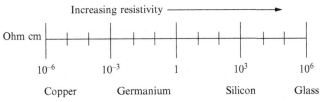

Fig. 3.17 Resistivity of typical conductors, semiconductors, and insulators.

and controlled. When the impurity atom has one more valence electron than the semiconductor atom, however, this extra electron cannot form an electron pair bond because no adjacent valence electron is available. The excess electron is then held very loosely by the atom and makes the material a better conductor; that is, its resistance to current flow is reduced.

Impurity elements, which are added to germanium and silicon crystals to provide excess electrons, include arsenic and antimony. When these elements are introduced, the resulting material is called n-type because the excess free electrons have a negative charge. A different effect is produced, when an impurity atom having one less valence electron than the semiconductor atom is substituted in the lattice structure. Although all the valence electrons of the impurity atom form electron pair bonds with electrons of neighboring semiconductor atoms, one of the bonds in the lattice structure cannot be completed because the impurity atom lacks the final valence electron. As a result, a vacancy or "hole" exists in the lattice. An electron from an adjacent electron pair bond may then absorb enough energy to break its bond and move through the lattice to fill the hole. As in the case of excess electrons, the presence of "holes" encourages the flow of electrons in the semiconductor material; consequently, the conductivity is increased and the resistivity is reduced.

The vacancy or hole in the crystal structure is considered to have a positive electrical charge because it represents the absence of an electron. Semiconductor material that contains these "holes" or positive charges is called p-type material. p-Type materials are formed by the addition of aluminum, gallium, or indium.

A slight difference in the chemical composition of n-type and p-type materials makes the electrical characteristics of the two types substantial and very important in the operation of semiconductor devices.

3.4.2 Photovoltaic Materials [1,11]

When the PV cell is exposed to sunlight, the photo electrons from solar light striking the doped semiconductor at light speed dislodges free electrons with part of its energy and moves the free electrons to flow with the balance of its kinetic energy due to its photoelectric effect. The physical process in a PV cell that converts photons of the absorbed sunlight into electricity is known as the PV effect. The PV cell or solar cell is made of a doped semiconductor material. PV cell materials are of two major types: crystalline and thin films. The types of materials vary in terms of light absorption efficiency, energy conversion efficiency, manufacturing technology, and cost of production.

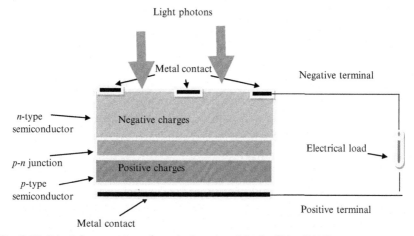

Fig. 3.18 Principle operation of producing electricity by PV cell [11].

Fig. 3.18 shows the principle operation of a PV cell. The current flow from the P layer, where the potential is high to the N layer with lower potential continues as long as the cell is exposed to solar light and the current flow depends on the strength of solar radiation.

The annual electrical power output of a PV plant depends on the following factors:

- Solar radiation incident on the installation site
- Inclination and orientation of the PV module
- Presence of shading or not
- Technical performances of the plant components, especially modules and inverters

The main applications of PV plants are for stand-alone off-grid with or without storage facilities, and on-grid connected to utility systems. A PV plant constitutes a PV generator, supporting framed structure to mount PV modules, inverter for power control and conditioning, energy storage systems, electrical switchboards and switchgear assemblies, protection equipment, and connection cables for interconnections of components. The system materials in a PV system other than the PV module is called balance of system (BOS).

3.4.2.1 Photovoltaic Technologies [11–13]

The solar cell manufacturing technologies are

- Single cell or monocrystalline
- Multicell or polycrystalline

- Bar crystalline silicon
- Thin film technology

The solar cells from crystalline silicon (Si) are made of thinly sliced wafers. A crystal of single silicon is monocrystalline and a whole block of silicon crystals are called multicrystalline. Their efficiency ranges between 12% and 19%.

Three PV technologies are monocrystalline or single crystal, polycrystalline, and thin film. Monocrystalline and polycrystalline represent the traditional technologies for solar panels and are grouped as crystalline silicon.

High-purity polycrystalline is melted in a quartz crucible. A single crystal seed is dipped into the molten mass of polycrystalline and pulls out a single crystal ingot. The ingots are sliced into thin wafers of 200–400 μm thickness, which are then polished, doped, coated, interconnected, and assembled as modules and arrays. A single crystalline cell has higher conversion efficiency, the ratio of power produced in a given area of exposure to the sunlight by the cell to that of available power from sunlight. The single crystal cells are more reliable and energy efficient in the range of 15%–20%.

Monocrystalline, or single crystal, is the original PV technology invented in 1955, and polycrystalline entered the market in 1981. They are similar in performance and reliability. Single crystal modules are composed of cells cut from a piece of continuous crystal. The material forms a cylinder that is sliced into thin circular wafers. To minimize waste, the cells may be fully round or they may be trimmed into other shapes, retaining more or less of the original circle. Because each cell is cut from a single crystal, it has a uniform color. Fig. 3.19 shows the typical crystalline silicon (c-Si) cell structure.

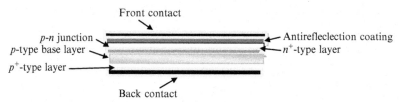

Fig. 3.19 Typical c-Si solar cell structure [9].

An absorber material is moderately doped with a *p*-type square wafer as a base layer of size 10×10 cm^2 with 300 µm thick. On both sides of the c-Si wafer, a highly doped *n*-type layer at the top and a *p*-type layer at the bottom are formed. These two high-doped layers help separate the photo-generated charge carriers from the bulk of the c-Si wafers. The metallic grid at the top and bottom collects the separated charge carriers and connects the cell to a load. A thin layer of antireflective coating at the top side of the cell is given to decrease the reflection of light from the cell. A glass sheet or another type of transparent encapsulant is attached to both sides of the cell to protect the cell against outer environment effects. The layers are deposited at low temperature on a substrate carrier such as glass sheet, metal, or polymer foil.

Polycrystalline cells are made from similar silicon material that is melted and poured into a mold. This forms a square block that can be cut into square wafers with less waste of space or material than round single-crystal wafers. As the material cools, it crystallizes in an imperfect manner, forming random crystal boundaries. The grain boundaries become a hindrance to the flow of electrons and reduce the power output. Thus, the efficiency of energy conversion is slightly lower, 10%–14%. The size of the finished module is slightly greater per watt than most single-crystal modules. The cells look different from single-crystal cells. The surface has a jumbled look with many variations of blue color. Alternatively, PV cells are developed by the ribbon growth method, where silicon is grown directly as thin ribbons or sheets at a lower manufacturing cost compared to wafers. Most crystalline silicon technologies yield similar results, with high durability.

A compound semiconductor made of two elements, gallium (Ga) and arsenic (As), has a crystal structure similar to silicon. GaAs has higher sunlight absorption efficiency with a lower thickness of material than for a crystalline material and has higher energy conversion efficiency than crystalline silicon in the range of 25%–30%. The GaAs cell material is more costly than crystalline cells and is mostly used in concentrator systems at a higher temperature.

The silicon used to produce crystalline modules is derived from sand. It is the second most common element on earth but purified expensively to an extremely high degree in order to produce the photovoltaic effect. Crystalline solar cells use a relatively large amount of silicon.

Thin Film Technologies

A PV cell is made with a microscopically thin deposit of silicon on a sheet of metal or glass, instead of a thick wafer. Thin film technology, also called amorphous, uses a more exotic material such as cadmium telluride. Thin film panels are made flexible and lightweight by using plastic glazing. Some of them perform slightly better than crystalline modules under low-light conditions. They are also less susceptible to power loss from partial shading of a module.

The disadvantages of thin film technology are lower energy conversion efficiency and uncertain durability. Thin film materials tend to be less stable than crystalline, causing degradation over time. The technology is greatly improved. Thin film cells will be strong where price is a critical factor.

PV materials used in thin film cells are

(a) Amorphous silicon (a-Si): This material, a noncrystalline form of silicon with irregular crystalline structure, has been used dominantly in thin film PV cells. Amorphous silicon has 40 times higher light absorption efficiency than that of single crystal silicon. Therefore, a thin layer of a-Si is deposited on low-cost substrates such as steel, glass, and plastics at low temperatures. All these factors lower the material costs and manufacturing costs than those of crystalline silicon cells. However, the cell energy conversion is very low in the range of 5%–9%, which further degrades with short life in outdoor exposure to sunlight.

(b) Cadmium telluride (CdTe): This is a semiconductor compound made of cadmium and tellurium. A micrometer thick CdTe coating has higher light absorption efficiency from 90% of the solar spectrum. The manufacturing process by high-rate evaporation, spraying, or printing is easy and cheap. The cell energy conversion efficiency is about 7%. The cell instability and toxic nature of the material are its disadvantages.

(c) Copper indium diselenide (CIS or $CuInSe_2$): A polycrystalline semiconductor compound of copper, indium, and selenium (CIS) has the highest energy conversion research efficiency of 17%, which is best among the existing materials in the thin film industry. It is close to that of 18% research efficiency of polycrystalline silicon cells. CIS has one of the highest light absorbing efficiencies with 0.5 μm thickness material that can absorb 90% of the solar spectrum. However, the manufacturing process is complex and the material has toxic gas. The material is now commercially unavailable.

Crystalline silicon materials are dominating PV cell materials due to their energy conversion efficiency, even though the cost is high. Thin film PV cells are focused on their cost even though they have poor energy conversion efficiency. Recent research advances by industry experts promise that thin film PV cells will dominate the industry by their low energy price and reliable energy source.

Table 3.5 PV cells and modules [10] (http://www.greenrhinoenergy.com/solar/technologies/pv_modules.php)

	Crystalline photovoltaics	Thin film
	(credit: Sharp)	(credit: sikod.com)
Description	Crystalline silicon wafers	A semiconductor is deposited directly on glass
Module efficiency	High	Low
Performance under heat	Performance degrades with higher temperatures	Up to 60% lower heat coefficient than crystalline silicon modules. A good choice in hot climates
Space required per kWp	Polycrystalline: 10–30 m^2 based on cell spacing Monocrystalline: >8 m^2	Glass-glass laminate ~25 m^2
Amount of photovoltaic material	Polysilicon: 8 g/W	CdTe: 0.22 g/W
Panel choice	Only 2 panel types	Many different panel types
Degradation	No known degradation	Depends on heat and material. Amorphous silicon can lose up to 30% within 3–6 months and are stable thereafter.
Direct or diffuse light	Direct light preferred, but diffuse light can be used too	Both direct and diffuse light

A lot of researchers are active the world over to bring out new PV materials and process methods with focus on total cost reduction of PV power systems. Table 3.5 gives the performance characteristics of PV cells and modules.

3.4.2.2 The Principles of Photovoltaics [10–12]

A solar module has a number of layers in the process of manufacturing. Fig. 3.20 shows the layers of a solar module. All PV modules contain a number of layers from the light-facing side to the back side of the cell:

1. Protection layer: Usually made from glass. Thin film modules are made of transparent plastic.

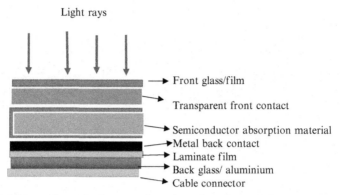

Light rays

Front glass/film

Transparent front contact

Semiconductor absorption material

Metal back contact

Laminate film

Back glass/ aluminium

Cable connector

Fig. 3.20 Different layers in the manufacturing of a solar module.

2. Front contact: The electric contact at the front has to be transparent to allow light into the cell.

3. Absorption material: Silicon is used as the single layer in most of the module, where the light is absorbed and converted into electric current. All materials used are semiconductors. However, to improve performance, there could be multiple layers of different materials. All layers are doped, split with some layers into an n-doped and some layers with a p-doped zone.

4. Metal back contact: A conducting material at the back completes the electric circuitry.

5. Laminate film: A laminate ensures that the structure is waterproof and insulated from heat.

6. Back glass: This layer made of glass, aluminum, or plastic gives protection on the back side of the module.

7. Connectors: Finally, the module is fitted and wired with connectors and cables.

3.5 ELECTRICAL CHARACTERISTICS [1,13–19]

The electrical characteristics of materials are important in materials used in electrical applications. The electrical conductivity of a substance is the ability of the substance to conduct electric current and is the inverse of the resistivity of the material. The conductivity varies on the order of 27 in magnitude between different materials as shown in Fig. 3.21.

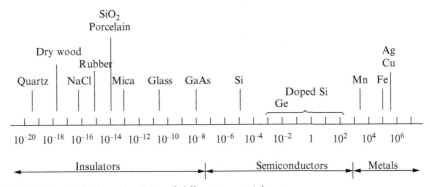

Fig. 3.21 Electrical conductivity of different materials.

3.5.1 Theory of *I-V* Characterization [1,18,19]

PV cell characterization involves measuring the cells electrical characteristics to determine the light conversion efficiency and critical equivalent circuit parameters. The performance of PV cells is compared by two simple tests to obtain necessary data for solar cell *I-V* characterization. These tests are the forward bias (illuminated) test and the reverse bias (dark) test. The SMUs (source measure units) are used for PV characterization of higher range PV cells as alternative cost-effective test systems. Measuring the electrical characteristics of a solar cell is critical for determining the output performance and efficiency of solar modules.

A typical voltage versus current characteristic is known as an *I/V* curve, of a PN diode without illumination. The applied voltage is in the forward bias direction. The curve shows the turn on and the buildup of the forward bias current in the diode. Without illumination, no current flows through the diode unless there is external potential applied. With incident sunlight, the *I/V* curve shifts up showing that there is external current flow from the solar cell to a resistive load.

3.5.1.1 Forward Bias (Illuminated) Test

The forward test gives the illuminated *I-V* curve for a PV cell. In a light test, an *I-V* sweep is conducted where the voltage is swept upward starting at $V = 0$, while measuring the sinking current. The following values are calculated using the forward bias (illuminated) test:

- Open circuit voltage (V_{OC})
- Short circuit current (I_{SC})
- Maximum power (P_{MAX}), current at P_{MAX} (I_{MP}), voltage at P_{MAX} (V_{MP})
- Fill factor (FF)

- Shunt resistance (R_{SH})
- Series resistance (R_S)
- Maximum efficiency (η_{MAX})

3.5.1.2 Reverse Bias (Dark) Test

The PV cell is tested as a passive diode element by blocking all light to prevent it from exciting a PV cell to determine its breakdown diode properties and internal resistances. The following I-V parameters are obtained via the reverse bias (dark) test.

- Shunt resistance (R_{SH})
- Series resistance (R_S)

PV cells are modeled as a current source in parallel with a diode. When there is no light present to generate any current, the PV cell behaves like a diode. As the intensity of incident light increases, current is generated by the PV cell, as illustrated in Fig. 3.22.

The total current I, in an ideal cell, is equal to the current I_0 generated by the photoelectric effect minus the diode current I_D, according to Eq. (3.15).

$$I = I_{SC} - I_0\left(e^{qV/kT} - 1\right) \tag{3.15}$$

$$I = I_{SC} - I_0\left(e^{38.9V} - 1\right) \tag{3.16}$$

where I_0 is the saturation current of the diode, q is the elementary charge 1.6×10^{-19} C, k is a constant of value 1.38×10^{-23} J/K, T is the cell temperature in Kelvin, and V is the cell voltage. A more accurate model will include two diode terms.

$$V_{OC} = \frac{kT}{q} \ln\left(\frac{I_{SC}}{I_0} + 1\right) \tag{3.17}$$

$$V_{OC} = 0.0257 \ln\left(I_{SC}/I_0 + 1\right) \tag{3.18}$$

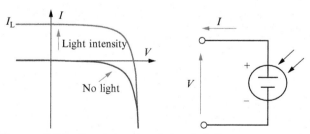

Fig. 3.22 *I-V* curve of PV cell and associated electrical diagram.

Fig. 3.23 shows the simplified circuit model and the following associated equation, where n is the diode ideality factor, typically between 1 and 2. R_S, is the series resistance that represents the ohmic loss in the front surface of the cell and R_{SH} is the shunt resistance that represents the loss due to diode leakage currents to account for the power dissipation in the internal resistance of the PV cells.

The PV equivalent circuit is modified to account for the losses due to series resistance (R_S) and shunt resistance (R_{SH}) as shown in Fig. 3.23. Accordingly, Eqs. (3.16), (3.18) become modified as below.

$$I = I_{SC} - I_0 \left(e^{38.9 V_d} - 1 \right) - V_d / R_{SH} \tag{3.19}$$

$$V_{OC} = 0.0257 \ln \left(I_{SC} / I_0 + 1 \right) \tag{3.20}$$

$$V_d = V + IR_S \tag{3.21}$$

$$V_{module} = n(V_d - IR_S) \tag{3.22}$$

where "n" are the number of cells in a module.

Fig. 3.24 shows the I-V curve of an illuminated PV cell. The voltage across the measuring load is swept from zero to V_{OC}, and many performance parameters for the cell is determined from this data.

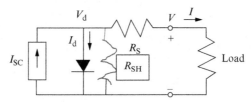

Fig. 3.23 Simplified equivalent circuit model for a PV cell.

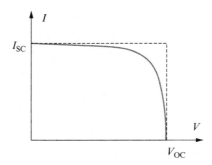

Fig. 3.24 Illuminated I-V sweep curve.

3.5.1.3 Short Circuit Current (I_{SC})

Short circuit current (I_{SC}) corresponds to the short circuit condition when the impedance is low and is calculated when the voltage equals 0.

$$I(\text{at } V = 0) = I_{SC} \qquad (3.23)$$

I_{SC} occurs at the beginning of the forward bias sweep and is the maximum current value in the power quadrant. For an ideal cell, this maximum current value is the total current produced in the solar cell by photon excitation.

$I_{SC} = I_{MAX} = I_0$ for forward bias power quadrant

3.5.1.4 Open Circuit Voltage (V_{OC})

Open circuit voltage (V_{OC}) occurs when there is no current passing through the cell.

$$V(\text{at } I = 0) = V_{OC} \qquad (3.24)$$

V_{OC} is also the maximum voltage difference across the cell for a forward bias sweep in the power quadrant.

$V_{OC} = V_{MAX}$ for forward bias power quadrant

3.5.1.5 Maximum Power (P_{MAX}), Current at P_{MAX} (I_{MP}), Voltage at P_{MAX} (V_{MP})

The power produced by the cell in watts can be easily calculated along the I-V sweep by the equation $P = IV$. The power will be zero at I_{SC} and V_{OC} points, and the maximum value for power will occur between the two. The voltage and current at this maximum power point are denoted as V_{MP} and I_{MP}, respectively, as shown in Fig. 3.25.

3.5.1.6 Fill Factor

FF is essentially a measure of the quality of the solar cell. It is calculated by comparing the maximum power to the theoretical power (P_T) that would be output at both the open circuit voltage and short circuit current together. FF can also be interpreted graphically as the ratio of the rectangular areas from Fig. 3.26 and Eq. (3.25).

A larger FF is desirable and corresponds to an I-V sweep that is more square-like. Typical FFs range from 0.50 to 0.82. FF is also often represented as a percentage.

$$\text{Fill factor} = P_{\max}/P_T$$
$$\text{FF} = V_{mp} \times I_{mp}/V_{OC} \times I_{SC} \qquad (3.25)$$

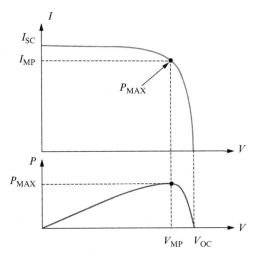

Fig. 3.25 Maximum power for an *I-V* sweep.

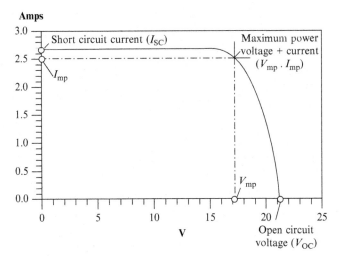

Fig. 3.26 The fill factor from the *I-V* sweep.

3.5.1.7 Efficiency (η)

The conversion efficiency is the ratio of the electrical power output, P_{out}, compared to the solar power input, P_{in}, into the PV cell. P_{out} can be taken to be P_{MAX} because the solar cell can be operated up to its maximum power output to get the maximum efficiency.

$$\eta = \frac{P_{\text{out}}}{P_{\text{in}}} \Rightarrow \eta_{\text{MAX}} = \frac{P_{\text{MAX}}}{P_{\text{in}}} \tag{3.26}$$

P_{in} is taken as the product of the irradiance of the incident light, measured in W/m^2 or in suns ($1000 \ W/m^2$), with the surface area of the solar cell in m^2. The maximum efficiency (η_{MAX}) found from a light test is not only an indication of the performance of the device but, like all of the I-V parameters, can also be affected by ambient temperature, the intensity, and spectrum of the incident light. Therefore, PV cells performance are compared with tests under standard test conditions using similar lighting and temperature conditions. These standard test conditions (STCs) are 25°C temperature, air mass (AM) $= 1.5$, and incident light of $1000 \ W/m^2$.

Example 3.6

The IV curve of a PV cell of $100 \ cm^2$ has reverse saturation current $I_0 = 10^{-12}$ A/cm^2. It produces a short circuit current $I_{SC} = 40 \ mA/cm^2$ at 25°C and cell voltage of 0.50 V. Calculate the open circuit voltage of the PV cell under full sun, where incident light is $1000 \ W/m^2$. Calculate the maximum power output, FF, and η at STC.

Solution

$$I_0 = 100 \times 10^{-12} = 10^{-10} \, A$$
$$I_{SC} = 0.040 \times 100 = 4.00 \, A$$

In Eq. (3.16),

$$I = I_{SC} - I_0 \left(e^{38.9V} - 1 \right)$$
$$I = 4.00 - 10^{-10} \left(e^{38.90 \times 0.50} - 1 \right) = 3.12 \, A$$

In Eq. (3.18),

$$V_{OC} = 0.0257 \, \ln \left(4.00 \times 10^{10} + 1 \right) = 0.627 \, V$$

Power output $(P_{out}) = VI = 0.50 \times 3.12 = 1.56 \, W$
Power input $(P_{in}) = 1000 \times 0.01 = 10 \, W$ for $100 \, cm^2$ cell area at 25°C
Maximum power $(P_T) = V_{OC} \times I_{SC} = 0.627 \times 4.00 = 2.5 \, W$

Therefore by Eq. (3.25), $FF = P/P_T = 1.56/2.5 = 0.624$
Efficiency $\eta = P_{out}/P_{in} = 1.56/10.00 = 0.156$ or 15.6%

3.5.1.8 Shunt Resistance (R_{SH}) and Series Resistance (R_S)

During operation, the efficiency of solar cells is reduced by the dissipation of power across internal resistances. These parasitic resistances are modeled as a

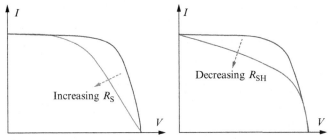

Fig. 3.27 Effect of diverging R_S and R_{SH}.

parallel shunt resistance (R_{SH}) and series resistance (R_S), as depicted in Fig. 3.27 For an ideal cell, R_{SH} would be infinite and would not provide an alternate path for current to flow, while R_S would be zero, resulting in no further voltage drop before the load.

Decreasing R_{SH} and increasing R_S will decrease the FF and P_{MAX} as shown in Fig. 3.27. If R_{SH} is decreased too much, V_{OC} will drop while increasing R_S excessively can cause I_{SC} to drop. It produces short circuit current instead.

It is possible to approximate the series and shunt resistances, R_S and R_{SH}, from the slopes of the I-V curve at V_{OC} and I_{SC}, respectively. The resistance at V_{OC}, however, is at best proportional to the series resistance but it is larger than the series resistance. R_{SH} is represented by the slope at I_{SC}. Typically, the resistances at I_{SC} and at V_{OC} will be measured as shown in Fig. 3.28.

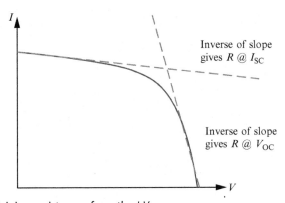

Inverse of slope gives R @ I_{SC}

Inverse of slope gives R @ V_{OC}

Fig. 3.28 Obtaining resistances from the I-V curve.

Example 3.7

A PV module is made out with 36 number cells connected in series. Each cell at STC produces short circuit current of 3.4 A and reverse saturation current is 6.6×10^{-10} A. Parallel resistance R_{SH} is 6.6 Ω and series resistance of $R_S = 0.005$ Ω. Each cell voltage is 0.50 V. Calculate the voltage, current, and power delivered by the module at full sun.

Solution

Number cells in the module $n = 36$. Each cell voltage $V_d = 0.50$ V, $I_{SC} = 3.4$ A, $I_0 = 6.6 \times 10^{-10}$ A, $R_{SH} = 6.60$ Ω, $R_S = 0.005$ Ω.

Using Eq. (3.19),

$$
\begin{aligned}
I &= I_{SC} - I_0 \left(e^{38.9 V_d} - 1 \right) - V_d / R_{SH} \\
&= 3.40 - 6.6 \times 10^{-10} \left(e^{38.9 \times 0.50} - 1 \right) - 0.50/6.60 \\
&= 3.16 \, \text{A}
\end{aligned}
$$

Using Eq. (3.22),

$$
V_{module} = 36(0.50 - 3.16 \times 0.005) = 17.43 \, \text{V}
$$

$$
\text{Power delivered} \quad P_{output} = V_{module} \, I = 17.43 \times 3.16 = 55 \, \text{W}
$$

If the incident light is prevented from exciting the solar cell, the I-V curve shown in Fig. 3.22 is obtained. This I-V curve is simply a reflection of the "no light" curve from Fig. 3.24 about the V-axis. The slope of the linear region of the curve in the third quadrant (reverse bias) is a continuation of the linear region in the first quadrant, which is the same linear region used to calculate R_{SH} in Fig. 3.29. It follows that R_{SH} can be derived from the I-V plot obtained with or without providing light excitation as in Fig. 3.29, even when power is sourced to the cell. These resistances are often a function of the light level, and differ in value between the light and dark tests.

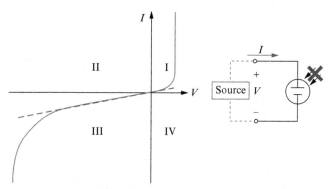

Fig. 3.29 I-V curve of solar cell without light excitation.

3.5.1.9 Temperature Measurement Considerations

The semiconductor crystals used in PV cells are sensitive to temperature. Fig. 3.30 depicts the effect of temperature on an I-V curve. When a PV cell is exposed to higher temperatures, I_{SC} increases slightly, while V_{OC} decreases more significantly.

Higher temperatures decrease the maximum power output P_{MAX}. As the I-V curve will vary according to temperature, it is beneficial to record the conditions under which the I-V sweep was conducted. Temperature is measured using sensors such as thermistors or thermocouples.

3.5.1.10 I-V Curves for Modules

The shape of the I-V curve does not change for a module or array of PV cells. However, it is scaled based on the number of cells connected in series and in parallel. Fig. 3.31 shows an I-V curve for a module with n number of cells connected in series and m number of cells connected in parallel, and I_{SC}

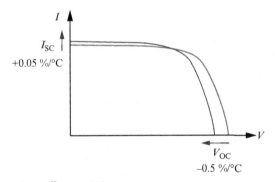

Fig. 3.30 Temperature effect on *I-V* curve.

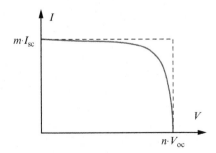

Fig. 3.31 *I-V* curve for modules and arrays.

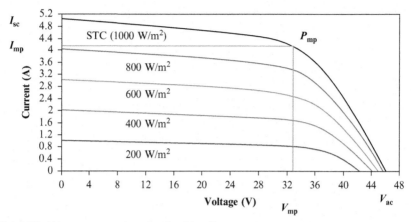

Fig. 3.32 *I-V* curves at various levels of irradiance.

and V_{OC} are valued for individual cells. Fig. 3.32 shows the influence of solar irradiance.

Solar cells produce more efficient photo conversion at cold temperatures and reduce as the temperature increases. The output power at the same irradiance level reduces as the temperature increases. V_{OC} and I_{SC} increase by mirror boosting of irradiance. The electrical parameters of the solar cell have been affected by the tilt angle of the cell, which is dependent on the location latitude angle of the cell and season of field exposure of the cell.

3.6 PV COMPONENTS AND STANDARDS [18–21]

Assembly of PV cells, mostly with 36 cells, connected in four parallel rows connected in series with an area ranging from 0.50 to 1.00 m^2 are called PV modules. Several modules assembled structurally and connected in series electrically are called panels. Assembly of several panels, electrically connected in series, are called arrays, and several arrays, electrically connected in parallel, to produce power is called PV generator. Fig. 3.33 shows the layout configuration.

PV cells are not exactly alike, so also the assembly of modules and assembly of arrays. These assemblies of inequalities of cells, modules, and arrays cause mismatch losses. Shades and faults in arrays and inequality of irradiation in cells cause voltage and current inequalities in the modules and arrays, causing overheating and damage to the modules. External factors such as temperature, shades, and imperfect cell surfaces cause deterioration of cells. Therefore, the cells are adequately encapsulated in a metal frame usually of aluminum and a rear supporting substratum of glass, metal, or plastic.

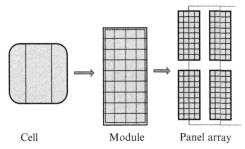

| Cell | Module | Panel array |

Fig. 3.33 PV generator assembly [18].

The PV generator produces direct current. The loads use alternating current. Therefore, an inverter is used to convert DC to AC with modulating waves to make the output to near sinusoidal AC current. The DC output current is fed through a diode static switching configuration to create a square wave of DC and an L-C filter, which equals to a sinusoidal waveform using pulse width modulation (PWM) technique.

The solar PV system is designed and installed as per the following National Electrical Code (NEC) standards and codes. The standards are under development as the technology is evolving rapidly.

- NEC 690—Electrical Codes Article 690-solar photovoltaic systems and NFPA
- NEC 230—Disconnects
- NEC 240—Overcurrent protection
- NEC 250—Grounding
- NEC 300—NEC 384—Wiring methods
- Building Codes: ICC, ASCE 7 UL standard 1701; Flat Plat PV modules and panels
- IEEE 1547, Standards for interconnecting distributed resources with electric power systems
- UL standard 1741, Standards for Inverter, Converters, Controllers and Interconnecting System Equipment for use with Distributed Energy Resources

3.7 PHOTOVOLTAIC POWER SYSTEMS AND TECHNOLOGIES [22–27]

The PV power system is a preferred and emerging technology that has been well understood and promoted for intensive use of renewable source of energy in a DG system. The large-scale PV system is well developed with interconnection systems to be used with a utility power system. A brief discussion on design practices of PV systems is presented in this section.

3.7.1 PV System Design Practices

The PV system is broadly classified into two major groups:

1. Stand-alone DC solar power system with and without battery back up

2. Grid-connected AC solar power system

PV systems comprise the following major equipment duly designed and wired to provide the required power loads. The DC power generated in solar modules is used instantly on the premises or by feeding it into the grid with the module wired up with a few other electrical components. The exact combination depends on the size of the installation and type of load, either stand-alone or grid-connected. Figs. 3.34 and 3.35 show a typical PV system component for stand-alone and a typical grid-connected, large-scale installation, respectively.

Grid-connected PV systems typically have the following basic components;

- Solar PV modules
- Array mounting racks and supporting structure
- Grounding equipment
- Combiner box
- Surge protection equipment
- Inverter to convert DC to AC
- Meters to measure system power and energy
- Disconnects:
  - Array disconnect
  - Inverter DC disconnect
  - Inverter AC disconnect
  - Exterior AC disconnect

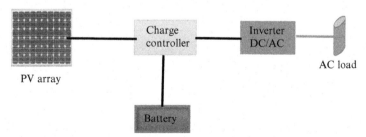

Fig. 3.34 Typical standalone PV system components.

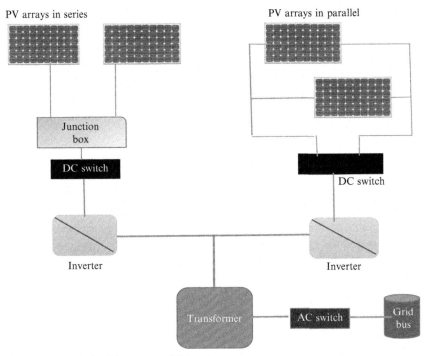

Fig. 3.35 A typical grid-connected PV system.

A grid-connected battery support system also includes the following components:

• Battery bank with cabling
• Charge controller
• Battery disconnect

Many individual solar cells are wired and connected in series to produce PV modules. The modules are wired in series at the site to forms strings. The strings of modules are wired together in parallel to form arrays. The rigid and flat framed modules are mostly either of single or multicrystalline or amorphous thin film types of solar cells. Other types of cells are either CdTe or CIS types. Buildings integrated with roof tiles PV cells (BIPV) are also used as modules.

Grid-connected PV systems have modules with rated power output of 100–300 W with PV system voltage of 235–600 V at standard temperature conditions of 25°C per module. The rated peak power of the module is expressed as kWp as the maximum power output from the panel with 1 kW/m^2 solar power at 25°C at air mass of AM = 1.5. Modules by different manufacturers are compared by factors on module cost per watts and module efficiency by watts per area in a square meter. Lower cost per watts and higher efficiency with higher watts in the lesser area are desired in module selection for PV systems. Array mounting racks may be used either for a building roof top or for direct on the ground support. The rack structure shall have adjustability to fit in the roof slope and to hold the automatic sun tracking system. Generally, one support bracket is used for a sun tracking system for every 100 W module.

The grounding equipment provides a well-defined low earth resistance path to surges from the panels to the earth as per standards for grounding. All system components including equipment metal boxes and appliance frames and PV equipment mountings are to be properly grounded for electrical safety as per codes and standards. A combiner box typically includes a safety fuse or breaker for each string with a surge protector for protection against lightning or surge voltage. Energy meters are used to measure energy delivered to or from the grid and a system meter either separately or part of a charge controller is used to display module performance.

An inverter is used for the following tasks:

- Converting the DC power from the panel or from the battery to AC power to the grid or to the AC load
- Ensuring the required frequency
- Reducing the voltage fluctuations
- Ensuring the AC wave shape suitable to utility sinewave

Inverters are carefully designed either for the grid system with and without a battery back-up system. Inverter design shall be strictly with reference to standards to match the utility system in terms of power quality, AC power output, and voltage input of 235–600 V for a system without a battery and 12–48 V for a system with a battery. Other important factors considered for good quality inverters are start-up surge capacity, frequency and voltage regulation, higher efficiency up to 92%, maximum power point tracking (MPPT), and integral safety disconnects. Inverter charge controllers for battery-based systems have factory integrated charge controllers, facilities to supply critical loads, and ensures load shedding during utility power outages.

Automatic and manual disconnects protect the wiring and other components from power surges and malfunctions of other components. The disconnects are used to safely shut down relevant equipment for maintenance and repairs. A disconnect is needed for each source of power or energy storage equipment. The disconnects ensure safety to equipment and operating personnel in case of fault isolation and during operation and maintenance in each live source in the power system.

Batteries are used to store surplus power generation for later use. The use of a battery bank increases the cost of a PV system and its complexity reduces system efficiency by ~10%. Two types of batteries are used in a PV system:
- Lead–acid batteries
  - Flooded
  - Sealed
    - Absorbent glass mat (AGM)
    - Gel cell
- Alkaline batteries
  - Nickel-cadmium
  - Nickel-iron

For grid-connected systems, the batteries are sized generally for short time periods, typically for 8 h. Solar arrays shall have higher voltage to fully charge these batteries. The rated voltage of the module should be V_{mpp}, also called maximum power point in the electrical specifications. For example, a 16.5–17.5 V module is required to fully charge a 12 V battery.

A charge controller in a PV system is also referred to as a PV controller or battery charger. The primary function of a charge controller is to prevent over-charging and overdischarging of batteries. Some controllers have a facility for MPPT, which optimizes the PV array's output for maximum energy. Two types of charge controllers are used: series and shunt controllers. Shunt controllers are inexpensive and bypass current when the battery is fully charged. Series controllers stop the current between the battery and the PV array when the battery is fully charged. A pulse controller, a type of shunt controller referred to as a multistage controller, optimizes the charging rate and thus increases the life of the battery. The charge controller is selected to deliver the charging current for the type of battery used in the system. The selection of charger depends on the charging requirement specified by the battery supplier.

3.7.2 Design of a PV Grid-Connected System

The following factors are important to consider when designing a perfect PV system for best performance. Site evaluation is first and foremost in the design a PV system. The site should have the following parameters.
- The site should have a good and excellent solar potential. Solar potential has been already evaluated, and published data is available with reference to Earth's latitude and longitude worldwide.

- The site should be suitable for mounting PV panels, ground mounted, roof mounted, and so on.
- PV arrays are affected badly by shading. The site should be free from shade by obstructions from trees and buildings. The site should be clear and unobstructed to sun rays for longer periods of days.
- PV modules should have good orientation, mostly true south, to attract more sunlight during the day. PV modules should be mounted on frames and tilted toward true south.
- The tilt of a PV array in a location achieves maximum power output during the year.
- The required area for installation of PV modules depends on the size of the power output. Five acres of land is required to generate 1 MWp of power in India. This area requirement depends on the location, the solar irradiance, and PV modules' capacity.

The AC energy output from the solar array is the energy delivered to the point of connection of the grid connect from the inverter to the grid. The output of the solar array is affected by the following factors:

- The average solar radiation from the published data available for the selected tilt angle and orientation and shading
- Manufacturing tolerance as specified by the supplier of modules
- Temperature effects on the modules
- Effects of dirt on the modules
- System losses such as power loss in cables
- Inverter efficiency

The system energy output over a whole year is called the energy yield and is obtained from the peak power rating (kWp) under STCs for a solar array and derated for the losses due to the factors discussed above. Solar radiation is typically expressed in $kW\,h/m^2$ or as daily peak sun hours (PSH), which is the number of hours of solar irradiance of $1\ kW/m^2$.

An example of designing a grid-connected PV system is illustrated for understanding the concept of system design. The design of a solar PV system sizing follows the following steps.

(a) Determine power consumption demands

The first step in designing a solar PV system is to find out the total power and energy consumption of all loads that need to be supplied by the solar PV system as follows:

Calculate the total watt-hours per day for each appliance used. Add the watt-hours needed for all appliances together to get the total watt-hours per day. Multiply the total appliances watt-hours per day times

1.3, accounting 30% for the energy lost in the system, to get the total watt-hours per day, which must be provided by the panels.

(b) Size the PV modules

Different sizes of PV modules will produce different amounts of power. The peak watt (Wp) produced depends on the size of the PV module and climate of the site location. We have to consider "panel generation factor," which is different in each site location. For example, the panel generation factor is 3.4 to a particular location. The sizing of PV modules is obtained by dividing the total watt-peak rating by the panel generation factor 3.4. The number of panels is arrived at by dividing the total watt-peak power divided by the panel generation factor and the rated output watt-peak of the modules approximated to the nearest full figure that gives the minimum number of PV panels. If more PV modules are installed, the system will perform better and battery life will be improved.

(c) Inverter sizing

An inverter is used in the system where AC power output is needed. The input rating of the inverter should never be lower than the total watts of appliances. For a stand-alone system, the inverter must have the same nominal voltage of the battery. The inverter size should be 25%–30% bigger than the total watts of appliances. If the appliance type is a motor or compressor then inverter size should be a minimum of three times the capacity of those appliances, and must be added to the inverter capacity to handle surge current during starting. For grid-tie systems or grid-connected systems, the input rating of the inverter should be the same as the PV array rating to allow for safe and efficient operation.

(d) Battery sizing

A deep cycle battery is specifically designed for a solar PV system. The battery should be large enough to store sufficient energy to operate the appliances at night and on cloudy days. The size of battery is calculated as follows:

The total watt-hours per day is accounted for by assuming battery loss at 0.85 and 0.60 for depth of battery discharge of the battery and by the nominal battery voltage, and multiplied by the days of autonomy to get the required ampere-hour capacity of the deep cycle battery. The days of autonomy are the number of days that the battery has to operate the loads when PV panels do not produce power. Battery capacity (Ah) = Total watt-hours per day used by appliances multiplied by days of autonomy $(0.85 \times 0.60 \times$ nominal battery voltage).

(e) Solar charge controller sizing

The solar charge controller is typically rated against amperage and voltage capacities. The solar charge controller must match the voltage of the PV array and batteries. The solar charge controller must have enough capacity to handle the current from the PV array. For the series charge controller type, the sizing of the controller depends on the total PV input current that is delivered to the controller and also depends on the PV panel configuration, series or parallel configuration. According to standard practice, the sizing of a solar charge controller is to take the short circuit current (I_{SC}) of the PV array and multiply it by 1.3. Solar charge controller rating = Total short circuit current of PV array × 1.3. For a MPPT charge controller, sizing will be different.

Example 3.8

A house has the following electrical appliance usage: Four 18-W fluorescent lamps with electronic ballast used 4 h/day. Two 60-W fans used for 2 h/day. One 75-W refrigerator that runs 24 h/day with the compressor run 12 h and off 12 h. The system will be powered by a 12 V_{dc}, 110 Wp PV module.

a. Determine power consumption demands

Total appliance use = $(4 \times 18 \text{ W} \times 4 \text{ h}) + (2 \times 60 \text{ W} \times 2 \text{ h}) + (75 \text{ W} \times 24 \times 0.5 \text{ h}) = 1428$ W h/day

Total PV panels energy needed = $1428 \times 1.3 = 1856.40$ W h/day.

b. Size the PV panel

Total Wp of PV panel capacity needed = $1856.40/3.40 = 546$ Wp

Number of PV panels needed = $546.00/110 = 4.96$ modules

Actual requirement = 5 modules

So this system should be powered by at least eight modules of 110 Wp PV module.

c. Inverter sizing

Total watts of all appliances = $72 + 120 + 75 = 267$ W

For safety, the inverter should be considered to be a 25%–30% bigger size.

The inverter size should be about 350 W or greater.

d. Battery sizing

Total appliances use = $(72 \text{ W} \times 4 \text{ h}) + (120 \text{ W} \times 2 \text{ h}) + (75 \text{ W} \times 12 \text{ h}) = 1428$ W h

Nominal battery voltage = 12 V

Days of autonomy = 3 days

Battery capacity = $1428 \times 3/(0.85 \times 0.6 \times 12)$

Total ampere-hours required = 700.00 Ah

So the battery should be rated 12 V, 700 Ah for 3-day autonomy.

e. Solar charge controller sizing
PV module specification
$P_m = 110$ Wp
$V_m = 16.7$ V_{dc}
$I_m = 6.6$ A
$V_{OC} = 20.7$ A
$I_{SC} = 7.5$ A
Solar charge controller rating $= (5$ strings $\times 7.5$ A$) \times 1.3 = 48.75$ A
So the solar charge controller should be rated at 50 A at 12 V or greater.

Example 3.9

Design a power system with 16 modules of 160 Wp rated monocrystalline PV modules in a place where the average PSH is 5 and ambient temperature is 30°C. Assume standard factors of derating modules/cells to account for losses and performance factors and calculate the average AC energy deliverable to the grid with its performance ratio (PR).

Solution

The derating of modules is assumed as follows:

- Derating of modules by manufacturers: 3%–5%
- Derating for dirt over modules: 5%
- Power loss due to temperature for every degree over standard temperature condition: 25°C: $= 0.5\%$
- Derating of module type over temperature (monocrystalline): 0.45% per degree over standard temperature condition
- DC system cable loss: 3%
- Efficiency of the inverter: 96%
- AC cable loss before delivering to grid: 1%

The solar power output from 160 Wp is derated on the above factors. The output power is reduced on derating. Ambient temperature plus the standard temperature condition is the cell temperature (55°C).

That is, DC power output from one module:

$$= 160 \times 0.95 \times 0.95 \times 30 \times 0.50 \times 30 \times 0.45$$
$$= 124.90\,W$$

The DC power output $=$ module output $\times$ number of modules in array $\times$ average daily PSH:

$$= 124.90 \times 16 \times 5$$
$$= 9992.00\,Wh$$

DC cable loss between array to inverter is 3%, and inverter efficiency is assumed to be 96%.

AC cable loss is between inverter and grid.

AC energy output from inverter to grid $= 9992.00 \times 0.97 \times 0.96 \times 0.99$

$$= 9.212\,\text{kWh/day}$$

Yearly energy yield $= 365 \times 9.212$

$$E_{\text{sys}} = 3362.40\,\text{kWh/year}$$

Power output from array at standard temperature condition, P_{array}
$= 16 \times 160 = 2.56\,\text{kW}$

$$\text{Specific energy yield} = E_{\text{sys}}/P_{\text{array}}$$
$$= 3362.40/2.56$$
$$= 1313.43\,\text{kWh/kWp}$$

The performance ratio PR is evaluated to assess the installation quality and to compare different types and sizes of PV systems. It is the ratio of yearly energy yield to that of ideal energy output of the array. E_{ideal} depends on the yearly average daily irradiance in kW h/m^2 for the specified tilt angle. This is expressed as PSH that is equivalent to 1 kW h/m^2. The H_{tilt} is the average daily irradiation at a specific tilt angle is expressed in PSH hours.

The yearly irradiation $= 365 \times H_{\text{tilt}}$
$$= 365 \times 5 = 1825\,\text{h}$$
The rated peak power output from array $= 16 \times 160 = 2.56$ kW
The ideal yearly energy from the array, $E_{\text{ideal}} = 2.56 \times 1825 = 4672$ kW h
The performance ratio (PR) $= E_{\text{sys}}/E_{\text{ideal}} = 3362/4672 = 0.72$

Example 3.10

Design an inverter for the solar PV array in Example 3.9 above.

Solution

Inverter design depends on the energy output from the array. The inverter size is decided based on the maximum DC input power and DC input current to the inverter and AC output power to the grid. The array voltage from PV modules must match with the minimum and maximum voltage at the maximum power point of the inverter.

The average peak power from the array $= 16 \times 160 = 2.56$ kW
The derating factor of the array $= 0.95 \times 0.95 \times 0.865 = 0.78$
Inverter must be rated for $= 2.56 \times 0.78 = 2$ kW

The inverter operating voltage must match the array voltage. The array voltage should not exceed the maximum inverter operating voltage. The maximum power point voltage V_{mp} of the array should not fall below the operating inverter voltage. Assuming the maximum cell temperature at 70°C, the voltage coefficient is assumed to be 0.177 per degree over standard temperature condition. The selected module voltage at maximum power point is 35.4 V. The effective cell temperature of 70°C is 45°C above standard temperature condition.

Therefore,

V_{mp} is reduced by $= 45 \times 0.177 = 7.97$ V

V_{mp} at 70°C $= 35.4 - 7.97 = 27.43$ V

Assuming a cable loss of 3% the V_{mp} at maximum power point is $= 0.97 \times 27.43 = 26.57$ V

This is the effective minimum input voltage to the inverter, assuming the minimum voltage for an inverter is 140 V and a safety factor of 10%.

The minimum inverter voltage is $1.10 \times 140 = 154$ V.

The minimum number of modules in a string $= 154/26.57 = 5.79 =$ say six modules

---

3.7.2.1 Maximum Voltage of Inverter

The lowest temperature of the day is used to determine the maximum V_{OC}. The effective minimum cell temperature is assumed as 15°C with V_{OC} of 43.20 V and a coefficient of 0.14°C. V_{OC} increases with low temperature at 10°C less than standard temperature condition by 1.40 V. That is, V_{OC} at 15°C is 44.60 V. The maximum voltage allowed by the inverter is 400 V. The maximum number of modules in the string is equal to (400/44.60) 8.96 modules, say eight modules. Since 16 modules are used, two parallel strings of 8 modules are considered to maintain the maximum voltage limit of the inverter.

3.8 MATERIALS FOR FUTURE PHOTOVOLTAIC SYSTEMS [28–41]

The development of new materials for future PV cells is towards reducing the cost per peak power output, improved electrical performance for longer life time, improved device stability under open environment, hazard and environment impact free material for module support system and to improve the conversion efficiency of light to electricity. A lot of researches on the following key areas are focused for developing;

- New absorber materials, better contact and barrier material.
- New electron/hole contact layers. combine molecular design using computational resources, develop new acceptors and donors to enhance device performance.

- Mechanisms of materials and device degradation. new materials and device architectures that mitigate degradation at different substrate temperatures.
- Environment impact free and health hazard free materials for future PV cells.

This section is devoted to discuss on various research developments for future PV systems.

3.8.1 Challenges and Opportunities [28–30]

The future of PV power systems is highly promising and lot of focus on promoting the technology is found the world over. There are great challenges and opportunities in developing this potential technology to harness solar energy and integrate into grid-connected utility systems. Therefore, a review of the challenges and opportunities with trends in ongoing research in materials, methods, performance procedures, and training for skills is presented.

Fig. 3.36 shows a new type of PV cell called organic photovoltaic (OPV) cells. OPV solar cells technology provides electricity at a lower cost than first- and second-generation solar technologies. Various absorbers can be used to create colored or transparent OPV devices. OPVs have achieved efficiencies near 11%.

OPV cells use molecular or polymeric absorbers, which results in a localized exciton. The absorber is used in conjunction with an electron acceptor, such as a fullerene, which has molecular orbital energy states that facilitate electron transfer. Upon absorbing a photon, the resulting exciton migrates to the interface between the absorber material and the electron acceptor material. At the interface, the energetic mismatch of the molecular orbitals

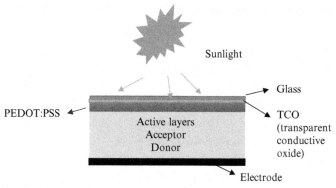

Fig. 3.36 Organic photovoltaic cell structure [28].

provides sufficient driving force to split the exciton and create free charge carriers. The low efficiencies of OPV cells are related to their small exciton diffusion lengths and low carrier mobilities. These two characteristics ultimately result in the use of thin active layers that affect overall device performance. Furthermore, the operational lifetime of OPV modules remains significantly lower than for inorganic devices.

Current research focuses on increasing device efficiency and lifetime. Substantial efficiency gains have been achieved already by improving the absorber material, and research is being done to further optimize the absorbers and develop an organic multijunction architecture. Improved encapsulation and alternative contact materials are being investigated to reduce cell degradation and increase cell lifetimes.

OPV solar cells have the following advantages:

- Low-cost manufacturing: Soluble organic molecules enable roll-to-roll processing techniques and allow for low-cost manufacturing.
- Abundant materials: Abundant availability of materials reduce price constraints.
- Flexible substrates: Flexible substrates permits a wide variety of uses.

OPV cells are categorized into two classes:

- Small-molecule OPV cells
- Polymer-based OPV cells

Small-molecule OPV cells use molecules with broad absorption in the visible and NIR portion of the electromagnetic spectrum. Highly conjugated systems are typically used for the electron-donating system such as phthalocyanines, polyacenes, and squarenes. Perylene dyes and fullerenes are often used as the electron-accepting systems. These devices are most commonly generated via vacuum deposition to create bilayer and tandem architectures. Recently, solution-processed small-molecule systems have been developed.

Polymer-based OPV cells use long-chained molecular systems for the electron-donating material (e.g., P3HT, MDMO-PPV), along with derivatized fullerenes as the electron-accepting system (e.g., $PC_{60}BM$, $PC_{70}BM$). Like small-molecule OPV cells, these systems have small exciton diffusion lengths. However, this limitation is circumvented by a high interface surface area within the active device.

Dye-sensitized solar cells (DSSC) are a hybrid organic-inorganic technology that uses small-molecule absorber dyes. These dyes adsorb onto a suitable electron-accepting material, such as titanium dioxide or zinc oxide, along with an electrolyte to regenerate the dye.

The promising economic and social opportunities are in the following key areas:

a. Waste disposal and new business opportunities
b. Cost reduction in PV systems
c. Training and skill development of the workforce and economic growth
d. Consultancy and evaluation of site performance of PV systems
e. R&D opportunities in manufacturing PV systems
f. Financing sector opportunities
g. Operation and maintenance opportunities for social and economic growth

3.8.1.1 End-of-Life Management [31]

Solar PV panels have a roughly 30-year lifetime. A large stock of raw materials and other valuable components are projected as PV panel wastes on end of life. These wastes may be recycled or used for repurposing solar PV panels. PV panel wastes comprise mostly of glass, estimated to total 78 million tonnes worth of USD 15 billion globally by 2050. This potential material influx could produce 2 billion new panels into global markets. This may increase the security of future PV supply or other raw material-dependent products. PV panel wastes create new environmental challenges, but also unprecedented opportunities to create and pursue new economic avenues to establish an industry on PV-specific waste regulation and waste management infrastructure. End-of-life PV management creates industry value creation in R&D organization, repair/reuse services, and the recycling treatment industry.

3.8.1.2 The International Renewable Energy Agency (IRENA) [32]

The cost reduction potential from solar PV technology could be by 59% between 2015 and 2025. The reductions in the cost of balance-of-system, operation and maintenance, and capital costs are becoming the important drivers for overall cost reduction.

3.8.1.3 Analytical Monitoring of PV Systems—IRENA [33]

The report provides detailed guidelines, methods and models for the analysis of the performance of PV systems. The best practices in PV monitoring have described and analyzed the energy flow in a grid–connected PV system with a selected collection of variables. Effects related to special PV technologies, namely CIGS and amorphous silicon PV may be different from experimental installations in the field. The specific behavior has to be compared to the expected performance of the designed performance. Variations in performance, on account of the sensitivity to the spectral composition and measurement of the incoming sunlight, have to be adjusted in site installations. These adjustments at site, based on meteorological variations, degradation, or other changes in the solar cell materials, have to be done periodically and by experts certified for the trade. Regarding system design decisions, the main factors of influence are mounting angle and row distance, related to irradiance gains and shading losses, inverter to module power ratio, and cabling optimizations. Several challenges on both shading losses and inverter to module power ratio are important to be addressed. These monitoring activities in emerging PV technologies provide great opportunities for service providers in the industry.

3.8.1.4 Workforce Challenges and Opportunities [34]
The potential economic benefits in employment generation and local business development associated with the solar PV industry with increased local employment and increased percentage of locally procured materials are significant. The local employment is in system installation, in component manufacturing and in operation and maintenance. Almost 50% of the local employment would be professional occupations, while the remaining are factory workers, field workers, and sales and marketing representatives. Lack of qualified and experienced installation technicians, project managers, and engineers with knowledge and background in the solar industry is posing a great challenge in workforce recruitment. There is the need to keep the industry sustainable by making sure an adequate workforce is available as demand increases. There are opportunities for training of PV manufacturing experts, skill development for PV design, and installation and service institutes.

3.8.1.5 Emerging Challenges and Opportunities for Manufacturing [35]
The global PV market has changed dramatically in recent years. Module prices have been reduced reasonably, and global deployment has grown strongly and shifted. China became a major demand market by consolidating its dominance in PV manufacturing. These include policy factors, financial factors, strong global market demand set against capacity shortages, and a resulting free flow of goods, services, and labor that accelerated the growth in China. The rapid and protracted capacity build-out is unlikely to be replicated. PV manufacturing faces challenges globally

on the relatively low price of incumbent electricity-generating sources in most large global PV markets, and the pricing achievable for PV products and services. This factor, combined with slowing rates of manufacturing cost reductions, may constrain profit opportunities for firms and poses a potential challenge to the sustainable operation and growth of the global PV manufacturing base. PV-specific characteristics provide opportunities to exploit global industry changes and accelerate manufacturing expansion. Cost reductions for standard PV modules appear to be slowing, and the path to continued reductions will require improved cell and module efficiencies typically found with innovative, advanced device architectures. A greater reliance on innovation could benefit PV manufacturers. The United States is a world leader in PV patents and research and development (R&D) expenditures. PV manufacturers already are pursuing diverse technological innovations.

3.8.1.6 Emerging Challenges and Opportunities in the Financing Sector [36]

Solar technology, solar markets, and the solar industry have changed dramatically. Cumulative solar deployment has increased more than 10-fold, while solar's levelized cost of energy (LCOE) has dropped by as much as 65%. New challenges and opportunities have emerged as solar technology has become much more affordable, and solar technologies have been deployed at increasing scale. Solar energy technologies have had high upfront costs and low operating costs, they provide long-term benefits, and financing is critically important to the solar market. A wide variety of federal, state, and local incentives have helped make solar more affordable and competitive but some of these incentives have also had a significant impact on financing solar projects. The important incentives are federal tax incentives for solar energy use, investment tax credits, and accelerated tax depreciation. The creation of complicated tax equity structures monetizes these tax benefits. These tax equity structures used in the utility-scale solar sector have also spurred the development of PV systems. Though these complicated structures have driven significant solar deployment, they are increasingly recognized as inefficient and costly, relying almost exclusively on the two most expensive sources of capital: sponsor equity and tax equity. Therefore, solar industry stakeholders have expended significant effort in developing lower-cost financing solutions for solar energy to reach long-term competitiveness with traditional sources of energy.

3.8.2 Research Trends on Materials and Technology Development [37–41]

Researchers are also developing other Earth–abundant materials, such as iron sulfide (FeS_2), lead sulfide (PbS), and tin sulfide (SnS). These materials have exhibited some solar-relevant properties and have similarities to other solar absorber materials.

3.8.2.1 Hybrid Organic-Inorganic Halide Perovskite Solar Cells [37]

The field of new materials for hybrid organic-inorganic perovskite solar cells has been dominated by absorber materials based on methyl ammonium lead halide perovskites. Perovskite solar cells have shown remarkable progress in recent years with rapid increases in conversion efficiency of 20%. Perovskite solar cells may offer the potential for an Earth–abundant and low-energy-production solution to truly large-scale manufacturing of PV modules. The application of a solid-state hole transport material and improvements in performance and stability with mixed halide perovskites, improved contact materials, new device architectures, and improved deposition processes has increased the efficiency from the initial 4% to 20%.

Hybrid organic-inorganic perovskite solar cells have proved to have higher competitive efficiencies and higher performance. Researchers are studying the relevant degradation mechanisms in both the perovskite materials and the contact layers, improved cell durability, and the development of commercial perovskite solar products.

Fig. 3.37A and B shows a hybrid OPV cell, thin film perovskite silicon cell and perovskite silicon tandem cell, respectively. The material has the potential environmental impacts related to the lead-based perovskite absorber. Current efforts are on lead-free perovskite structures to eliminate potential environmental concerns. Perovskite solar cells have very low energy losses, allowing for high open-circuit voltage, and serve as an excellent wide gap absorber in tandem devices with low gap absorbers, such as crystalline or multicrystalline silicon cells. A silicon-based tandem device architecture using a low-cost, high-performance, wide band gap perovskite cell might offer a cost-effective path toward high-efficiency modules. Researchers are presently evaluating wider band gap perovskite absorbers and contact materials for the tunnel junction between the subcells, as well as mechanical tandem architectures. A final challenge lies in scale-up and optimization of the deposition processes for reproducible perovskite solar cell performance. The benefits are

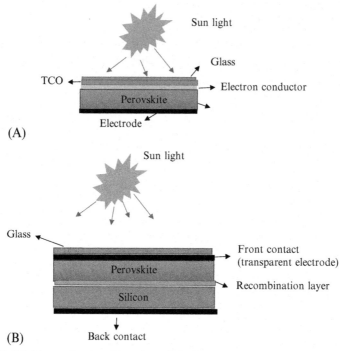

Fig. 3.37 Hybrid organic perovskite solar cells [37] (http://www.energy.gov/eere/sunshot/hybrid-organic-inorganic-halide-perovskite-solar-cells). (A) Thin film perovskite solar cell. (B) Perovskite on silicon tandem solar cell.

- Low cost technology and ease of manufacturing.
- Perovskites are stable and available in abundant resource materials, such as methyl ammonia, lead, and iodine.
- The band gap of perovskite solar cells can be modified by modifying the composition of the perovskite material, enabling for higher efficiency tandem PV applications.

Table 3.6 shows a comparison of performance characteristics with different types of solar cells.

3.8.2.2 Multijunction III–V PVs [38]

Fig. 3.38 shows a typical multijunction III–V solar cell. Research drives down the costs of the materials, manufacturing, tracking techniques, and concentration methods used with this technology. High-efficiency multijunction devices use multiple band gaps, or junctions, that are tuned to absorb a specific region of the solar spectrum to create solar cells having record efficiencies over

Table 3.6 Comparison of performance of Solar cells

Characteristics	CdTe	CIGS	c-Si	Perovskite
Raw materials cost	Low	Medium	Low	Low
Finished materials cost	Low	High	High	Low
Fabrication cost	Medium	Medium	High	Low
Energy payback period	Medium	High	High	Low
LCOE	Medium	High	High	Low
Efficiency	Medium	Medium	High	High

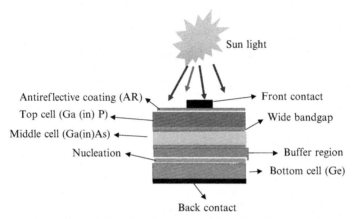

Fig. 3.38 Multijunction III–V PV cell structure [38] (http://www.nrel.gov/pv/high-concentration-iii-v-multijunction-solar-cells.html).

45%. The maximum theoretical efficiency that a single-band gap solar cell can achieve with nonconcentrated sunlight is about 33.5%, primarily because of the broad distribution of solar-emitted photons. This limiting efficiency, known as the Shockley-Queisser limit, arises from the fact that the open-circuit voltage (V_{OC}) of a solar cell is limited by the band gap of the absorbing material and that photons with energies below the band gap are not absorbed. Photons that have energies greater than the band gap are absorbed, but the energy greater than the band gap is lost as heat.

Multijunction devices use a high-band gap top cell to absorb high-energy photons while allowing the lower-energy photons to pass through. A material with a slightly lower band gap is placed below the high-band gap junction to absorb photons with slightly less energy with longer wavelengths. Typical multijunction cells use two or more absorbing junctions, and the theoretical maximum efficiency increases with the number of junctions. Multijunction devices have leveraged the properties of semiconductors comprised from elements in the III and V columns of the periodic table, such as gallium indium phosphate (GaInP), gallium indium arsenide (GaInAs), and gallium arsenide (GaAs). Three-junction devices using III–V semiconductors have reached efficiencies of >45% using concentrated sunlight. Multijunction cells made from CIGS, CdSe, silicon, organic molecules, and other materials are under investigation. The use of concentrating optics such as Fresnel lenses with semiconductor substrates and a dual-axis sun-tracking system may reduce the high costs. The concentrating optics increase the amount of light incident on the solar cell, thus leading to more power production. Multijunction III–V cells have higher efficiencies than competing technologies but are costlier because of fabrication techniques and materials. Therefore, active research efforts are on developing new substrate materials, absorber materials, and fabrication techniques; increasing efficiency; and extending the multijunction concept to other PV technologies. The benefits of multijunction III–V solar cells are

- Spectrum matching: High-efficiency cells (>45%) are fabricated with specific absorber layers having specific band gaps to match the solar spectrum.
- Crystal structure: The various combinations of III–V semiconductors have similar crystal structures and ideal properties for solar cells, including long exciton diffusion lengths, carrier mobility, and compatible absorption spectra.

3.8.3 New Materials for Future Photovoltaic Systems [39–41]

3.8.3.1 High-Efficiency DASH Solar Cell [39]

A new mix of materials known as dopant-free asymmetric heterocontacts (DASH) has been reported by researchers for use in solar cells. Fig. 3.39 shows the details. The new mixer of materials is amorphous silicon that is applied as layers over the crystalline silicon core and coated with ultrathin coatings of molybdenum oxide (moly) at the sun-facing surface of the solar cell and lithium fluoride at the bottom surface. This process eliminates the doping process and increases the efficiency to 20% over the earlier 14% efficiency. Molybdenum oxide and lithium fluoride have properties, transparent and complementary electronic structures, that make dopant-free electrical contacts for holes and electrons, respectively.

Lithium fluoride has been found good for electron contacts to crystalline silicon coated with a thin amorphous layer. That layer complements the moly oxide layer for hole contacts. The thermal evaporation technique at

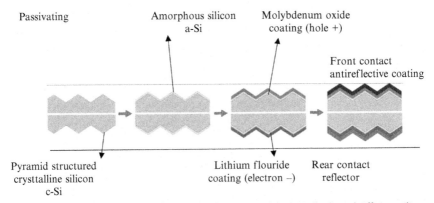

Passivating   Amorphous silicon   Molybdenum oxide
                          a-Si   coating (hole +)

                                                     Front contact
                                                     antiref lective coating

Pyramid structured   Lithium f louride   Rear contact
crysttalline silicon   coating (electron –)   reflector
    c-Si

Fig. 3.39 DASH solar cell. *(Reproduced with permission from J. Bullock et al. Efficient silicon solar cells with dopant-free asymmetric heterocontacts, Nat. Energy (2016). http://nano. eecs.berkeley.edu/publications/NatureEnergy_2016_SiDASH.pdf. Macmillan Publishers Ltd: Nature Energy, 15031/Doi:10.1038/NENERGY.2015.31, copyright ©2016.)*

room temperature has been used to deposit the layers of lithium fluoride and moly oxide.

Silicon commonly would be the dominant solar material, but a promising class of materials called perovskites are being developed. Perovskites have a chemical structure with unique electronic properties. This material combines the light-absorbing characteristics of silicon with those of a perovskite material and achieves 25% efficient performance with a silicon-perovskite tandem device.

High-performance tandem devices made of semiconductor materials other than perovskite have been reported to have a lab efficiency over 40%, but they are extremely expensive due to technically complex manufacturing processes. Making perovskite solar cells as a tandem device with silicon is much simpler and cheaper in manufacturing and integrating with existing processes. Perovskite with the substitution of cesium ions maintaining the materials' structural stability achieves the desired PV properties and enhances the absorption range of the solar spectrum for high performance by developing silicon-perovskite tandem devices.

Perovskites are materials with a particular crystalline structure. The perovskite contains relatively abundant and cheap materials including ammonia, iodine, and lead. Perovskites also convert certain parts of the solar spectrum into electricity more efficiently than silicon. The researchers took a cheap silicon solar cell with an efficiency of 11.4% and increased it to 17% by adding the perovskite cell with a transparent electrode made of silicon nano-wires. The researchers believe that perovskite-silicon cells will convert over 30% of the energy in sunlight into electricity, which would reduce the cost of panels and installation.

Multiferroic films are increasingly considered for applications in solar energy conversion because of their efficient ferroelectric polarization-driven carrier separation and above-band gap generated photovoltages with energy conversion efficiencies beyond the maximum value ($\sim 34\%$). A new approach is to effectively tune the band gap of double perovskite multiferroic oxides.

3.8.3.2 National Renewable Energy Laboratory (NREL) [40]

NREL researches in developing OPV cells, transparent conducting oxides (TCOs), combinatorial (combi) methods, and atmospheric processing. OPV is a rapidly emerging PV technology with improving cell efficiency (currently >8%), encouraging an initial lifetime of more than 5000 h, and potential for roll-to-roll manufacturing processes. OPV cells may be emerging as attractive because of the availability of absorbers in several different colors and the ability to make efficient transparent devices. OPV's great strength lies in the diversity of organic materials that can be designed and synthesized for the absorber, acceptor, and interfaces. Fig. 3.40 shows the details of an OPV cell.

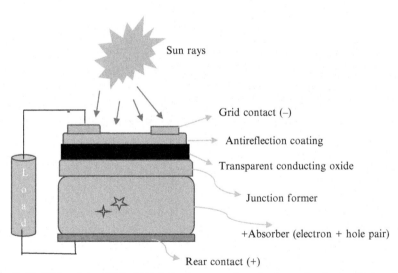

Fig. 3.40 OPV cell by NREL [40] (http://www.nrel.gov/pv/organic-photovoltaic-solar-cells.html).

3.8.4 Research and Development Challenges

The key research and development (R&D) challenges are in the following areas:

- New absorber, contact, and barrier materials. New high-performance absorber materials for improved performance and lifetime, focusing on improving photovoltage, and stability to photo-oxidation.
- Better contact materials and device architectures. New or improved TCOs, including robust indium-free TCOs for PV applications. Development of inverted OPV devices.
- New electron/hole contact layers. Development of new acceptors and donors to enhance device performance and lifetime.
- Mechanisms of materials and device degradation. New materials and device architectures that mitigate degradation, leading to improved device stability.

Atmospheric processing can reduce overall cost and scalability issues of PV production. Focus on developing "inks" and ink-conversion processes to create materials with desired properties. Research is performed across all PV technologies, that is, silicon, thin film, and OPV. Atmospherically processable materials developed include metals (Ag, Ni, Cu, Al), semiconductors (CuInGa[S,Se], CuZnSn[S,Se], CdTe), and oxides (ZnO, In_2O_3, SiO_x, $Ba_xSr_{1-x}TiO_3$).

The National Renewable Energy Laboratory (NREL) continues R&D in PV materials and devices, measurement and characterization, reliability testing and analysis, and development of new materials and devices.

Material technologies include crystalline silicon and thin film materials with high-efficiency concepts such as polymer solar cells and nanostructured solar cells. Device processing uses many different techniques to deposit PV materials and process PV devices.

3.8.4.1 NREL Research Contributions

- Polymer solar cells into a high vacuum deposition chamber, where metal contacts are deposited onto the solar cells.
- A pulsed laser deposition chamber to grow solar cell samples.
- Analytical devices—custom laser equipment for assessing semiconductor crystal structure.
- Measure, image and characterize properties of PV and electronic materials, devices, and interfaces. This includes such properties as optical, electrical, material, surface, chemical, and structural performance.

Researchers develop new materials that can be used for module encapsulants and moisture barrier coatings under a variety of conditions to ensure materials work at the highest efficiency.

3.8.4.2 Development of New Materials and Devices [40,41]

Thermophotovoltaics (TPV) research focuses on developing semiconductor materials and devices that convert radiation from heat energy to electricity, with emphasis on alloys from gallium-indium arsenide and indium-phosphide arsenide, which achieve low-energy band gaps (0.40–0.70 eV).

Basic materials science develops materials to improve the efficiency, economy, and environmental acceptability of energy generation, transmission, and use. NREL's research emphasizes semiconductor and condensed matter physics, high-temperature superconductors, and electrochromics. NREL conducts research on the synthesis and characterization of quantum dots and third-generation solar cells.

Solar cells made out of new material based on perovskite absorb light and screens emit light when electricity passes through it. It is a high-quality material, and very durable under light exposure, where it can capture light particles and convert them to electricity, or vice versa. Crystalline silicon comprises over 85% of PV materials, but silicon converts <15% of captured solar energy into watt hours. It is also bulky and expensive. Solid perovskite solar cells made out of liquid mineral of the crystal structure is five times cheaper than silicon cells and improves efficiency to 20%. The mineral with its crystalline structure causes electrons generated by sunlight to travel a long way within the material, producing electricity more efficiently per mass and purity than silicon. There are significant advances in using perovskite to convert electricity into light.

3.9 SUMMARY

The solar PV power system has promising potential as an alternative energy source and as a distributed energy system for the compelling need to preserve fossil-based power plants and systems.

The photoelectric effect exhibited by semiconductor material has been used for converting the abundantly available and free source of solar light energy into electrical energy as a stand-alone power system or grid-connected large-scale power system interconnected to utility systems.

The ongoing research to develop PV materials for improved performance of solar cells with efficient electrical characteristics are in focus.

The market for solar power systems is promising and a lot of attention is given to developing new and various PV materials to get efficient and cheap systems. PV technology is evolving fast, and outstanding achievements are reported by manufacturers and researchers toward highly efficient methods of manufacturing PV cells of various types of materials. Therefore, a review has been made on the following important and interesting topics of materials science and the solar energy sources that are pertinent to this chapter.

- Solar energy sources and its system to harness light energy for conversion by PV cells into electrical energy sources.
- Materials science and basics of semiconductor principles with a focus on its photoelectric properties,
- Types of PV materials and types of PV power systems.
- PV characterization of electrical characteristics of PV materials, and testing and measurements of electrical parameters for comparison of various PV cells and modules.
- PV system components and PV technology for solar power systems.
- Principles and practices in designing of a solar PV system.
- Review of R&D on new materials and future trends to harness the solar energy power in large-scale and integrating the distributed systems with utilities and smart grids.
- Challenges and opportunities in R&D, e-waste management, and cost reductions on PV systems, innovative industry on-site performance monitoring and evaluation, skill development and creation of jobs in operation and maintenance services, regulatory policies, and financing investment in manufacturing.

The materials included in this chapter, with brief notes on solar energy sources, give the basics involved in PV materials and electrical characteristics. A brief on the design of a simple PV solar power system has also been discussed. These are very useful to graduate students, practicing engineers and consultants, manufacturers and researchers on the important and emerging topic of the PV power system to meet the challenging and compelling environmental pressure for a renewable solar PV power system as an alternative future energy source.

REFERENCES

[1] J. Nelson, Imperial College, UK, The Physics of Solar Cells (Chapters 1 and 2). http://www.uccs.edu/~rtirado/PES_1600_SolarEnergy/fotovoltaic_effect.pdf.
[2] IEA PVPS TRENDS 2015 in photovoltaic applications—executive summary, Report IEA-PVPST 1-27, October 2015. http://www.iea-pvps.org/fileadmin/dam/public/report/national/IEA-PVPS_-_Trends_2015_-_MedRes.pdf.

[3] Solar radiation. http://www.kippzonen.com/Knowledge-Center/Theoretical-info/ Solar-Radiation.

[4] Basic research needs for solar energy utilisation, Report of the basic energy sources workshop, April 18, 2005. http://.sc.doe.gov/bes/reports/files/SEU_rpt.pdf.

[5] The University Corporation for Atmospheric Research (UCAR), Layers of earth's atmosphere, http://scied.ucar.edu/atmosphere-layers, 2015.

[6] P. Hofmann, Solid State Physics: An Introduction, WILEY-VCH Verlag GmbH & Co. KGaA, Weinheim, 2008, ISBN 978-3-527-40861-0.

[7] O. Mah, Fundamentals of Photovoltaic Materials, National Solar Power Research Institute, Inc., California, 1998.

[8] W.D. Callister Jr., Materials Science and Engineering: An Introduction, University of Utah, John Wiley & Sons Inc., New York, 2007.

[9] A chapter on "Semiconductor Materials for Solar Cells" (Chapter 3). https://ocw. tudelft.nl/wp-content/uploads/Solar-Cells-R3-CH3_Solar_cell_materials.pdf.

[10] Green Rhio Energy, Photovoltaics. http://www.greenrhinoenergy.com/solar/ technologies/pv_modules.php.

[11] Istrian Regional Energy Agency (IRENA), Photovoltaic systems, (2012). https://mail. google.com/mail/u/0/#sent/157dddb189f2b2ac?projector=1.

[12] M. Zeman, Photovoltaic Systems, Solar Cells, Delft University of Technology, Netherlands, 2004 (Chapter 9).

[13] Asian Brown Bowri (ABB), Technical application papers no. 10—photovoltaic plants.

[14] M. Petkov, et al., Modelling of electrical characteristics of photovoltaic power supply sources, Contemporary Materials (Renewable Energy Sources), II-2, Technical University of Gabrovo, Bulgaria, 2011, pp. 173–177.

[15] J. Merten, et al., Improved equivalent circuit and analytical model for amorphous silicon solar cells and modules, IEEE Trans. Electron Devices 45 (2) (1998) 423–429.

[16] L.A. Dobrzariski, et al., Comparison of electrical characteristics of silicon solar cells, J. Achiev. Mater. Manuf. Eng. 18 (1–2) (2006) 215–218.

[17] A. Ibrahim, Analysis of electrical characteristics of photovoltaic single crystal silicon solar cells at outdoor measurements, Smart Grid Renew. Energy 2 (2011) 169–175. http://www.SciRP.org/journal/sgre.

[18] G.M. Masters, Renewable and Efficient Electric Power Systems, John Wiley & Sons, Inc., New Jersey, 2004 (Chapters 7 and 8).

[19] National Instruments paper on, Part II—Photovoltaic cell I–V characterization theory and LabVIEW analysis code, May 2012. http://www.ni.com/white-paper/7230/en/.

[20] J.H. Wohlgemuth, Standards for PV Modules and Components-Recent Developments and Challenges, National Renewable Energy Laboratory, Golden, CO, 2012.

[21] K. Nguyen, SEMI PV standards. http://www.semi.org/standards.

[22] A Working Paper on, Solar Photovoltaics, Vol. 1. Power Sector, Issue 4/5, International Renewable Energy Agency (IRNEA), 2012. http://www.irena.org/ Publications.

[23] A report by California Energy Commission on, a guide to photovoltaic (PV) system design and installation, 2001.

[24] C. Roos, Solar Electric System Design, Operation and Installation, An Overview for Builders in the Pacific Northwest, Washington State University, Olympia, WA, 2009.http://www.energy.wsu.edu.

[25] A publication by Microgeneration Certification Scheme (MCS), London on, Guide to the installation of photovoltaic systems, 2012. http://www. microgenerationcertification.org.

[26] P. Tiwana, et al., Electron Mobility and Injection Dynamics in Mesoporous ZnO, SnO_2, and TiO_2 Films Used in Dye-Sensitized Solar Cells, vol. 5, no. 6, University of Oxford, United Kingdom/American Chemical Society, 2011, pp. 5158–5166.

[27] A Publication on, A strategic Research Agenda for Photovoltaic Solar Energy Technology, Science and Technology and Applications of the EU PV Technology Platform, European Union, 2007.

[28] Organic photovoltaic solar cells. http://www.nrel.gov/pv/organic-photovoltaic-solar-cells.html.

[29] Photovoltaics research on new materials for PV technology by National Renewable Energy Laboratories (NREL). http://www.nrel.gov/pv/advanced_concepts.html.

[30] J.P. Connolly et al., III–V Solar Cells, Nanophotonic Technology Centre, Spain. http://arxiv.org/ftp/arxiv/papers/1301/1301.1278.pdf.

[31] End-of-life management: solar photovoltaic panels, IRENA 2016 and IEA-PVPS 2016, June 2016. www.irena.org, www.iea-pvps.org.

[32] IRENA report on, The power to change: solar and wind cost reduction potential to 2025. http://bit.ly/233POFQ.

[33] Analytical monitoring of grid-connected photovoltaic systems, good practices for monitoring and performance analysis, report IEA-PVPST13-03:2014, March 2014. http://www.iea-pvps.org.

[34] Workforce challenges and opportunities in the solar photovoltaic industry in Toronto—a study report. http://www.greencollarcareers.ca/students/.

[35] Emerging opportunities and challenges in US solar manufacturing—a report from the National Renewable Energy Laboratory (NREL), May 2016. www.nrel.gov/publications.

[36] Emerging opportunities and challenges in financing solar—a report from the National Renewable Energy Laboratory (NREL), May 2016. www.nrel.gov/publications.

[37] Hybrid organic-inorganic halide perovskite solar cells, US Department of Energy. http://www.energy.gov/cere/sunshot/hybrid-organic-inorganic-halide-perovskite-solar-cells.

[38] High-concentration III–V multijunction junction solar cells by NREL. http://www.nrel.gov/pv/high-concentration-iii-v-multijunction-solar-cells.html.

[39] J. Bullock, et al., Efficient silicon solar cells with dopant-free asymmetric heterocontacts, Nat. Energy (2016)http://nano.eecs.berkeley.edu/publications/NatureEnergy_2016_SiDASH.pdf.

[40] Photovoltaic research materials and devices by NREL. http://www.nrel.gov/pv/perovskite-organic-photovoltaics.html.

[41] Hybrid tandem solar cells by NREL. http://www.nrel.gov/pv/hybrid-tandem-solar-cells.html.

CHAPTER 4

Microturbine Generation Power Systems

R. Noroozian, P. Asgharian
University of Zanjan, Zanjan, Iran

4.1 INTRODUCTION

There are various methods to produce energy such as wind turbine, solar cell, microturbine (MT), and fuel cell but MT offers several potential advantages compared to other technologies for small-scale power generation.

Generally, MT is a small gas turbine, which has similar cycles and in many cases its components are made of the same materials as a heavy gas turbine. The MT power-to-weight ratio is superior to the heavy gas turbine because based on aerodynamic equations, reduction of turbine diameters cause an increase in shaft rotational speed, so output power is improved. Heavy gas turbine generators are too large and too expensive for distributed power applications, so, MTs are developed for small-scale power like electrical power generation alone or as combined cooling, heating, and power (CCHP) systems.

High power-to-weight ratio and efficiency are not the only advantages of microturbine generation (MTG) systems. Advantages that make MTs superior to other distributed generation (DG) systems are compact size, high reliability, light weight, low operation and maintenance costs, simple and remote control, low environmental emissions, fewer moving parts, low noise level, and fuel flexibility [1]. MTs can operate on a variety of gaseous or liquid fuels including natural gas, associated gas, LPG/propane, flare gas, landfill gas, digester gas, diesel, biogas, and kerosene, some of which are renewable energy.

The interest in MTGs is increasing due to technical, economic, and reliability merits. Potential applications of MTs include peak shaving, premium power, remote power, cogeneration applications, and as grid supporter.

This chapter presents applications, configurations, and operational studies of MTs as DG. First, the MT and its operation cycle are introduced and power electronic interface and type of MT are discussed. Finally, dynamic modeling of MT is considered and simulation study is carried out in PSCAD/EMTDC software for grid-connected and isolated operation modes.

Distributed Generation Systems
http://dx.doi.org/10.1016/B978-0-12-804208-3.00004-2
149

4.2 HISTORY

The idea of using a turbine was hatched many years ago, and Leonardo da Vinci was the first man who drew an illustration of his idea, which he called a "chimney jack." Early models were based on the steam turbine. The first true gas turbine was invented by John Barber in 1791, and his invention had most of the elements present in modern gas turbines. Over the years, many gas turbine designs were based on John Barber's model. In fact, gas turbines were involved in many military and aerospace industrial projects done during the 1950s through the 1970s. Using the gas turbine as a constant power producer was done in the 1980s and 1990s and the concept of DG was accelerated.

In 1950s and 1960s, the Boeing Company was a pioneer in using gas turbines for helicopters, aircraft, and army tanks. The first cogeneration based on a gas turbine was installed by California Gas Company in 1962. The Boeing Company developed a 100 kW outboard motor that had positive effects like reduction of weight and fuel of gas turbines, but the company failed to keep its products and, finally, this product line was sold to Caterpillar. In 1960s, big companies like BMW, Rover, Nissan, Ford, and Benz produced gas turbine-based automobiles and Rover pioneered them.

In 1980, a program entitled Advanced Energy Systems supported by the Gas Research Institute was initiated to develop a small gas turbine rated at 50 kW and including heat recovery. In 1990, this program was abandoned due to problems of product final cost. Since then, Gas Research Institute has supported new projects with other companies like Northern Research & Engineering Energy System and NoMac Energy Systems.

In 1988, Robin Mackay and Jim Noe, former members of Allied Signal Company and Garrett, respectively, decided to establish a new company, NoMac Energy Systems, for developing a small gas turbine in the automotive industry. In fact, NoMac was the first manufacturer of MTs in the world, under the support of the Gas Research Institute. In 1993, the company name was changed to Capstone Turbine Corporation, which is one of the biggest MT manufacturers in the world.

Generally, research on turbochargers in the transportation industry and the aerospace industry as well as improvement of gas turbine efficiency were the main reasons to develop MTs. Recently, output power of MTs has reached up to 500 kW and improvement of efficiency and size is considered in new models.

4.3 STATISTICAL INFORMATION

In this section, MT statistical aspects are considered and charts are collected from credible sources. There are various methods to generate electricity, which are subsets of renewable sources or fossil fuels. Fossil fuels are the traditional method and now the international focus is on renewable energies. There are several factors to consider when choosing a power source such as pollutant emission level, fuel cost, the cost of construction, and availability. Global warming and greenhouse gasses have driven trends toward renewable energy, but these energies (e.g., wind power or solar power) are not reliable resources and they are not the main sources for power generation. Fig. 4.1 shows that along with the increase of energy demand, the total share of fossil fuels will not be affected by a drastic reduction. With respect to growth of renewable energies, there is a gap between them and fossil fuels to be the main resource of power generation.

The use of fossil fuels from 1965 to 2035 by percentage is shown in Fig. 4.2.

Fossil fuels share of the primary energy mix will decrease only modestly while the new renewable share will increase. Fossil fuels will remain the

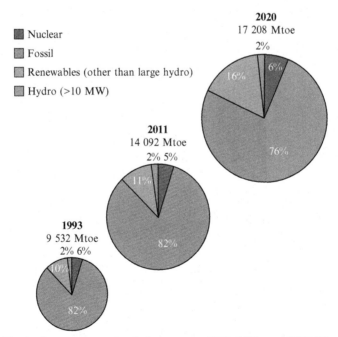

Fig. 4.1 Total primary energy supply by resource 1993, 2011, and 2020 [2].

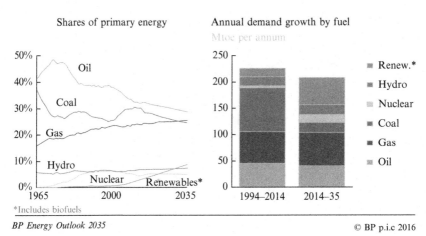

Fig. 4.2 Evolution of world primary energy consumption [3]. *Used with permission from 2016 BP Energy outlook 2035.*

dominant source of energy powering the global economy, providing around 60% of the growth in energy and accounting for almost 80% of total energy supply in 2035. Gas is the fastest growing fossil fuel (1.8% p.a.), with its share in primary energy gradually increasing [3].

Fig. 4.3 presents percentage shares of energy sources with greater clarity. Global interest in natural gas is obviously increasing during the 1975–2035 period.

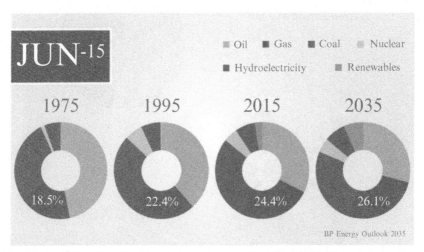

Fig. 4.3 Evolution of natural gas in power generation [3]. *Used with permission from http://www.bp.com/en/global/corporate/bp-magazine/observations/today-and-tomorrow-the-rise-of-gas.html.*

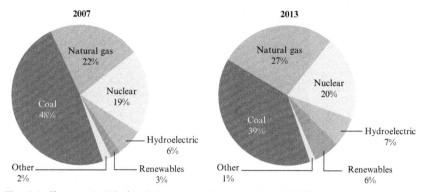

Fig. 4.4 Changes in US electricity generation mix, 2007–13 [4].

The United States is one of the highest power consuming countries in the world that has accordant growth in the use of natural gas. Fig. 4.4 shows percentages of US electricity generation in different years. As coal generation declines, it has been replaced by generation from a combination of cleaner power sources led by natural gas [4].

The use of natural gas is increasing due to high reliability and clean fuel. The chart in Fig. 4.5 compares the amounts of carbon dioxide emitted per unit (p.u.) of heat content generated. As shown, natural gas is the cleanest fossil fuel.

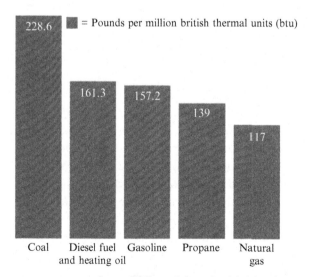

Source: US Energy Information Administration

Fig. 4.5 Amounts of carbon dioxide emitted by fossil fuels [3]. *Used with permission from http://www.bp.com/en/global/corporate/bp-magazine/observations/today-and-tomorrow-the-rise-of-gas.html.*

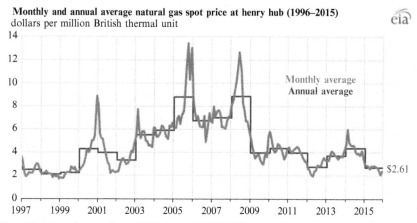

Monthly and annual average natural gas spot price at henry hub (1996–2015) dollars per million British thermal unit

Fig. 4.6 Natural gas prices, 1997 through 2015 [4].

The natural gas industry has a well-documented history of price volatility. Due to the nature of its production, storage, and transmission; the fact that it supplies many end uses; and its susceptibility to extreme weather events, the price of natural gas fluctuates more than does the price of its competitors. For example, the natural gas market experienced a number of dramatic price swings between 2000 and 2014, which is shown in Fig. 4.6.

In 2005, the spot market price spiked to nearly $14 per million British thermal unit (MMBtu) in response to hurricane activity in the Gulf of Mexico, where many gas wells are located. In contrast, prices dipped to about $2 per MMBtu in 2012 in response to decreased demand resulting from the economic recession and from a series of warm winters, which reduced competition between heating and electricity suppliers for limited natural gas pipeline resources [5].

Many experts believe that low natural gas prices are not sustainable over the long term. For example, the US Energy Information Administration's (EIA) Annual Energy Outlook projects that spot prices will significantly increase from the recent low point of $2.75 per MMBtu in 2012 to $6.03 per MMBtu in 2030 and $7.65 per MMBtu in 2040. Factors that contribute to upward pressure on prices and the risk of price volatility include uncertain available supply and potentially increasing demand for natural gas from electric utilities, other competing domestic users, and exporters [5].

The market for new MT installations is expected to surpass $1 billion in annual revenue by 2020. Over the next decade, Navigant Research projects more than 4 GW of cumulative MT capacity will be installed globally. Despite this healthy growth, the crowded DG technology landscape is expected to temper upside growth for MT installations over the forecast period. North America and Asia Pacific are expected to account for 63% of the cumulative revenue generated from global MT installments between 2015 and 2024. Substantial growth in shale gas production and an improving combined heat and power (CHP) environment point to the North American market leading all regions with $5.2 billion in cumulative revenue. This represents nearly half of the cumulative revenue generated globally. Annual MT revenue in North America is expected to experience a CAGR[1] of 14.8%. Asia Pacific is a distant second to the North American market with $1.4 billion in cumulative revenue generated from MT installments, but the region is expected to experience higher annual revenue growth between 2015 and 2024 (CAGR of 17.7%). Western Europe is third with $1.3 billion in cumulative revenue, followed by Eastern Europe at $1.0 billion [6]. Fig. 4.7 shows annual MT revenue

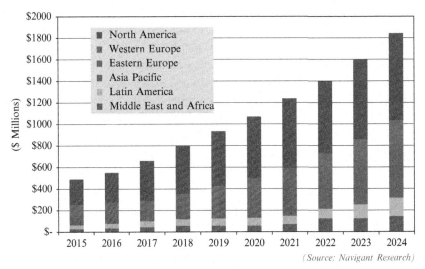

(Source: Navigant Research)

Fig. 4.7 Annual microturbine revenue by region, world markets: 2015–24 [6]. *From Microturbines: Global Market Analysis and Forecasts for Residential, Commercial, and Industrial Applications. Navigant research. From http://www.prnewswire.com/news-releases/microturbines-global-market-analysis-and-forecasts-for-residential-commercial-and-industrial-applications-300105652.html.*

[1] *CAGR refers to compound average annual growth rate, using the formula:* $CAGR = (End\ Year\ Value \div Start\ Year\ Value)(1/steps) - 1$. CAGRs presented in the tables are for the entire timeframe in the title. Where data for fewer years are given, the CAGR is for the range presented. Where relevant, CAGRs for shorter timeframes may be given as well.

by region and represents increasing global MTG market size as a reliable resource. MTs can operate with various fuels in addition to natural gas. Fuel flexibility is one of the favorable factors influencing the choice of MTs.

4.4 GAS TURBINE

Today, gas turbines play an important role in electric power systems. Use of the gas turbine has increased in aircraft propulsion, the automotive industry, and the oil and gas industry for driving pumps and other uses. There are many reasons that distinguish gas turbines from other sources. One of most striking features of a gas power plant is fast startup time, which is suitable for peak shaving and used as a base load in electric power systems. High power to weight ratio, long life, high reliability, and remote control are other advantages of gas turbines. Compared with other plants such as thermal plants, this system needs low water cooling and therefore water is not a restriction on the construction of a gas plant.

A gas turbine is a rotating engine that extracts energy from a flow of combustion gases (resulting from the ignition of compressed air and a fuel-like natural gas). Gas turbines consist of three general parts: a compressor, a combustor, and a turbine. A basic condition for turbines' power generation is the difference between inlet pressure and outlet pressure, so the compressor is used to increase inlet pressure and the compressor is located before the turbine. Compressed air cannot produce enough mechanical power for driving a turbine because the compressor consumes a large part of turbine output power. For driving a turbine, more energy is required and a fuel like natural gas in needed. Compressed air and fuels are mixed in the combustor due to more power production. Therefore, the combustor or combustion chamber is located between the compressor and the turbine. Fig. 4.8 shows a gas turbine diagram. Gas turbine operation is based on the Brayton cycle that will be described in detail below.

Output efficiency of gas turbine depends on compressor and turbine efficiency as well as environmental conditions like temperature. For improving output efficiency, high-temperature inlet gas is necessary. In general, the electrical efficiencies of modern gas turbines range from 30% to 40% and output power is about 1–500 MW.

There are several type of gas turbine such as turboprop, MT, and turbofan and the purpose of this chapter is the detailed description of MTs.

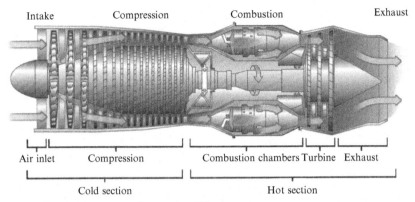

Fig. 4.8 Gas turbine generation system [7].

4.4.1 The Brayton Cycle

Large gas turbines operations are based on the Brayton thermodynamic cycle. Understanding of the Brayton cycle concept is an essential issue because MTs are a subset of gas turbines and they have a similar cycle.

This cycle was proposed by George Brayton (1830–92), who was an American engineer, in 1872. He applied this cycle for his patent that was for a reciprocating oil-burning engine. Today, gas turbines and jet engines based on the Brayton cycle are widely used in industry.

The Brayton cycle or Joule cycle is a thermodynamic cycle and needs to have a working fluid. Working fluid in a thermodynamic system is a liquid or gas that absorbs or transmits. Air is the working fluid of the Brayton cycle. This system is based on four steps: compression, heat addition, expansion, and heat rejection, and each cycle consists of two isobaric processes and two adiabatic processes. An isobaric process is a thermodynamic process in which pressure stays constant and an adiabatic process is a thermodynamic process without heat transferring between two systems.

In a Brayton open cycle, inlet fresh air at ambient conditions is running through a compressor and after that, the resulting compressed air enters the combustion chamber. In the combustor, fuel is added to air and ignited, which causes an increase of gases volume. Expanded gases run through a turbine and provide mechanical force for driving. Finally, combustion products are discharged to the atmosphere. Fig. 4.9 diagrams the open cycle that discharges air flow to the atmosphere and occurs in each cycle. The compressor

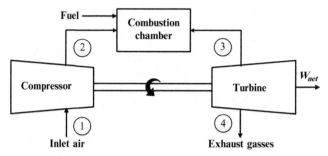

Fig. 4.9 An open cycle gas turbine engine.

power requirement is about 40%–80% of turbine power output, which is a large amount, so power loss reduction is necessary.

A typical close cycle and T–S and P–V diagram of an ideal Brayton cycle are represented in Fig. 4.10.

A close cycle consists of four main stages, as follows:

- *Stage 1 to 2: Air compression through adiabatic process*—fresh air enters the compressor in stage 1. Through stage 1 to 2, air is compressing and increasing temperature whereas entropy is constant and volume of the air is decreasing. At stage 2, high pressure and high temperature air is discharged from the compressor.
- *Stage 2 to 3: Air combustion through isobaric processes*—in this stage, compressed air enters the combustion chamber. Fuel and air are mixed and burned at constant pressure. After the combustion process, gas volumetric expansion occurs and at stage 3, there is a very high temperature gas.
- *Stage 3 to 4: Air expansion through adiabatic process*—at stage 3, working fluid enters the turbine to drive it. When hot gas runs through the turbine, volume is larger than before and gas pressure is decreasing, so the turbine is called an expander.
- *Stage 4 to 1: Air heat transfer through isobaric process*—in stage 4, turbine exhaust gas is not proper for stage 1, thus heat transfer to the surroundingis necessary. After the heat transfer, there is a gas at atmospheric pressure as well as appropriate volume and temperature, which is ready to enter the compressor again.

The Brayton cycle network with respect to Fig. 4.10B is equal to area of a 1-2-3-4 close loop. Output efficiency of this cycle is an important parameter because it represents a reasonable operation or inappropriate performance of the gas turbine. The following equations are according to Fig. 4.10.

Added heat to cycle is equal to [8]:

$$Q_{in} = h_3 - h_2 = C_p(T_3 - T_2) \tag{4.1}$$

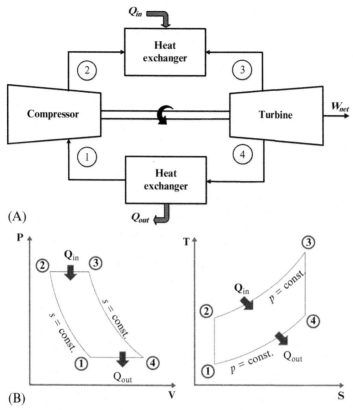

Fig. 4.10 Brayton cycle: (A) A close cycle gas turbine engine and (B) P-V and T-S diagram.

where C_p is heat capacity in constant pressure, T_2 and T_3 are temperatures in stages 2 and 3, respectively, and h is enthalpy. Heat rejection from turbine is as follows:

$$Q_{out} = h_4 - h_1 = C_p(T_4 - T_1) \tag{4.2}$$

From Eqs. (4.1) (4.2), net work is

$$W_{net} = Q_{in} - Q_{out} = C_p[(T_3 - T_2) - (T_4 - T_1)] \tag{4.3}$$

Brayton cycle efficiency is equal to the ratio of net work to added heat. Efficiency is as follows:

$$\eta = \frac{W_{net}}{Q_{in}} = \frac{C_p[(T_3 - T_2) - (T_4 - T_1)]}{C_p[(T_3 - T_2)]} = 1 - \frac{T_4 - T_1}{T_3 - T_2} \tag{4.4}$$

The isentropic process is an idealized thermodynamic process in which there is no heat transfer from the system to its surrounding. In the isentropic process, the following equation is true.

$$\frac{T_2}{T_1} = \left(\frac{P_2}{P_1}\right)^{(\gamma-1/\gamma)} = \left(\frac{V_1}{V_2}\right)^{(\gamma-1)}, \quad \gamma = \frac{C_p}{C_v} = \frac{c_p}{c_v} \tag{4.5}$$

where γ is heat capacity ratio, C is heat capacity, and c is specific heat capacity that index p and v are related to constant pressure and constant volume, respectively. Similarly,

$$\frac{T_3}{T_4} = \left(\frac{P_3}{P_4}\right)^{(\gamma-1/\gamma)} \tag{4.6}$$

According to Fig. 4.3, because $P_2 = P_3$ and $P_1 = P_4$, thus

$$\frac{T_2}{T_1} = \frac{T_3}{T_4} = \left(\frac{P_2}{P_1}\right)^{(\gamma-1/\gamma)} \tag{4.7}$$

$$\frac{T_4 - T_1}{T_3 - T_2} = \frac{T_1}{T_2} = \frac{T_3}{T_4} \tag{4.8}$$

From Eqs. (4.8) (4.4), efficiency is as follows:

$$\eta = 1 - \frac{T_1}{T_2} = 1 - \frac{T_3}{T_4} \tag{4.9}$$

Pressure ratio is called r_p and Eq. (4.10) is rewritten Eq. (4.7).

$$\frac{T_2}{T_1} = \left(\frac{P_2}{P_1}\right)^{(\gamma-1/\gamma)} = \left(r_p\right)^{(\gamma-1/\gamma)} \tag{4.10}$$

Finally, the efficiency of the Brayton cycle is as follows:

$$\eta = 1 - \frac{1}{\left(r_p\right)^{\left(\frac{\gamma-1}{\gamma}\right)}} = 1 - \left(r_p\right)^{\left(\frac{1-\gamma}{\gamma}\right)} \tag{4.11}$$

Note, all mentioned equations are based on ignorance of kinetic energy and potential energy changes. According to Eq. (4.11), efficiency depends on temperature and pressure ratio but maximum and minimum temperatures are limited. Maximum temperature is limited by materials because turbine blades cannot withstand high temperatures and minimum temperature is defined by the temperature of inlet air to the engine.

Example 4.1

A gas turbine power plant operating on an ideal Brayton cycle has a pressure ratio of 6. The gas temperature is 350°K at the compressor inlet and 1500°K at the turbine inlet. Utilizing the air-standard assumptions, determine (a) the gas temperature at the exits of the compressor and the turbine, (b) the back work ratio, and (c) the thermal efficiency.

Solution

The T–S diagram of the ideal Brayton cycle described is shown in Fig. 4.11. Note that the components involved in the Brayton cycle are steady flow devices.

(a) The air temperatures at the compressor and turbine exits are determined from isentropic relations in Appendix B:

Process 1-2: $T_1 = 350\,\mathrm{K} \rightarrow h_1 = 350.49\,\mathrm{kJ/kg}, P_{r1} = 2.379$

(At compressor exit) $P_{r2} = \dfrac{P_2}{P_1} P_{r1} = 6 \times 2.379 = 14.274 \rightarrow T_2 = 580\,\mathrm{K},$

$$h_2 = 586.04\,\mathrm{kJ/kg}$$

Process 3-4 : $T_3 = 1500\,\mathrm{K} \rightarrow h_3 = 1635.97\,\mathrm{kJ/kg}, P_{r3} = 601.9$

(At turbine exit) $P_{r4} = \dfrac{P_4}{P_3} P_{r3} = \left(\dfrac{1}{6}\right) \times 601.9 = 100.32 \rightarrow T_4 = 970\,\mathrm{K},$

$$h_4 = 1011.9\,\mathrm{kJ/kg}$$

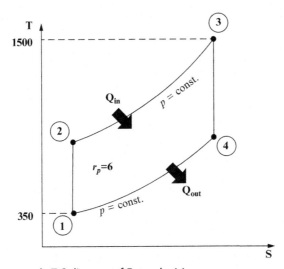

Fig. 4.11 Brayton cycle T-S diagram of Example 4.1.

(b) To find the back work ratio, we need to find the work input to the compressor and the work output of the turbine:

$$\begin{cases} w_{comp,\,in} = h_2 - h_1 = 586.04 - 350.49 = 235.55\,kJ/kg \\ w_{turb,\,out} = h_3 - h_4 = 1635.97 - 1011.9 = 624.07\,kJ/kg \end{cases} \rightarrow r_{bw} = \frac{w_{comp,\,in}}{w_{turb,\,out}} = 0.377$$

That is, 33% of the turbine work output is used just to drive the compressor.

(c) The thermal efficiency of the cycle is the ratio of the net power output to the total heat input:

$$\begin{cases} Q_{in} = h_3 - h_2 = 1635.97 - 586.04 = 1049.93 \\ w_{net} = w_{out} - w_{in} = 624.04 - 235.55 = 388.49 \end{cases} \rightarrow \eta_{th} = \frac{w_{net}}{Q_{in}} = 0.37\,or\,37\%$$

Example 4.2

In a Brayton cycle-based power plant, the air at the inlet is at 27°C, 0.1 MPa. The pressure ratio is 6.25 and the maximum temperature is 800°C. Find (a) the compressor work per kilograms of air, (b) the turbine work per kilograms of air, (c) the heat supplied per kilograms of air, and (d) the cycle efficiency. (Consider $\gamma = 1.4$, which is the specific heat ratio value of air at room temperature)

Solution

Based on mentioned values: $T_1 = 27°C = 300$ K, $P_1 = 100$ kPa, $r_p = 6.25$, $T_3 = 800°C = 1073$ K, and thus

(a) Process 1–2:

$$\frac{T_2}{T_1} = r_p^{\left(\frac{\gamma-1}{\gamma}\right)} = 6.25^{\left(\frac{1.4-1}{1.4}\right)} = 1.689 \rightarrow T_2 = 506.69\,K\,,\ W_{comp}$$
$$= c_p(T_2 - T_1) = 1.005(506.69 - 300) = 207.72\,kJ/kg$$

(b) Process 3–4:

$$\frac{T_3}{T_4} = r_p^{\left(\frac{\gamma-1}{\gamma}\right)} = 6.25^{\left(\frac{1.4-1}{1.4}\right)} = 1.689 \rightarrow T_4 = 635.29\,K\,,\ W_{turb}$$
$$= c_p(T_3 - T_4) = 1.005(1073 - 635.29) = 439.89\,kJ/kg$$

(c) Heat input,

$$Q_{in} = c_p(T_3 - T_2) = 1.005(1073 - 506.69) = 569.14\,kJ/kg$$

(d) Cycle efficiency,

$$\eta_{th} = \frac{\left(W_{turb} - W_{comp}\right)}{Q_{in}} = \frac{439.89 - 207.72}{569.14} = 0.408 \text{ or } 40.8\%$$

There are three general methods for improving Brayton cycle efficiency: regeneration, reheat. and intercooling. Fig. 4.12 shows a Brayton cycle that simultaneously uses reheating, intercooling, and regeneration methods.

T-S and P-V diagrams for improved Brayton cycle are given in Fig. 4.13. According to these diagrams, surface increasing causes more net work and higher efficiency.

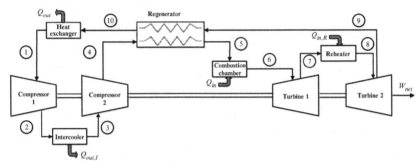

Fig. 4.12 Improved Brayton cycle.

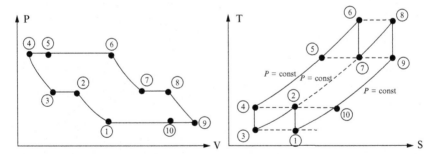

Fig. 4.13 Diagrams of improved Brayton cycle.

Example 4.3
Solve Example 4.2 if a regenerator of 80% effectiveness is added to the plant.

Solution
With respect to Fig. 4.14,

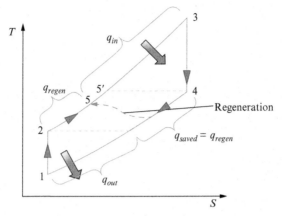

Fig. 4.14 T-S diagram with a regenerator of Example 4.3.

$$\varepsilon = \frac{T_5 - T_2}{T_4 - T_2} = 0.8 \rightarrow 0.8 = \frac{T_5 - 506.69}{635.29 - 506.69} \rightarrow T_5 = 603.14 \text{K}$$

and T_4, W_{comp}, W_{turb} remain unchanged.
The new input heat is equal to:

$$Q_{in} = c_p(T_3 - T_5) = 472.2 \text{kJ/kg}$$

so,

$$\eta_{th} = \frac{\left(W_{turb} - W_{comp}\right)}{Q_{in}} = \frac{439.89 - 207.72}{472.2} = 0.492\% \text{ or } 49.2\%$$

4.5 MT COMPONENTS

MTG basic components are a compressor for creating appropriate pressure and a combustor and turbine that are coupled to a generator. An induction generator or synchronous generator is used in MTGs but a permanent magnet synchronous generator (PMSG) is superior. The field winding is replaced by a permanent magnet in PMSGs, which cause lower power losses, higher mechanical consistency, structure simplicity, and lower maintenance; so PMSGs are mostly used in MTGs. To get higher efficiency, a recuperator, which is a heat exchanger, is added to other components. In a single-shaft design, all components are mounted on the same shaft. Fig. 4.15 shows a MTG configuration in detail. This section provides a brief description of MTG components.

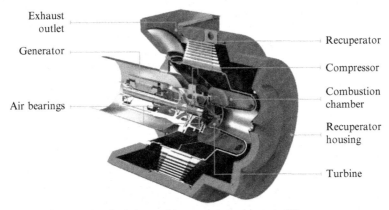

Fig. 4.15 Cross section of a microturbine generation system [9].

4.5.1 Compressor

MTs require high-pressure gases for appropriate combustion, thus a compressor is used. A gas compressor is a mechanical device that increases the pressure of a gas by reducing its volume, and an air compressor is a specific type of gas compressor. An air compressor is used to increase the pressure of inlet air at ambient temperature by using some external energy. The compressors are divided into dynamic and positive displacement, which differ in energy transfer. In a dynamic compressor, energy transfer is done continuously but in a positive displacement type transfer is periodic.

There are two main types of positive displacement compressors: reciprocating and rotary. In the reciprocating compressor, air volume is drawn into a cylinder and compression is done in it. Airflow is controlled by a cylinder valves piston and this valve acts as a check valve. In a rotary compressor, gases volume is decreased by lobes, screws, and vanes.

A dynamic type of compressor is used in gas turbines. Dynamic type includes a radial flow compressor and an axial flow compressor. In a radial or centrifugal compressor, air is compressed through an impeller or rotating disk by use of centrifugal force. After the impeller, flow enters a diffuser and static pressure is increased by reducing velocity. In other words, a centrifugal compressor accelerates the gasses velocity (increase in kinetic energy), which is then converted into pressure as the airflow leaves the volute and enters the discharge pipe. A centrifugal compressor delivers higher flow rates than a positive displacement compressor. An axial compressor is composed of a rotor that has rows of fans as blades where rotating blades are attached to a shaft to push air over stationary blades called stators. The stator blades

PMSG   Turbine

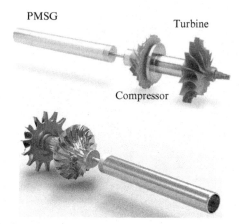

Compressor

Fig. 4.16 Single-shaft microturbine components [9].

are attached to the casing. Gas or working fluid flows parallel to the rotating shaft and rotating blades increase air velocity, which is down by the stationary blades. The decreasing air velocity converts kinetic energy to pressure. Centrifugal type compressors are simpler and lower cost than axial type and are favorable for MTs. Most MTGs are single-stage centrifugal compressors, and the axial flow compressor is more suitable for a heavy gas turbine. Fig. 4.16 shows a centrifugal compressor in MTs that is coupled with a turbine.

Compressor efficiency has a direct influence on MTG overall efficiency because output mechanical power of the turbine is consumed by a compressor that are coupled together in a same shaft. Recently, compressor manufacturers have used advanced computational fluid dynamics to obtain higher efficiency.

Compressor ratio is defined as the ratio of air volume before compression to air volume after compression, and pressure ratio between 3:1 or 4:1 is best suitable for a single-stage centrifugal compressor. Fig. 4.17 shows electrical efficiency of MT as a function of compressor pressure ratio for various turbine inlet temperatures and optimum performance exists in the pressure ratio range of 2.5–4.5.

Specific power of the MT as a function of compressor pressure ratio for the same range of firing temperatures as last figure is shown in Fig. 4.18.

Greater specific power can be obtained with higher pressure ratios but higher pressure ratio is depending on higher rotational speed. Rotational speed is limited by turbine blade materials, so current pressure ratio is appropriate in cost and operation.

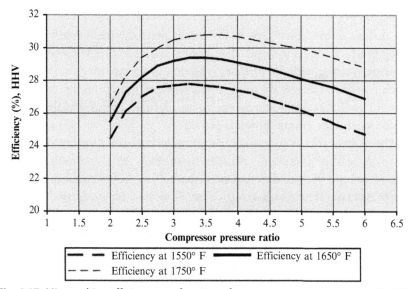

Fig. 4.17 Microturbine efficiency as a function of compressor pressure ratio and turbine inlet temperature [10].

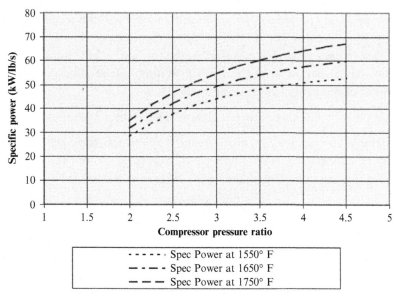

Fig. 4.18 Microturbine specific power as a function of compressor pressure ratio and turbine inlet temperature [10].

4.5.2 Turbine

The turbine is the heart of MTs because it plays the main role in electric power production and in single-shaft design, where a compressor and a generator are rotating with the turbine. A turbine is a rotary mechanical device, which is used to convert the energy of a working fluid to useful work. The working fluid contains potential energy and kinetic energy and the turbine extracts energy by several physical principles.

The main factors in choosing turbines include manufacturing costs, cycle flow path, leakage problems, and stability of shaft. Axial and radial are two types of turbine that are different in air path flow. In the radial turbine the flow of working fluid is radial to the shaft. The axial types are used in heavy gas turbines but radial types are more efficient in micro gas turbines. The advantages of the use of radial turbines in MTs are higher single-stage ratio, small volume, simple structure. and higher efficiency. Fig. 4.16 shows a radial turbine that is coupled to a compressor.

4.5.3 Combustor

Combustion means the act of burning, and the combustion process is a high-temperature chemical reaction between the fuel and oxygen in which heat is released.

A combustion chamber is a part of an internal combustion engine in which fuel and compressed air are mixed and burned; the resulting high-temperature exhaust gas is used to drive the turbine. The combustion chamber in agas turbine is called a combustor, and a MT combustor has continuous operation unlike in a reciprocating engine. The combustion equation of the natural gas is given by [11]

$$CH_4 + 4O \rightarrow CO_2 + 2H_2O + \text{heat} \tag{4.12}$$

Because air contains 78% nitrogen and 22% oxygen, there are four molecules of nitrogen for every molecule of oxygen in air. So, the complete combustion reaction of methane can be written as follows [11]:

$$\underset{\text{(Methane + Air)}}{CH_4 + 2(O_2 + 4N_2)} \rightarrow \underset{\text{(Carbon dioxide + Nitrogen + Water + Heat)}}{CO_2 + 8N_2 + 2H_2O + \text{heat}} \tag{4.13}$$

The combustor's purpose is to produce enough energy for the system. T main factors in a combustor's design include complete combustion, lower pressure losses, smaller size, and lower environmental emissions. There are three main types of combustor: annular type, can type, and can-annular type, which are briefly described below.

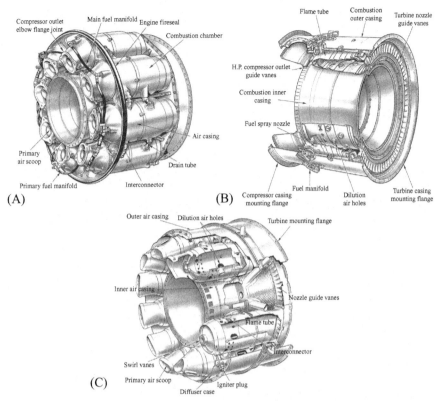

Fig. 4.19 Types of gas turbine combustors: (A) can, (B) annular, and (C) can-annular [11].

- *Can type*: This type is the earliest one and consists of separate cans. Each "can" has its own fuel injector, liner, interconnectors, and casing. Fig. 4.19A shows a can type of combustor.
- *Annular type*: This type is a modern combustor that is shown in Fig. 4.19B. Annular combustors have separate combustion zones, a continuous liner, and casing in a ring (the annulus). Annular designs have more efficient combustion in which nearly all of the fuel is burned completely.
- *Can-annular type*: This type is combination of the two abovementioned types, with discrete combustion zones contained in separate liners with their own fuel injectors and all the combustion zones share a common air casing. Can-annular designs have less-efficient combustion but their modular design is easy to repair or replace. Fig. 4.19C shows can-annular type combustor.

The annular type of combustor is usually used in MTs and integrated design. Advantages of the annular type include shorter size, more uniform combustion, less surface area, permitting better mixing of the fuel and air as well as simple structure. With respect to recent discussions, features of annular combustor are suitable for MT.

Example 4.4

With respect to Table 4.1, find the higher heating value (HHV) of methane CH_4 in kJ/mol and kJ/kg when it is oxidized to CO^2 and liquid H_2O [12].

Solution

First, HHV and lower heating value (LHV) are defined. The HHV, also known as the gross heat of combustion, includes latent heat, while the LHV, or net heat of combustion, does not.

The reaction is written below, and beneath it are enthalpies taken from Table 4.1. Notice that the equation must be balanced so that we know how many moles of each constituent are involved.

$$CH_4(g) + 2O_2(g) \rightarrow CO_2(g) + 2H_2O(l)$$
$$\underset{(-74.9)}{} \quad \underset{(2\times0)}{} \quad \underset{(-393.5)}{} \quad \underset{(2\times-282.8)}{}$$

Notice that we have used the enthalpy of liquid water to find the HHV. The difference between the total enthalpy of the reaction products and the reactants is

$$\Delta H = [(-393.5) + 2 \times (-285.8)] - [(-74.9) + 2 \times 0]$$
$$= -890.2 \text{kJ/mol of } CH_4$$

Because the result is negative, heat is released during combustion; that is, it is exothermic. The HHV is the absolute value of ΔH, which is 890.2 kJ/mol. As there are $12.011 + 4 \times 1.008 = 16.043$ g/mol of CH_4, the HHV can also be written as

$$HHV = \frac{890.2 \text{kJ/mol}}{16.043 \text{g/mol}} \times 1000 \text{g/kg} = 55,490 \text{kJ/kg}$$

Table 4.1 Enthalpy of formation $H°$ at 1 atm, 25°C for selected substances

Substance	State	$H°$ (kJ/mol)
H	Gas	217.9
O	Gas	247.5
O_2	Gas	0
H_2O	Liquid	−285.8
H_2O	Gas	−241.8
CH_4	Gas	−74.9
CO_2	Gas	−393.5

4.5.4 Recuperator

Heat exchangers are used to transfer heat from the turbine exhaust flow to the cycle of power generation. Recuperator and regenerator are two main types of heat exchanger. A recuperator operates based on direct heat exchange and there are separate flow paths for each fluid, which flow simultaneously through the heat exchanger. A regenerator is on the basis of an intermediate storage in which there is one flow path that hot and cold fluids alternately pass through. In other words, a recuperator is a heat exchanger for transferring heat from the hot turbine discharges to the colder compressed air.

MTs are required to have a recuperator to achieve desirable thermodynamic efficiency. Adding heat to compressed air causes a reduction in the amount of fuel required to raise the temperature of gases used by the turbine, so a recuperator in MT is necessary for fuel and cost reduction and improving efficiency. However, the recuperator causes additional pressure losses leading to reduction of overall efficiency. Therefore, the recuperator has positive and negative effects on output efficiency and then balancing is required. There are connections between the recuperator and the compressor discharge, the expansion turbine discharge, the combustor inlet, and the system exhaust. Thus, design of recuperator configuration is a great challenge because it is required to lower pressure losses, lower manufacturing costs, and increase reliability in high temperatures. Main factors of the recuperator include high effectiveness, low pressure losses, minimum volume and weight, high reliability and durability, and low cost. The recuperator impact on efficiency of the MT is shown in Fig. 4.20. The inclusion of a high effectiveness recuperator increases the electric efficiency of a MT [10].

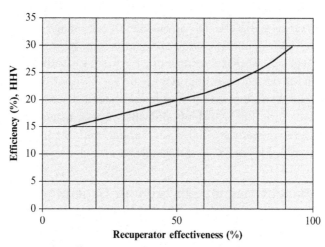

Fig. 4.20 Microturbine efficiency as a function of recuperator effectiveness [10].

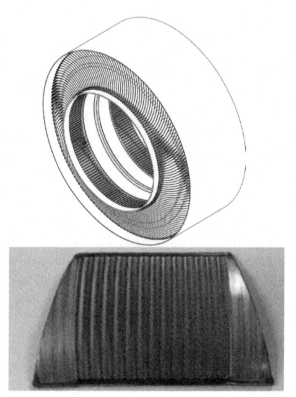

Fig. 4.21 Annular primary surface stainless steel recuperator.

Recuperators are divided into annular wraparound and rear-mounted cube types. The advantages of the annular type are compact design, good aerodynamic gas flow path, low pressure drop, no need for external ducts, and lower acoustic signature. The advantages of cube shape are simplicity of hot gas bypass for cogeneration and an external combustor for a variety of dirty fuels. Fig. 4.21 shows a recuperator for small gas turbine.

Recuperator design is an important issue for limiting deformation and thermal fatigue. Stainless steel is the typical material for a recuperator and it does provide long life with high efficiency. Some materials used in the structure of a recuperator are 300 series stainless steel, Inconel 625, Inconel 803, Haynes 120, Haynes 214, and PM2000.

4.5.5 Permanent Magnet Synchronous Generator

The term PMSG refers to two facts about its configuration. Permanent magnet means a generator without excitation coils (DC field winding), which are replaced by a permanent magnet. Synchronous means rotor and magnetic field, which are rotating with the same speed and frequency of induced voltage in the stator directly proportional to revolutions per minute (rpm). Today, PMSGs are widely used in steam turbines, gas turbines, hydro turbines, and wind turbine for converting mechanical to electrical output power for users. Generally, in a MT generator a PMSG is used instead of an induction generator because of following advantages [13]:

- No significant losses generated in the rotor
- Lower maintenance cost
- Design flexibility
- Very high torque is achieved at low speed
- Eliminates the need for separate excitation system
- Higher power density and higher reliability

However, the disadvantages of PMSG are

- Higher initial cost
- Flexibility loss of field flux control
- Possible demagnetization and thermal stress

Two general types of PMSGs are salient pole or nonsalient pole on average 70% efficiency. Construction materials typically are neodymium-iron-boron (NdBFe) or samarium-cobalt magnets, which are very suitable for such high-speed electrical machines. Fig. 4.9 shows a permanent magnet that is used in MTs, and a cross section of PMSG is shown in Fig. 4.22.

Fig. 4.22 Cross section of a permanent magnet synchronous generator.

4.6 MT APPLICATIONS

Recently, power sources with high-efficiency, reliability, and low pollutant emission level have been considered. MTs have several factors that are acceptable and appropriate for use in various applications. In the following sections, the most important and popular applications of MT are discussed.

4.6.1 Peak Shaving

One of the most important applications of MTs is using them in peak load periods. In peak demand, MTs are used to reduce costs and improve overall efficiency as well as reduce investments in bulk transmission. In general, peaking power plants are run only when there is peak demand for electricity. The MT start-up time is about 1–2 min. Moreover, appropriate output power allows using it in periods of high demand for peak shaving, which reduces overall costs. Peak shaving in combination with emergency power or standby power is economically advantageous.

4.6.2 Combined Heat and Power

Cogeneration or CHP is using wasted heat to generate electricity and useful heat (or cooling (CCHP)) at the same time. During normal operation, MTs produce significant amounts of high-temperature exhaust gases that with proper management and use of them, system overall efficiency reaches up to 80%. Wasted heat is used in forms of warm water, steam, direct use like heating or drying, driving thermally activated equipment, as well as chilling water. Fig. 4.23 shows MTG-based CHP.

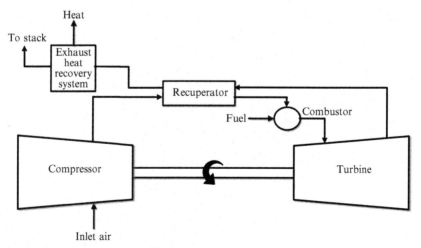

Fig. 4.23 Microturbine-based combined heat and power system.

Generally, the advantages of CHPs, which have created a global trend toward their use, include competitive capital investment, high efficiency, high power quality, environmental friendliness (low pollutant emissions level of NO_X, CO, and CO2), and operational flexibility. These CHP advantages are in addition to MT benefits such as simple installation, light weight, and low maintenance; hence, this combination is suitable. Typically the range of MT, which is used in CHP, is around 100 kW and they are known as micro CHPs.

Example 4.5

The Elliott TA 100A at its full 105 kW output burns 1.24×106 Btu/h of natural gas. Its waste heat is used to supplement a boiler used for water and space heating in an apartment house. The design calls for water from the boiler at 120°F to be heated to 140°F and returned to the boiler. The system operates in this mode for 8000 h per year [12].

(a) If 47% of the fuel energy is transferred to the boiler water, what should the water flow rate be?

(b) If the boiler is 75% efficient, and it is fueled with natural gas costing $6 per million Btu, how much money will the MT save in displaced boiler fuel?

(c) If utility electricity costs $0.08 kWh, how much will the MT save in avoided utility electricity?

(d) If operation and maintenance is $1500 year, what is the net annual savings for the MTe?

(e) If the MT costs $220,000, what is the ratio of annual savings to initial investment (called the initial rate of return)?

Solution

(a) The heat Q required to raise a substance with specific heat c and mass flow rate $\dot{m}$ by a temperature difference of ΔT is:

$$Q = \dot{m}c\Delta T$$

Because it takes 1 Btu to raise 1 lb of water by 1°F, and 1 gal of water weighs 8.34 lb, we can write

$$\text{Water flow rate } \dot{m} = \frac{0.47 \times 1.24 \times 10^6 \text{Btu/h}}{1\text{Btu/lb°F} \times 20°\text{F} \times 8.34\text{lb/gal} \times 60\text{min/h}}$$
$$= 58 \text{gpm}$$

(b) The fuel displaced by not using the 75% efficient boiler is worth

$$\text{Fuel saving} = \frac{0.47 \times 1.24 \times 10^{6\text{Btu/h}}}{0.75} \times \frac{\$6.00}{10^6\text{Btu}} \times 8000\text{h/year}$$
$$= \$37,300 \text{year}$$

(c) The utility electricity savings is

$$\text{Electric utility savings} = 105\,\text{kW} \times 8000\,\text{h/year} \times \$0.08/\text{kWh}$$
$$= \$67,200/\text{year}$$

(d) The cost of fuel for the MT is

$$\text{Microturbine fuel cost} = 1.24 \times 10^6 \text{Btu/h} \times \frac{\$6.00}{10^6 \text{Btu}} \times 8000\,\text{h/year}$$
$$= \$59,520/\text{year}$$

So the net annual savings of the MT, including $1500 year in operation and maintenance, is MT savings = ($37,300 + $67,200) − $59,520 −$1500 = $43,480 year

(e) The initial rate of return on this investment would be:

$$\text{Initial rate of return} = \text{Annual saving/Initial investment} =$$
$$\frac{\$43,480/\text{year}}{\$220,000} = 0.198 = 19.8\%/\text{year}$$

4.6.3 Backup Generation

In this application, MTs are used as standby power or uninterruptible power supply (UPS). A standby generator is used in event of an outage, as a backup electrical system that operates automatically. Similarly, an UPS is an electrical device, which produces emergency power to load. The MTs are an ideal choice to use as an UPS and standby power because of their low initial cost, high level of reliability, and low maintenance requirement. In this situation, grid reliability for supplying of sensitive loads is increased due to the high level of support.

4.6.4 Resource Recovery

Selective extraction of disposed materials for a specific next use to extract maximum benefits from products is called resource recovery. In this application, MTs burn waste gasses that are flared or released directly to the atmosphere. MTs can convert unprocessed gas that contain up to 7% H_2S to electricity and this reduces fuel consumption and costs. In addition to its fuel flexibility, the MT was introduced as a more efficient and most economical power source.

4.6.5 Transportation Application

Recently, the use of MTs has increased in hybrid electric vehicles (HEVs) because of their high efficiency and low pollutant emissions. MT application in transportation includes transit busses, heavy trucks, HEVs, and super cars. The MTs enabled increased mileage without refueling (reduction of fuel consumption reaches up to 40%). The MTGs with low levels of emissions and noise and high efficiency are appropriate cases for a transportation system.

4.6.6 Base Load

MTs are used as base load to increase power quality and reliability. Base load power sources can consistently generate the electrical power needed to satisfy minimum demand (minimum amount of electric power delivered or required over a given period at a constant rate). Generally, continuous operation of the MTs in rated capacity cause its efficiency to improve.

4.6.7 Hybrid System

Recently, research has focused on the combination of two or more DGs that are known as hybrid systems. To improve the reliability of renewable DGs, a MT is an appropriate choice and the combination of a fuel cell and a MT is a desired and popular hybrid system with overall efficiency that reaches up to 50%. Advantages of hybrid system include reduction of investment, lower pollutant emission, higher reliability, fuel consumption reduction, and higher efficiency.

4.6.8 Remote Power

In locations where network construction is impossible or it is not economically beneficial, MTs are useful. Typically, cost of grid construction is more than a DG. The MTs fuel flexibility is a determining factor for use in remote applications, which causes significant reduction of investment costs.

Mentioned applications are most important in MTGs but there are other applications, which are not included in this section. Generally, the main purpose is to achieve high efficiency, reliability, and economic power source.

4.7 TYPES OF MT

The main parts of a MT are the compressor, combustor, and turbine. As mentioned earlier, the compressor is located before the turbine and the combustion chamber is between them. Inlet air at ambient temperature is compressed by the compressor and then compressed air is mixed with fuels in the combustor. High temperature and pressure gases drive the turbine and

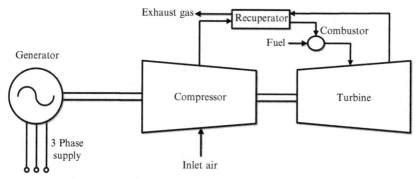

Fig. 4.24 Single-shaft microturbine generation system.

produce mechanical power. A compressor and a generator are required turbine output mechanical power for rotating. This cycle is true in gas turbines and its subsets.

The shaft structure defines important characteristics of the MT that eventually influence the required power electronics and control system. There are two types of MT design: single-shaft design and split-shaft or two-shaft design. Single-shaft design of a MT is represented in Fig. 4.24.

In this model, all components are mounted on same shaft and generator and compressor operation are dependent on the turbine. In the shingle-shaft design, the speed of the shaft is about 50,000–120,000 rpm, which causes high-frequency output voltages. High-frequency voltage is needed to power electronic interfaces for converting to the desired frequency for customers, which is discussed later.

Split-shaft design is not a required power electronic interface and Fig. 4.25 shows this type of MT. In the split-shaft design, there is a turbine for driving the compressor and another turbine for driving generators, which are coupled together through a gearbox. Speed of the shaft is about 3600 rpm and the output voltage frequency is 60 Hz (or 50 Hz) without using any converters. More moving parts cause a negative effect on efficiency and maintenance costs.

In general, single-shaft MTs are more popular because of higher flexibility, reliability, and lack of required lubrication.

4.8 TYPES OF POWER ELECTRONIC INTERFACE TOPOLOGIES

Power electronic interfaces are required in single-shaft MT designs because MT output power is high frequency, which is not appropriate for customers.

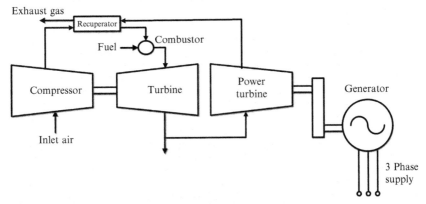

Fig. 4.25 Split-shaft microturbine generation system.

There are a variety of methods for converting of the MTs output power to desired power that is based on AC/DC/AC or AC/AC structures and there is not absolute excellency for any of them.

In an AC/DC/AC design based on a DC-link capacitor, high-frequency voltage is converted to DC voltage and then back to the desired frequency through an inverter. A controlled rectifier or active rectifier and uncontrolled rectifier or passive rectifier are considered as types of rectification. A passive rectifier is applied by diodes and this structure is simple and low cost but the advantages of an active rectifier are more flexibility and providing MTs starting-up features. The disadvantages of DC-link converters are large physical dimensions, lower lifetime, heavy weight, and lower reliability. Fig. 4.26 shows two typical structures of DC-link converters. However, a high-frequency link converter (HFLC) is another method of AC/DC/AC, which is shown in Fig. 4.27. One advantage of the HFLC is robust isolation, but it is more expensive with higher power electronics parts and losses that lead to less use in industry. Generally, power electronic interfaces based on a passive rectifier and inverter combination are the most common method because of their potential merits.

Recently, attention to AC/AC interface topology has increased. Two typical types of this structure are cycloconverters and matrix converters, which are shown in Fig. 4.28. Input AC voltage with arbitrary amplitude and frequency is converted to the desired frequency. These converters are without a DC-link capacitor that causes higher lifetime and efficiency. Other advantages of AC/AC structure include lower thermal stress on semiconductor switches, higher reliability, compact size, bidirectional power flow, and controllable input power factor. A major drawback of

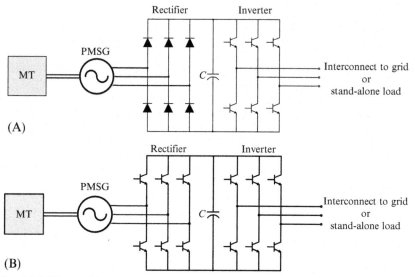

Fig. 4.26 Microturbine generation system power electronic interface: (A) passive rectifier and (B) active rectifier.

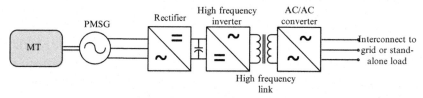

Fig. 4.27 Microturbine generation system with a high-frequency link power converter.

AC/AC converters is more switches lead to more power losses. Moreover, using of battery or other power sources and energy storage is not possible. These converters are sensitive to disturbance of input voltage and input fluctuation directly influences the output and vice versa. Control complexity and lower voltage transfer ratio are other disadvantages of these topologies.

None of the abovementioned structurea is superior. The simplest, least costly, and most common method for power electronic interface is the AC/DC/AC structure based on diode rectification, which is the simulation model of this chapter. In the following, converter simulation and dynamic model of MT with controls block diagrams will be considered.

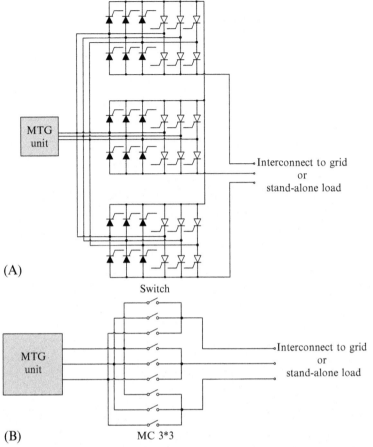

Fig. **4.28** Microturbine generation system power electronic interface: (A) cycloconverter and (B) matrix converter.

Example 4.6

A MT has a natural gas input of 13,700 Btu (LHV) per kWh of electricity generated. Find its LHV efficiency and its HHV efficiency [12] (Table 4.2).

Solution

Generally, the thermal efficiency of power plants is often expressed as a heat rate, which is the thermal input (Btu or kJ) required to deliver 1 kWh of electrical output (1 Btu/kWh = 1.055 kJ/kWh). The smaller the heat rate, the higher the efficiency. Heat rates are usually expressed in Btu/kWh, which results in the following relationship between it and thermal efficiency,

Table 4.2 Higher heating value and lower heating value for various fuels

Fuel	Higher heating value (HHV)		Lower heating value (LHV/HHV)		LHV/HHV
	Btu/lbm	kJ/kg	Btu/lbm	kJ/kg	
Methane	23,875	55,533	21,495	49,997	0.9003
Propane	21,669	50,402	19,937	46,373	0.9201
Natural gas	22,500	52,335	20,273	47,153	0.9010

$$\text{Heat rate (Btu/kWh)} = \frac{3412\,\text{Btu/kWh}}{\eta}$$

Using the LHV for fuel gives the LHV efficiency:

$$\text{Efficiency } (LVH) = \frac{3412\,\text{Btu/kWh}}{13,700\,\text{Btu/kWh}} = 0.2491 = 24.91\%$$

Using the LHV/HHV ratio of 0.9010 for natural gas in the above equation, gives the HHV efficiency for this turbine: $\text{Efficiency } (HHV) = 24.91\% \times 0.901 = 22.44\%$.

4.9 CONSTRUCTION AND OPERATION OF MTs

In this section, different aspects of the MT along with several diagrams are presented. The information is about overall efficiency, output power, and pollutant emissions, which are main factors in the construction and operation of MTs.

The ambient conditions at the inlet of the MT affect both power output and efficiency. At inlet air temperatures above 59°F (15°C), both the power and efficiency decrease. The power decreases due to the decreased air density with increasing temperature and the efficiency decreases because the compressor requires more power to compress higher-temperature air. Conversely, the power and efficiency increase when the inlet air temperature is below 59°F. Fig. 4.29 shows the variation in power and efficiency for a MT as a function of ambient temperature relative to the reference International Organization for Standards (ISO) condition of sea level and 59°F.

It should be noted that some manufacturers may place limitations on maximum power output below a certain ambient temperature due to maximum power limitations of the gearbox, generator, or power electronics, and this may modify the shape of the curves shown in Fig. 4.29 [10].

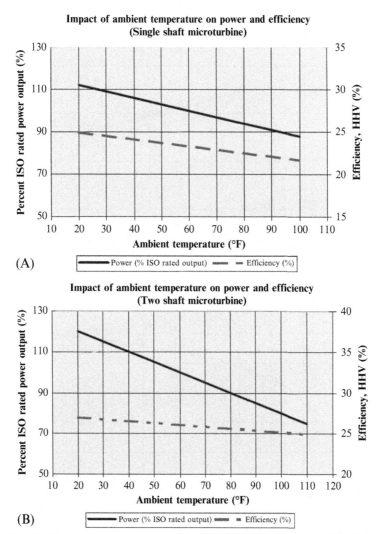

Fig. 4.29 Ambient temperature effects on microturbine performance: (A) single-shaft and (B) split-shaft [10].

Elevation has influence on the MT operation because air density is lower in altitude. Based on a rule of thumb, for each 305 m above sea level, 3% performance losses occur at full load rating. Fig. 4.30 shows the typical effect of altitude on MTG output power.

When less than full power is required from a MT, the output is reduced by reducing rotational speed, which reduces temperature rise and pressure ratio through the compressor and temperature drop through the turbine,

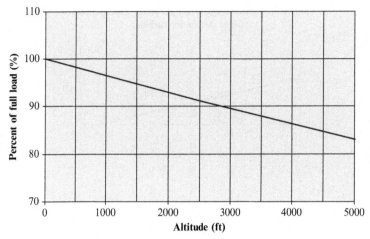

Fig. 4.30 Altitude effects on microturbine output power [10].

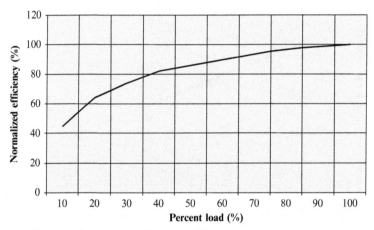

Fig. 4.31 Microturbine part-load efficiency [10].

and by reducing turbine inlet temperature so that the recuperator inlet temperature does not rise. In addition to reducing power, this change in operating conditions also reduces efficiency. The efficiency decrease is minimized by the reduction in mass flow (through speed reduction) at the same time as the turbine inlet temperature is reduced. Fig. 4.31 shows a typical part-load efficiency curve based on a 30 kW MT [10]. Experimental results show that when the MT is operated 50% below its rated capacity, pollutant emissions tend to increase.

MT operation in a partial load leads to lower efficiency. Efficiency as a function of output electrical power for two typical types of the MT is represented is Fig. 4.32.

Efficiency and output power as a function of ambient temperature for Capstone as the largest manufacture of MTs is shown in Figs. 4.33 through 4.35. Figs. 4.33–4.35 show C30, C65, and C200 Capstone MTs, respectively.

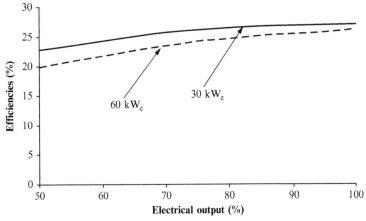

Fig. 4.32 Electrical efficiency for a 30 kW and a 60 kW microturbine [14].

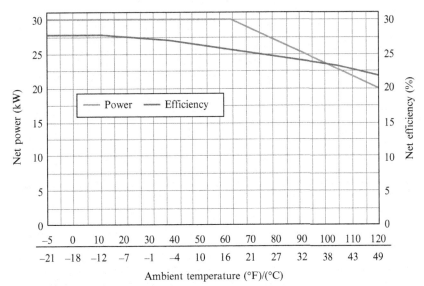

Fig. 4.33 The electrical power and net efficiency of a C30 MT as a function of temperature at sea level [9].

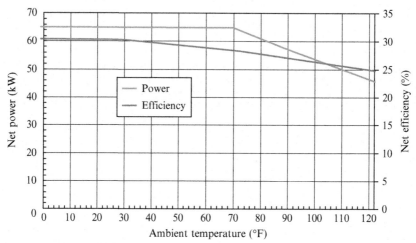

Fig. 4.34 The electrical power and net efficiency of a C65 MT as a function of temperature at sea level [9].

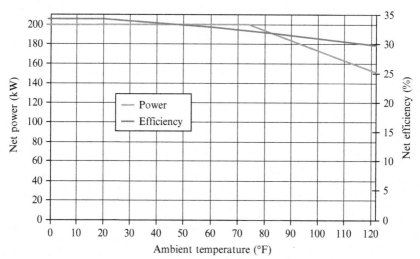

Fig. 4.35 The electrical power and net efficiency of a C200 MT as a function of temperature at sea level [9].

Evaluation of pollutant emission level is important issue and the MT with lower pollutant is more favorable. Manufacture and researchers are focused on more clean energy sources to prevent greenhouse gases emission. The US Environmental Protection Agency (EPA) lists six criteria air pollutants for which ambient air limits have been set [15]. These air pollutants include NO_X, CO, SO_2, lead (Pb), ozone (O_3) and particulates. And between them, NO_X, CO, and SO_2 are the most relevant for the operation of the MT. In the following, pollutant emission of the MTG is considered at various conditions.

Fig. 4.36 shows the pollutant emission characteristic of a 30 kW MT at steady state. MT operations at partial load release more pollutant compared to full load. Moreover, Fig. 4.37 shows the pollutant characteristic of a 60 kW MT at steady state.

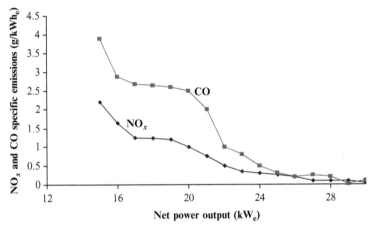

Fig. 4.36 NO_X and CO emission characteristic of the 30 kW MT [14].

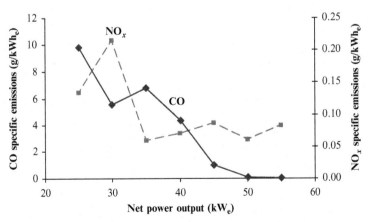

Fig. 4.37 NO_X and CO emission characteristic of the 60 kW MT [16].

Most studies about MT pollutant emissions are at steady state and full load conditions. The transient pollutant emissions during MT startup are often neglected. In fact, at partial loads and transient processes, the pollutant emission characteristics can be very different from rated conditions [12]. The following are from [12], and they reveal some results of the MT transient emissions startup at both partial and full load for various output powers. This study considers nitrogen oxide (NO_X), carbon monoxide (CO), and sulfur dioxide (SO_2) the most hazardous emissions for human health. In [12], a Turbec T100P MT fueled by natural gas used in the experimental studies of transient emissions. According to the manufacturer's catalog, output power of the unit at ISO conditions is 100 kW, and the efficiency is about 30%. The rated concentrations of CO and NO_X emissions at 15% oxygen are under 15 ppm.

Experimental results show that CO concentration of the MT startup increases rapidly to a rather high level (up to 5–100 times higher than that of the steady condition) and then drops down to a stable value [12]. There is no significant difference for various power outputs.

Moreover, the NO_X startup transient pollutant is similar to CO emission, and there is no significant difference in various powers. Generally speaking, the NO_X concentration at steady state condition is lower than 15 ppm, as stated in the manufacturer's catalog.

At higher power ratings, SO_2 emission is lower than 4 ppm, and it may be deduced that no significant damage occurs.

Finally, results of MT start-up pollutant emission show that [12]:
- Normal operation of the microturbine at full load power output produces the lowest emissions of air pollutants (CO, NO_X, and SO_2).
- The CO startup emissions have significant effect on the total amount of emission.
- Although the impact of NO_X and SO_2 startup emission is not as serious as that of CO, it also should be considered.

For comprehensive pollutant analysis, MT emissions compare with other DG technologies. The values are given for a variety of emissions, which are nitrogen oxide (NO_X), sulfur dioxide (SO_2), carbon dioxide (CO_2), and particulate matter (PM-10), and they are characterized in terms of pounds of emissions per unit of electrical output. The values are based on typical operation condition. Fig. 4.38 shows pollutant emission of the MT, which has favorable conditions in air emission. MT pollutant emission is desirable and catalytic exhaust gas cleaning is not required but natural gas fired reciprocating engines or diesel reciprocating engines are required.

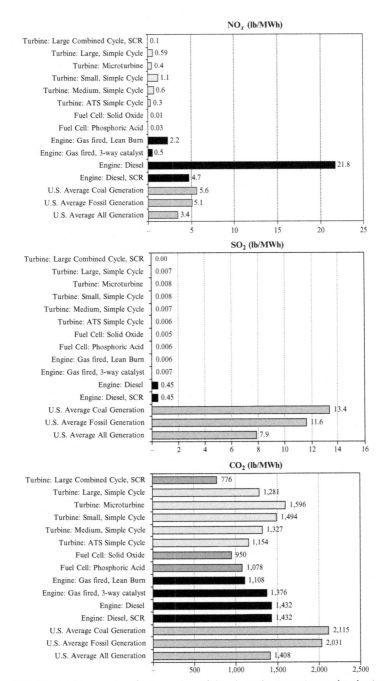

Fig. 4.38 Air emissions values for a number of distributed generation technologies [17].

(Continued)

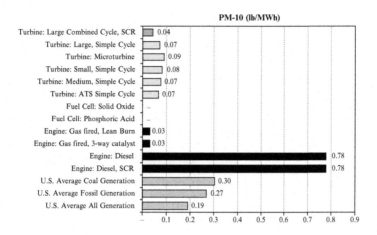

Fig. 4.38, cont'd

Although noise of the MT can be a problem, new technologies have solved this issue. Internal noises of the MT are from flow vortices, blading, magnetic field, and combustion. However, the level of noise is low and measurements show an average of 65 dB or less on a 10-m radius.

A single MT produces power in the range of 25–1000 kW, but research on higher output power is continuing and output power of MTs with parallel connections can reach up to 30 MW. Output power and efficiency of MTs are acceptable and appropriate in many cases. The efficiency of a non-recuperated MT is about 15%–20%, that for a recuperated MT is about 20%–35%, and in a cogeneration application or use as CHP efficiency can reach up to 80% [1].

Table 4.3 contains appropriate comparisons between various DGs. Based on Table 4.3 the MT as a reliable DG has a suitable rankings in many respects.

Table 4.3 Rankings for distributed power generating technologies [18]

Category	Most positive → least positive
Efficiency	Fuel cells—reciprocating—stirling—microturbine—rankine
Technology status	Reciprocating—microturbine—fuel cells—stirling—rankine
Cost	Reciprocating—microturbine—fuel cells—stirling—rankine
Emissions	Fuel cells—stirling—microturbine—rankine—reciprocating
Noise	Fuel cells—stirling—microturbine—rankine—reciprocating
Load matching flexibility	Reciprocating—fuel cells—microturbine—stirling—rankine

4.10 MT MODELING

The MT model includes temperature control, a fuel system, a speed governor, acceleration control, and turbine dynamic blocks. Outputs of the speed control, temperature control, and acceleration control are inputs of a low value selector (LVS), where the LVS output is the lowest value of all inputs and the fuel system is supplied by output of the LVS. A dynamic model of a single-shaft MT is presented in Fig. 4.39, and step-by-step modeling is done in the following sections. Note that parameters values are given in Appendix A to this chapter.

4.10.1 Speed and Acceleration Control

A speed controller operates on an error between reference speed and PMSG rotor speed, which is its output as it enters the LVS and usually a lead–lag transfer function or a proportional–integral–derivative (PID) controller is used for speed governor modeling. Fig. 4.40 shows a block diagram of the speed and acceleration control.

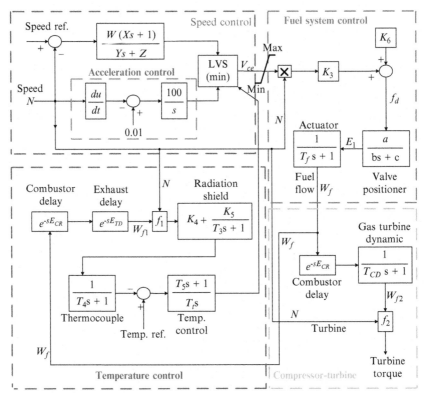

Fig. 4.39 Dynamic model of the microturbine.

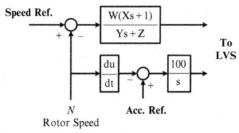

Fig. 4.40 Speed controller of a microturbine.

In this figure, W is controller gain, X (Y) is the governor lead (lag) time constant, and Z is a constant representing the governor mode in droop or isochronous [19].

Acceleration control is used for start-up time of the MT to limit the rate of rotor speed increasing before reaching operating speed. The system operating speed close to rated speed cause the elimination of the acceleration control in the modeling, but acceleration control is used in this study.

Example 4.7

Speed controller (governor) of a MT is modeled by a lead-lag transfer function with three parameters, which are X, Y, and Z. X and Y are the governor lead and lag time constant, respectively, and Z is a constant. Please illustrate the role of Z in the transfer function and the range of numbers that can be chosen for Z?

Solution

Z is a constant that represents the governor mode in droop or isochronous. A droop governor is a straight proportional speed controller in which the output is proportional to the speed error. An isochronous speed controller is a proportional-plus-reset speed controller in which the rate of change of the output is proportional to the speed error. Therefore, the output of an isochronous governor will integrate in a corrective direction until the speed error is zero [20]. $Z=1$ represents a droop governor and $Z=0$ represents an isochronous governor.

4.10.2 Temperature Control

The input signals to the temperature control system are fuel demand signal and turbine speed, which output is a temperature control signal to the LVS. A temperature control block diagram is shown in Fig. 4.41.

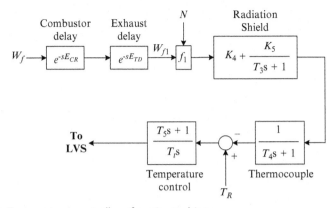

Fig. 4.41 Temperature controller of a microturbine.

The fuel and compressed air are burned in the combustor results in turbine torque. The exhaust temperature characteristic is as the following function:

$$f_1 = T_R - a_{f1} \cdot (1 - W_{f1}) + b_{f1} \cdot N \tag{4.14}$$

where f_1 is a function with inputs of fuel flow and turbine speed to produce a value of turbine exhaust temperature, a_{f1} and b_{f1} are constant values, and T_R is temperature reference. The exhaust temperature is measured using a series of thermocouples incorporating radiation shields [13]. According to Fig. 4.41, K_4 and K_5 are constant in radiation shield transfer function, T_3 and T_4 are the time constant of the radiation shield and thermocouple transfer function, respectively, and T_5 with T_t are the time constant of the temperature control transfer function. The output of the thermocouple is compared with a reference value and desirable performance is in high temperature. The reference temperature is normally higher than the thermocouple output but when thermocouple temperature is higher than reference value the result is a negative error, which is as input of the LVS and temperature control start decreasing to reach the former temperature.

4.10.3 Fuel System

Fuel system controls are a series block of the fuel valve and actuator. Fig. 4.42 shows a fuel control system for the MT. The output of the LVS, Vce, is scaled by the gain K_3 and is offset by K_6 that is representing fuel flow at no load condition. The valve positioner transfer function is as follows:

$$E_l = \frac{a}{bs + c} f_d \tag{4.15}$$

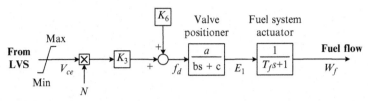

Fig. 4.42 Fuel control system of a microturbine.

Where a is the valve positioner gain, b is the valve positioner time constant, and f_d is input of the valve positioner. The actuator transfer function is

$$W_f = \frac{1}{T_f s + 1} E_l \tag{4.16}$$

where T_f is actuator time constant, E_l is output of the valve positioner, and W_f is the fuel demand signal per unit.

Example 4.8
Input of the MT fuel system is supply by output of the LVS block. The outputs of the speed controller, acceleration controller, and temperature controller enter the LVS and a minimum of them are passed through the LVS. Why is this mechanism chosen for the fuel system? What is the role of the limiter after the LVS?

Solution
These three control functions (speed governor, temperature control, and acceleration control) are all inputs to a LVS. The output of the LVS, which is called V_{ce}, is the lowest of the three inputs, whichever requires the least fuel. In other words, desirable operation of the MT depends on lower fuel consumption with higher output temperature and with respect to this fact, the LVS block and higher reference temperature are considered in modeling. Transfer from one control to another is bumpless and without any time lags. The output of the LVS is compared with maximum and minimum limits. Of the two, the maximum limit acts as a backup to temperature control and is not encountered in normal operation; the minimum limit is the more important, dynamically. This is because the minimum limit is chosen to maintain adequate fuel flow to ensure that a flame is maintained within the gas turbine combustion system [20].

Example 4.9

Output of the LVS in a MT fuel system is scaled by the gain K_3 and is offset by K_6. What is the role of these parameters in the fuel system? Why are K_3 and K_6 equal to 0.77 and 0.23, respectively?

Solution

In terms of dynamic response, gas turbines have many differences from steam turbines. One of the more obvious differences is the need for a significant fraction of rated fuel to support self-sustaining, no-load conditions. This amounts to ~23% and is one of the economic driving forces to minimize operating time at full-speed, no-load conditions [20]. To allow the use of governor parameters comparable to steam turbines and hydroturbines, the governor operates from 0%–100% of the active load range. This range must then be corrected to 23%–100% of the fuel flow range to be compatible with the thermodynamics of the turbine. This relationship between governor output signal and fuel flow is shown in Fig. 4.43. With respect to the mentioned explanation, K_6 is equal to 0.23 and that represents fuel flow at a no-load condition and $K_3 = 1 - K_6 = 1 - 0.23 = 0.77$.

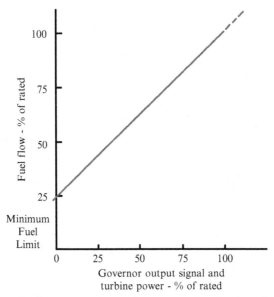

Fig. 4.43 Relationship between governor output signal and turbine fuel flow.

4.10.4 Compressor-Turbine

The compressor-turbine package is an important part of a MT and they are considered as a package because they are mounted on the same shaft. The input signals to the gas turbine are the fuel flow W_f signal that is achieved from the fuel control system and the rotor speed deviation ΔN. The compressor-turbine package is represented in Fig. 4.44. The output signal is the turbine torque. The gas turbine dynamic transfer function is

$$W_{f2} = \frac{1}{T_{CD}s + 1} W_f \tag{4.17}$$

where T_{CD} is the gas turbine dynamic time constant.

The torque characteristic is as following equation:

$$f_2 = a_{f2} + b_{f2} \cdot W_{f2} + c_{f2} \cdot N \tag{4.18}$$

where a_{f2}, b_{f2}, and c_{f2} are constant values and f_2 is a function whose inputs are fuel flow and turbine speed to produce mechanical power.

Note that all mentioned controllers are per unit except the temperature controller. Temperature control mostly is based on English units with $T_R = 950°F$.

4.10.5 PMSG Model

Electric power is produced by using a high-speed generator, which is driven by a turbine. A PMSG is similar to the conventional alternator configuration, and modeling of the PMSG is expressed in rotor reference frame (d-q frame). The following equations are applied for transformation of three-phase abc to rotor reference frame dq0 and vice versa.

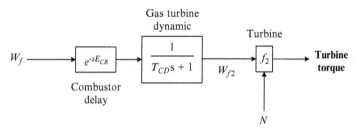

Fig. 4.44 Compressor–turbine package of a microturbine.

$$\begin{bmatrix} v_q \\ v_d \\ v_o \end{bmatrix} = \frac{2}{3} \begin{bmatrix} \cos\theta_r & \cos(\theta_r - 120) & \cos(\theta_r + 120) \\ -\sin\theta_r & -\sin(\theta_r - 120) & -\sin(\theta_r + 120) \\ \frac{1}{2} & \frac{1}{2} & \frac{1}{2} \end{bmatrix} \begin{bmatrix} v_a \\ v_b \\ v_c \end{bmatrix} \tag{4.19}$$

$$\begin{bmatrix} v_a \\ v_b \\ v_c \end{bmatrix} = \begin{bmatrix} \cos\theta_r & -\sin\theta_r & 1 \\ \cos(\theta_r - 120) & -\sin(\theta_r - 120) & 1 \\ \cos(\theta_r + 120) & -\sin(\theta_r + 120) & 1 \end{bmatrix} \begin{bmatrix} v_q \\ v_d \\ v_o \end{bmatrix} \tag{4.20}$$

And then PMSG modeling in d-q form are given by Eqs. (4.21)–(4.23).

- Electrical equations

$$v_d = R_s i_d + L_d \frac{di_d}{dt} - L_q p \omega_r i_q \tag{4.21}$$

$$v_q = R_s i_q + L_q \frac{di_q}{dt} + L_d p \omega_r i_d + \lambda p \omega_r \tag{4.22}$$

$$T_e = 1.5P\left[\lambda i_d + \left(L_q - L_d\right) i_d i_q\right] \tag{4.23}$$

where, L_d and L_q are the d-axis and q-axis inductance, respectively; R_s is the stator winding resistance; i_d and i_q are the d-axis and q-axis current, respectively; v_d and v_q are the d-axis, q-axis voltage, respectively; ω_r is the angular velocity of the rotor; λ is the flux linkage; P is the number of poles; and T_e is the electromagnetic torque.

- Mechanical equations

$$\frac{d\omega_r}{dt} = \frac{1}{J}\left(T_e - T_{shaft} - F\omega_r\right) \tag{4.24}$$

$$\frac{d\theta_r}{dt} = \omega_r \tag{4.25}$$

where J is rotor and load combined inertia, T_{shaft} is mechanical torque, F is rotor and load combined viscous friction, and θ_r is the rotor angular position. Fig. 4.45 shows the rotor reference frame equivalent circuit of a PMSG model.

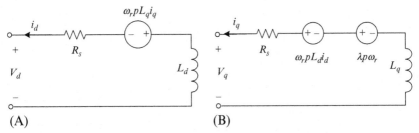

Fig. 4.45 Permanent magnet synchronous generator equivalent circuit model: (A) d-axis and (B) q-axis.

Example 4.10

A 30 kW, three-phase, PMSG produces an open-circuit voltage of 400 V line-to-line, 60 Hz, when driven at a speed of 96,000 rpm. When operating at rated speed and supplying a resistive load, its terminal voltage is observed to be 320 V line-to-line for a power output of 15 kW.

(a) Calculate the generator phase current under this operating condition.

(b) Assuming the generator armature resistance to be negligible, calculate the generator 60 Hz synchronous reactance.

(c) Write the torque in normal three-phase (abc) frame and compare with Eq. (4.23).

(d) What is the difference between permanent magnet synchronous motor and generator?

Solution

(a) Because the load is resistive,

$$I_a = \frac{P}{3V_a} = \frac{15{,}000}{\sqrt{3} \times 320} = 27.06 \text{ A}$$

(b)

$$E_{af} = 400/\sqrt{3} = 230.9 \text{V}$$

So we have,

$$E_{af} = \sqrt{V_a^2 - (X_s I_a)^2} \rightarrow X_s = \frac{\sqrt{E_{af}^2 + V_a^2}}{I_a} = 14.58 \ \Omega$$

(c) We assume that linkage flux is sinusoidal:

$$\lambda_m = \lambda_{m\ 0} \begin{bmatrix} \sin\theta_r \\ \sin\left(\theta_r - \frac{2\pi}{3}\right) \\ \sin\left(\theta_r + \frac{2\pi}{3}\right) \end{bmatrix}$$

where λ_{m0} is amplitude of λ_m. According to the mentioned equation for linkage flux, the stator-induced voltages are sinusoidal. The torque standard equation in electrical machines is

$$T_e = \frac{p}{2} \frac{\partial W_c}{\partial \theta_r}$$

where W_c is coenergy and p is the number of poles. The coenergy equation is as follows: $W_c = \frac{1}{2} i_{abcs}^T L_s i_{abcs} + i_{abcs}^T \lambda_m + W_{pm}$

where L_s is stator inductance and W_{pm} is the energy of the coupling field. From the previously mentioned torque equation and coenergy equations we have

$$T_e = \pm \left(\frac{p}{2}\right)$$

$$\left\{ \frac{\left(L_{md} - L_{mq}\right)}{3} \left[\left(i_{as}^2 - 0.5i_{bs}^2 - i_{as}i_{cs} + 2i_{bs}i_{cs}\right) \sin 2\theta_r + \frac{\sqrt{3}}{2}\left(i_{bs}^2 - i_{cs}^2 - 2i_{as}i_{bs} + 2i_{as}i_{cs}\right) \cos 2\theta_r \right] + \right.$$
$$\left. \lambda_{m0}\left[\left(i_{as} - 5i_{bs} - 0.5i_{cs}\right) \cos\theta_r + \frac{\sqrt{3}}{2}\left(i_{bs} - i_{cs}\right) \sin\theta_r \right] \right\}$$

where $L_{md} = 3/2(L_A - L_B)$ and $L_{mq} = 3/2(L_A + L_B)$

(d) To illustrate the PMSG it is easier to focus on motor type. The equations are similar but there are two great differences. Current direction in the generator type is outgoing and the electromagnetic torque sign is negative. So, equations and equivalent circuit are based on permanent magnet generator and permanent magnet motor is similar with the mentioned difference. You can see [21] for better understanding.

4.11 CONTROL CIRCUITS

A power conditioning unit consists of a rectifier–inverter system with a DC link. It is a general configuration of a power electronic interface in MT units (MTUs). The purposes of this configuration are simple structure, simple control circuit, and ease of understanding. The high-frequency electric power of a PMSG must be converted to DC and inverted back to 60 or 50 Hz AC. An insulated gate bipolar transistor (IGBT)-based pulse width modulation (PWM) inverter is used with a 2 kHz carrier frequency. The inverter injects AC power from a DC link of the MTU to the AC distribution system [22,23]. Grid-connected mode allows the MTU to operate parallel to the grid, providing base loading and peak shaving and grid support. Moreover, stand-alone mode allows the MTU to operate completely isolated from the grid. There are two different control strategies for each operation mode [24,25]:

- P-Q control strategy for grid-connected (power control)
- V-f control strategy for stand-alone (voltage-frequency control)

4.11.1 Control Strategy for Grid-Connected Operation

In this situation, voltage amplitude and frequency of the MTG is determined by the grid. The grid is like an infinite bus with constant frequency and voltage in which appropriate power flow is an important issue. In other words, for grid-connected operation mode, a P-Q control strategy is used where delivering desired active and reactive instantaneous power to load is

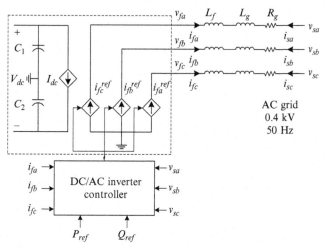

Fig. 4.46 Equivalent circuit of the inverter in grid-connected mode.

considered and the power deficit is compensated by the grid. The equivalent circuit of a DC/AC inverter in grid-connected operating condition is shown in Fig. 4.46. This converter is represented with three ideal current sources, which are i_{fa}^{ref}, i_{fb}^{ref}, and i_{fc}^{reff}. The converter manages the amount of the current injected to the AC grid from the DC bus, and the inverter controller uses the hysteresis current control (HCC) switching technique. The input signals to the inverter control circuit are three-phase grid voltages, which are called v_{sa}, v_{sb}, and v_{sc}, three-phase output currents of the converter, which are called i_{fa}, i_{fb}, and i_{fc}, reference of the active power P_{ref} and the reference of the reactive power Q_{ref}. With respect to Fig. 4.46, L_f is inductance of the converter filter, and R_g and L_g are the resistance and inductance of the grid, respectively.

Fig. 4.47 shows the inverter control strategy in grid-connected mode. The required power that must be injected to the grid is determined by P_{ref} and Q_{ref} signals, which are set by power management units. Based on Fig. 4.47, equations are expressed in Clarke transformation or α-β frame:

$$\begin{bmatrix} i_{sa} \\ i_{sb} \\ i_{sc} \end{bmatrix} = - \begin{bmatrix} i_{fa} \\ i_{fb} \\ i_{fc} \end{bmatrix} \tag{4.26}$$

$$\left\{ \begin{array}{c} v_{s\alpha} \\ v_{s\beta} \\ v_{s0} \end{array} \right\} = T_{\alpha\beta0} \left\{ \begin{array}{c} v_{sa} \\ v_{sb} \\ v_{sc} \end{array} \right\}, \quad T_{\alpha\beta} = \frac{2}{3} \begin{bmatrix} 1 & -1/2 & -1/2 \\ 0 & \sqrt{3}/2 & -\sqrt{3}/2 \\ 1/2 & 1/2 & 1/2 \end{bmatrix} \tag{4.27}$$

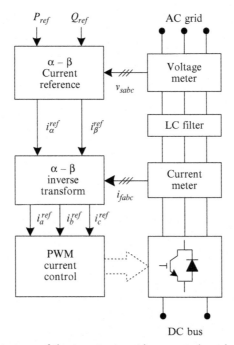

Fig. 4.47 Control strategy of the inverter in grid-connected mode.

With Eq. (4.27), the voltages and currents converts to α-β frame. Note that in a balanced three-phase system, the third equation is equal to zero. The instantaneous active and reactive power are as in Eqs. (4.28) and (4.29), respectively:

$$P = \frac{3}{2}\left(v_\alpha i_\alpha + v_\beta i_\beta\right) \tag{4.28}$$

$$Q = \frac{3}{2}\left(v_\beta i_\alpha - v_\alpha i_\beta\right) \tag{4.29}$$

The α-β reference currents are related to the reference power by Eq. (4.30).

$$\begin{bmatrix} i_{s\alpha}^{ref} \\ i_{s\beta}^{ref} \end{bmatrix} = \frac{2/3}{v_{s\alpha}^2 + v_{s\beta}^2} \begin{bmatrix} v_{s\alpha} & v_{s\beta} \\ v_{s\beta} & -v_{s\alpha}^f \end{bmatrix} \begin{bmatrix} P_{ref} \\ Q_{ref} \end{bmatrix} \tag{4.30}$$

HCC requires three-phase current references, which are achieved by α-β inverse transformation.

$$
\begin{bmatrix} i_{sa}^{ref} \\ i_{sb}^{ref} \\ i_{sc}^{ref} \end{bmatrix} = T_{\alpha\beta0}^{-1} \begin{bmatrix} i_{s\alpha}^{ref} \\ i_{s\beta}^{ref} \end{bmatrix}, \quad T_{\alpha\beta0}^{-1} = \begin{bmatrix} 1 & 0 \\ -\dfrac{1}{2} & \dfrac{\sqrt{3}}{2} \\ -\dfrac{1}{2} & -\dfrac{\sqrt{3}}{2} \end{bmatrix} \tag{4.31}
$$

$$
\begin{bmatrix} i_{fa}^{ref} \\ i_{fb}^{ref} \\ i_{fc}^{ref} \end{bmatrix} = - \begin{bmatrix} i_{sa}^{ref} \\ i_{sb}^{ref} \\ i_{sc}^{ref} \end{bmatrix} \tag{4.32}
$$

Finally, inverter switching is done through comparison between calculated reference currents and inverter actual measurement currents.

4.11.2 Control Strategy for Stand-Alone Operation

In this situation, voltage amplitude and frequency are important because the MTG cannot automatically set these parameters. The equivalent circuit of the inverter in islanding mode is represented in Fig. 4.48.

The inverter is represented with three ideal voltage sources, which are called v_{fa}^{ref}, v_{fb}^{ref}, and v_{fc}^{ref}, to regulate load voltages. Input signals to the inverter control circuit are load voltages and desired frequency.

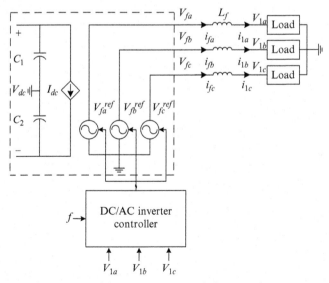

Fig. 4.48 Equivalent circuit of the inverter in stand-alone mode.

The following equation describes inverter voltages and currents:

$$\begin{bmatrix} v_{fa} \\ v_{fb} \\ v_{fc} \end{bmatrix} = \begin{bmatrix} v_{la} \\ v_{lb} \\ v_{lc} \end{bmatrix} + \begin{bmatrix} L_f & 0 & 0 \\ 0 & L_f & 0 \\ 0 & 0 & L_f \end{bmatrix} \frac{d}{dt} \begin{bmatrix} i_{fa} \\ i_{fb} \\ i_{fc} \end{bmatrix} \tag{4.33}$$

where v_{fa}, v_{fb}, and v_{fc} are three-phase output voltage of the inverter; i_{fa}, i_{fb}, and i_{fc} are three-phase output currents; v_{la}, v_{lb}, and v_{lc} are three-phase load voltages; and L_f is inductance of the output filter. Based on Eq. (4.19), the voltage formula in Eq. (4.33) converts to Park transformation or $dq0$ reference frame as follows:

$$\begin{bmatrix} v_{fd} \\ v_{fq} \\ v_{f0} \end{bmatrix} = \begin{bmatrix} v_{ld} \\ v_{lq} \\ v_{l0} \end{bmatrix} + \begin{bmatrix} L_f & 0 & 0 \\ 0 & L_f & 0 \\ 0 & 0 & L_f \end{bmatrix} \frac{d}{dt} \begin{bmatrix} i_{fd} \\ i_{fq} \\ i_{f0} \end{bmatrix} + \begin{bmatrix} 0 & -\omega L_f & 0 \\ \omega L_f & 0 & 0 \\ 0 & 0 & 0 \end{bmatrix} \begin{bmatrix} i_{fd} \\ i_{fq} \\ i_{f0} \end{bmatrix} \tag{4.34}$$

The circuit configuration and control scheme for the inverter in islanding mode is depicted in Fig. 4.49. In this mode, V-f control strategy is used so that voltage magnitude and frequency are adjustable parameters. In the V-f controller, desirable frequency (e.g., 50 Hz) can be obtained by a phase-locked loop (PLL) as well as desired load voltages set by a proportional integral (PI) controller to eliminate steady state errors.

The output filter is used to eliminate harmonics and ripples to achieve sinusoidal output. At first, load voltages measure and convert to $dq0$ frame as follows:

$$\begin{bmatrix} v_{ld} \\ v_{lq} \\ v_{l0} \end{bmatrix} = T_{dq0} \begin{bmatrix} v_{la} \\ v_{lb} \\ v_{lc} \end{bmatrix}, \quad T_{dq0} = \frac{2}{3} \begin{bmatrix} \cos(\omega t) & \cos(\omega t - 120) & \cos(\omega t + 120) \\ -\sin(\omega t) & -\sin(\omega t - 120) & -\sin(\omega t + 120) \\ \frac{1}{2} & \frac{1}{2} & \frac{1}{2} \end{bmatrix} \tag{4.35}$$

Note that in balanced three-phase voltages, v_{l0} is equal to zero. After converting to dq0 frame, it is required to achieve the difference between expected voltage and converted measured voltage because the load phase voltage should be kept balanced and sinusoidal with constant amplitude and frequency. So, the expected load voltages per unit are considered as the following values:

$$\begin{bmatrix} v_{ldp} \\ v_{lqp} \\ v_{l0p} \end{bmatrix} = \begin{bmatrix} 1 \\ 0 \\ 0 \end{bmatrix} \tag{4.36}$$

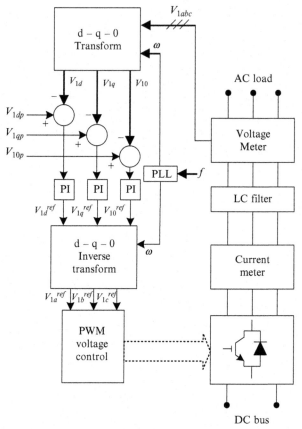

Fig. 4.49 Control strategy of the inverter in stand-alone mode.

Reference voltages of the inverter conclude from the difference between expected values and measured values after passing through a PI controller. The reference voltages in $dq0$ rotating frame is as in Eq. (4.37):

$$\begin{bmatrix} v_{ld}^{ref} \\ v_{lq}^{ref} \\ v_{l0}^{ref} \end{bmatrix} = \begin{bmatrix} PI(v_{ldp} - v_{ld}) \\ PI(v_{lqp} - v_{lq}) \\ PI(v_{l0p} - v_{l0}) \end{bmatrix} \tag{4.37}$$

Inverse transform of the inverter voltages cause the desired voltage for switching to be achieved.

$$\begin{bmatrix} v_{la}^{ref} \\ v_{lb}^{ref} \\ v_{lc}^{ref} \end{bmatrix} = T_{dq0}^{-1} \begin{bmatrix} v_{ld}^{ref} \\ v_{lq}^{ref} \\ v_{l0}^{ref} \end{bmatrix}, \quad T_{dq0}^{-1} = \begin{bmatrix} \cos(\omega t) & -\sin \omega t & 1 \\ \cos(\omega t - 120) & -\sin(\omega t - 120) & 1 \\ \cos(\omega t + 120) & -\sin(\omega t + 120) & 1 \end{bmatrix}$$

$$\tag{4.38}$$

The reference output voltages for the inverter are transformed to the abc by using Eq. (4.38), and then the available voltages in the abc coordinate are compared with the triangular wave provided by the PWM voltage control block and then apply suitable switching pattern to the inverter.

4.12 SIMULATION RESULTS

To analyze the previously mentioned structure as well as the impacts of a MTG unit on the distribution system, simulation is carried out in PSCAD/EMTDC software. The MTG system that is considered in the simulation is shown in Fig. 4.50. The power electronic interface is based on AC/DC/AC through a passive rectifier and the system parameters are given in Appendix A to this chapter. The simulation scenarios are focused on suitable operation with supplying unbalanced passive loads in the grid-connected and islanding modes. All time functions are in seconds and speed reference is kept constant at 1 p.u.

4.12.1 Grid-Connected Operation

In this case, the P-Q control scheme is applied. The reference values of P_{ref} and Q_{ref} are changed from 8 to 16 kW and from 1.4 to 3.5 kVar, at $t = 10$ s, and then back to 8 kW and 1.4 kVar, at $t = 20$ s, respectively.

Fig. 4.51 shows P and Q delivered to the grid and the system output power is varied according to the reference points.

Fig. 4.52 shows the rotor speed variations (N) and fuel demand signal (W_f). Rotor speed set at 0.988 p.u. after transient state variations, and a negative change occurs along with power variations. When power is increased, speed is decreased to 0.977 p.u. and then back again to initial value with power decreasing. In Fig. 4.52B, when the power reference variations are applied,

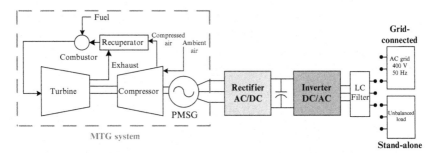

Fig. 4.50 Microturbine generation system.

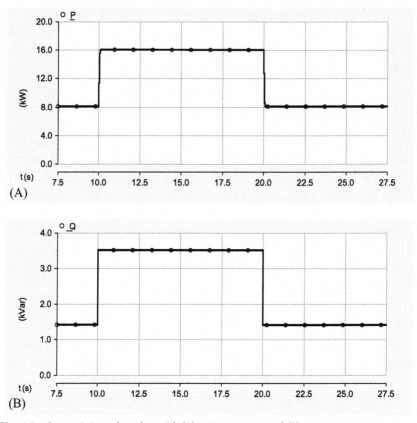

Fig. 4.51 Power injected to the grid: (A) active power and (B) reactive power.

fuel demand for combustion process is increased rapidly from 0.45 p.u. at 8 kW to 0.67 p.u. at 16 kW and vice versa. According to Fig. 4.52, fuel demand variations are the opposite of rotor speed variations.

Fig. 4.53A shows the shaft torque (T_{shaft}) and the developed electromagnetic torque (T_e). The shaft torque and electromagnetic torque are about equal at steady state. Fig. 4.53B shows the turbine mechanical torque with delay caused by time constants, which are considered in the model of the turbine and it follows the variations of the reference values.

Fig. 4.54A shows the power output from the MT and DC bus, which are equal at steady state. The DC bus voltage of the diode rectifier is shown in Fig. 4.54B. Note that the DC bus voltage drops when the load is increased and returns to its initial value, which is about 785 V when power is at its initial level.

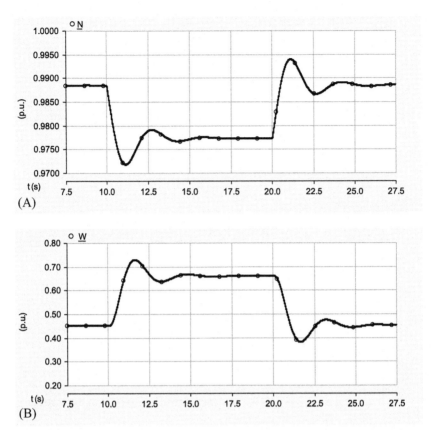

Fig. 4.52 (A) Rotor speed and (B) fuel demand signal of the microturbine in grid-connected mode.

Fig. 4.55 shows grid-side phase voltages and line currents of the inverter output. The line currents change with power reference variations. However, the inverter can deliver the MTGs power to the grid with low harmonic current. This verifies the effectiveness of the P-Q control strategy.

4.12.2 Stand-Alone Operation

In this mode, MTG response to unbalanced load in the stand-alone operation mode is studied. Load parameters are given in Appendix A to this chapter and V-f control strategy is used. The unbalanced load No. 1 is changed to the unbalanced load No. 2 at $t = 10$ s and the load is reverse changed to its initial value at $t = 20$ s.

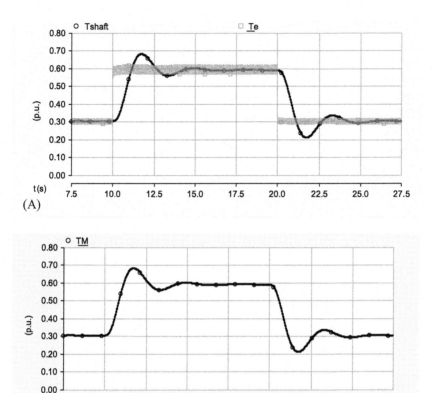

Fig. 4.53 (A) Shaft torque and electric torque and (B) turbine mechanical torque of the microturbine in grid-connected mode.

Fig. 4.56 shows the active and reactive power consumed by unbalanced AC loads. When the load is increased, power output is increased, too, and vice versa. Fig. 4.57 shows the MTG rotor speed variations and fuel signal in supplying unbalanced loads. Similar to what was mentioned earlier, when demand is increased this causes speed to drop and the fuel consumed to increase, and when demand is decreased it leads to an increase of rotor speed and a decrease of fuel demand signal. When the MT is supplied the unbalanced AC loads No. 1 and No. 2, the rotor speed is equal to 0.988 and 0.977 p.u. respectively. When at $t=10$ s, load No. 2 is applied, the fuel demand for combustion process is increased rapidly to 0.661 p.u. from 0.458 p.u. and vice versa.

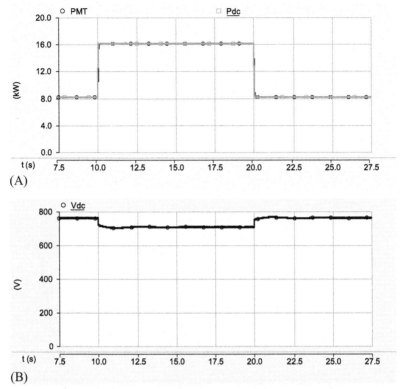

Fig. 4.54 (A) Microturbine and DC bus power output and (B) DC bus voltage in grid-connected mode.

The shaft torque (T_{shaft}) and electromagnetic torque (T_e) are shown in Fig. 4.58A. The shaft torque and electromagnetic torque are equal together at steady state. Fig. 4.58B shows the turbine mechanical output torque, which is matched according to load variations.

Fig. 4.59A shows the power output from the MT and DC bus, which is matched with unbalanced loading conditions. Fig. 4.59B shows the DC bus voltage of the uncontrolled rectifier, when the unbalanced loads are connected to the MT system. The DC bus voltage fluctuates and drops back to initial value, when No. 2 is disconnected from the MTG.

Fig. 4.60 shows the phase voltages and line currents at unbalanced load terminals. The inverter maintains the output at the desired level irrespective of the load applied on the system. The AC voltage level across the load

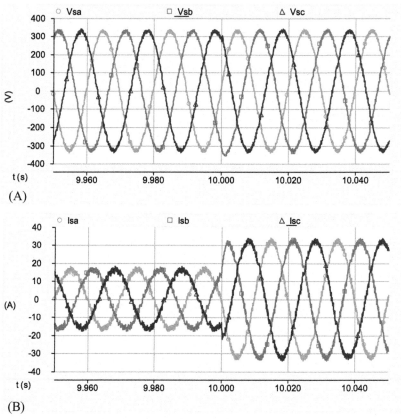

Fig. 4.55 (A) Grid-side phase voltages and (B) line currents of the inverter in grid-connected mode.

remains unchanged with the load variation switching and staying at the reference value the whole time of simulation. The balanced voltages are provided for unbalanced AC loads, while the load phase currents are not sinusoidal and balanced. To quantify level of the voltage unbalance, the percentage of unbalance is expressed by "degree of unbalance in three-phase system" [26]. In this case, the negative sequence unbalance is lower than 1.1%, which is acceptable. It should be noted that international standards admit unbalances lower than 2% [26]. The three-phase line currents at load terminals change with the load variation switching. It should be noted that the frequency of output voltage of the PMSG is over 1.6 kHz, while that at the output of the inverter, at the load terminals,

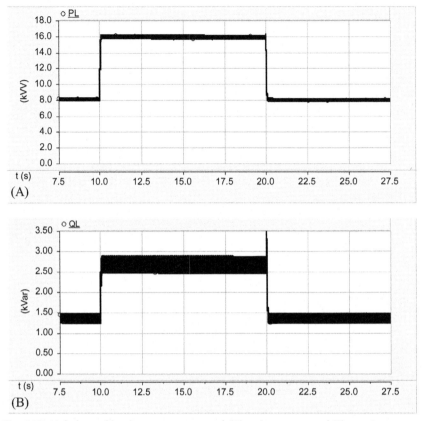

Fig. 4.56 Unbalanced loads power consumed: (A) active power and (B) reactive power.

it is 50 Hz. This verifies the effectiveness of the V-f control strategy for the stand-alone operation.

4.13 FUTURE TRENDS

Considering MTs potential advantages, a bright future can be expected for them. MTGs are suitable for DG systems because of their low maintenance and high reliability. One characteristic that drives manufacturers to focus on them is efficiency improvement. A MT with higher efficiency and fuel flexibility is an ideal choice for power generation. Higher efficiency of the MT in renewable fuels like biogas is another subject that will be considered. Further size reduction is another issue that is considered by manufacturers and

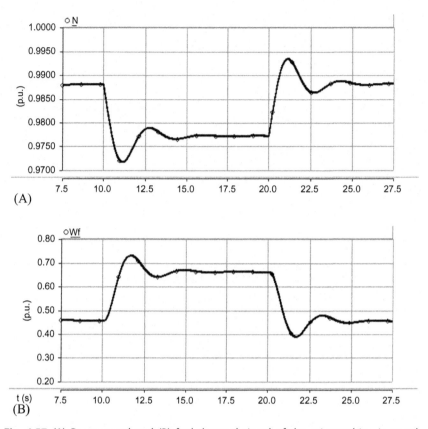

Fig. 4.57 (A) Rotor speed and (B) fuel demand signal of the microturbine in stand-alone mode.

extensive research is being done. Higher efficiency and smaller size cause more global attention to MTGs.

Recently, a combination of two or more DGs is considered for maximum use of sources and MT has an important role in this regard. The combination of a MT with a fuel cell is one of the best sources; however, other combinations such as CCHP with PV are also considered. With the advancement of technology, using a single DG is not desirable.

Global attention on reduction of emissions like CO_2 and NO_X and achieving higher efficiency are major reasons for expanding the use of hybrid vehicles. Using MTs in hybrid vehicles and transportation as well as aircraft, turbojets, turboprops, and unmanned vehicles are developing.

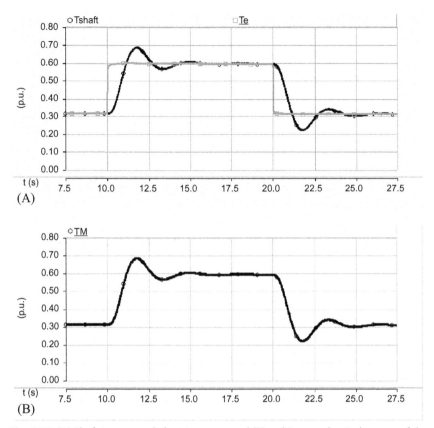

Fig. 4.58 (A) Shaft torque and electric torque and (B) turbine mechanical torque of the microturbine in stand-alone mode.

Remote monitoring, quiet operation, oxygen rich exhaust, and so on are also favorite issues to MTG manufacturers.

In general, manufacturers focus on MT configuration to produce higher overall efficiency and smaller size for customers. MTs will be a desirable, appropriate, and acceptable source for producing power for years to come.

4.14 SUMMARY

Recently, the development of DG has been a main issue for governments, and high efficiency and reliable sources have been considered. MTs are suitable with respect to their potential merits and they are favorable for

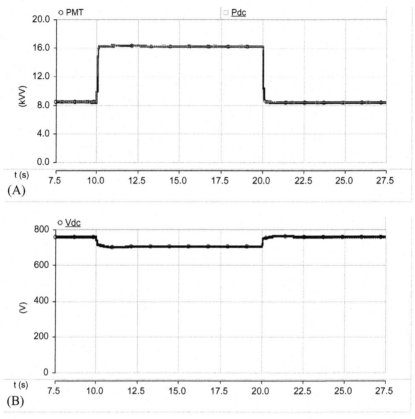

Fig. 4.59 (A) Microturbine and DC bus power output and (B) DC bus voltage in stand-alone mode.

various applications. The MTs' operation is independent of choose of fuel type and they are suitable for CHP. Moreover, start-up time of MTs is short, thus they are serious sources for grid supporting and load supplying.

To begin this chapter, MT components and applications were illustrated and statistical information was presented showing that MTs are a main source for power systems in the future. Modeling of the MT was presented in grid-connected and stand-alone modes and grid-connected control strategy based on $\alpha\beta$ transformation and $dq0$ transformation was demonstrated in islanding mode. Control circuits were described completely. Simulations were carried out in PSCAD/EMTDC software for two mentioned modes. Simulation results approved modeling and the MT performance as a suitable power source.

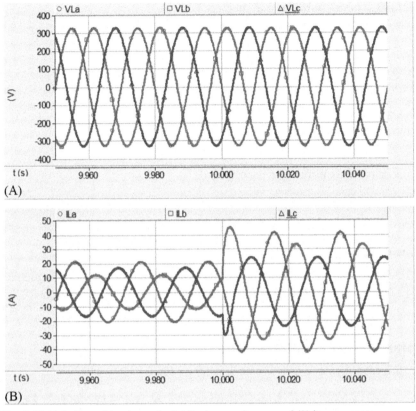

Fig. 4.60 Unbalanced load terminal: (A) phase voltages and (B) line currents.

Finally, in the future MTGs should receive more global attention and enjoy more progress because they are appropriate for the distribution grid and are a clean source of power.

APPENDIX A

Parameters of the MT

- Speed governor parameters:

$W = 25$, $X = 0.4$, $Y = 0.05$, $Z = 1$, speed ref. $= 1$ p.u., acceleration ref. $= 0.01$ p.u.

- Fuel system controller parameters:

$$K_6 = 0.23, \ K_3 = 0.77, \ a = 1, \ b = 0.05,$$
$$c = 1, \ T_f = 0.4, \ E_{CR} = 0.01, \ E_{TD} = 0.04,$$
$$Max\,limit = 1.5 \text{ p.u.}, \ Min\ limit = -0.1 \text{ p.u.}$$

- Compressor–turbine combination parameters:

$$T_{CD} = 0.2, \ a_{f2} = -0.3, \ b_{f2} = 1.3, \ c_{f2} = 0.5$$

- Temperature controller parameters:

$$K_4 = 0.8, \ K_5 = 0.2, \ T_3 = 15, \ T_4 = 2.5, \ T_R = 950°F, \ T_5 = 3.3, \ T_t$$
$$= 450°F, \ a_{f1} = 700, \ b_{f1} = 550$$

Parameters of the PMSG

Rated output $= 30$ kW, Rated voltage $= 0.320$ kV, Base Angular Frequency $= 10,053.1$ rad/s, $R_s = 0.25\ \Omega$, $p = 2$, $L_d = L_q = 0.0687\ \mu H$, $\lambda = 0.0534$ Wb, $J = 0.011$ kg m^2, $F = 0$.

Parameters of the Unbalanced AC Loads

- RL load No. 1:

$$z_{la1} = 15.62\angle 11.61°\Omega, \ z_{lb1} = 27.48\angle 6.57°\Omega, \ z_{lc1} = 19.56\angle 9.25°\Omega$$

- RL load No. 2:

$$z_{la2} = 7.86\angle 11.53°\Omega, \ z_{lb2} = 9.83\angle 9.21°\Omega, \ Z_{lc2} = 13.79\angle 6.54°\Omega$$

APPENDIX B

Ideal gas properties of air [8]

T (K)	h (kJ/kg)	Pr	T (K)	h (kJ/kg)	Pr	T (K)	h (kJ/kg)	Pr
200	199.97	0.3363	510	513.32	9.031	1000	1046	114.0
210	209.97	0.3987	520	523.63	9.684	1020	1068.89	123.4
220	219.97	0.4690	530	533.98	10.37	1040	1091.85	133.3
230	230.02	0.5477	540	544.35	11.10	1060	1114.86	143.9
240	240.02	0.6355	550	555.74	11.86	1080	1137.89	155.2
250	250.05	0.7329	560	565.17	12.66	1100	1161.07	167.1
260	260.09	0.8405	570	575.59	13.50	1120	1184.28	179.7
270	270.11	0.9590	580	586.04	14.38	1140	1207.57	193.1
280	280.13	1.0889	590	596.52	15.31	1160	1230.92	207.2
285	285.14	1.1584	600	607.02	16.28	1180	1254.34	222.2
290	290.16	1.2311	610	617.53	17.30	1200	1277.79	238.0
295	295.17	1.3068	620	628.07	18.36	1220	1301.31	254.7
298	298.18	1.3543	630	638.07	18.36	1240	1324.93	272.3
300	300.19	1.3860	640	649.22	20.64	1260	1348.55	290.8
305	305.22	1.4686	650	659.84	21.86	1280	1372.24	310.4
310	310.24	1.5546	660	670.47	23.13	1300	1395.97	330.9
315	315.27	1.6442	670	681.14	24.46	1320	1419.76	352.5
320	320.29	1.7375	680	691.82	25.85	1340	1443.60	375.3
325	325.31	1.8345	690	702.52	27.29	1360	1467.49	399.1
330	330.34	1.9352	700	713.27	28.80	1380	1491.44	424.2
340	340.42	2.149	710	724.04	30.38	1400	1515.42	450.5
350	350.49	2.379	720	734.82	32.02	1420	1539.44	478.0
360	360.58	2.626	730	745.62	33.72	1440	1563.51	506.9
370	370.67	2.892	740	756.44	35.50	1460	1587.63	537.1
380	380.77	3.176	750	767.29	37.35	1480	1611.79	568.8
390	390.88	3.481	760	778.18	39.27	1500	1635.97	601.9
400	400.98	3.806	780	800.03	43.35	1520	1660.23	636.5
410	411.12	4.153	800	821.95	47.75	1540	1684.51	672.8
420	421.26	4.522	820	843.98	52.59	1560	1708.82	710.5
430	431.43	4.915	840	866.08	57.60	1580	1733.17	750.0
440	441.61	5.332	860	888.27	63.09	1600	1757.57	791.2
450	451.80	5.775	880	910.56	68.98			
460	462.02	6.245	900	932.93	75.29			
470	472.24	6.742	920	955.38	82.05			
480	482.49	7.268	940	977.92	89.28			
490	492.74	7.824	960	1000.55	97.00			
500	503.02	8.411	980	1023.25	105.2			

REFERENCES

[1] P. Asgharian, R. Noroozian, Modeling and simulation of microturbine generation system for simultaneous grid-connected/islanding operation, in: 24th Iranian Conference on Electrical Engineering (ICEE), (pp. 1528-1533). Shiraz: IEEE, 2016.

[2] World Energy Council, World Energy Resources, Summary, World Energy Resources, 2013.

[3] BP, BP Energy Outlook 2035, 2015. Retrieved from BP global, http://www.bp.com/.

[4] EIA, EIA's Annual Energy Outlook 2014, 2014. Retrieved from U.S. Energy Information Administration, http://www.eia.gov/.

[5] J. Deyette, S. Clemmer, R. Cleetus, S. Sattler, A. Bailie, M. Rising, The Natural Gas Gamble, Union of Concerned Scientists (2015).

[6] Navigant Research, Microturbines: Global Market Analysis and Forecasts for Residential, Commercial, and Industrial Applications, Navigant Research, 2015.

[7] I. Gurrappa, I.V.S. Yashwanth, I. Mounika, H. Murakami, S. Kuroda, The importance of hot corrosion and its effective prevention for enhanced efficiency of gas turbines, in: Gas Turbines: Materials, Modeling and Performance, InTech, Rijeka, Croatia, 2015, pp. 55–102.

[8] Y. Cengel, M. Boles, Thermodynamics: An Engineering Approach, 8th ed., McGraw-Hill Education, New York, USA, 2014.

[9] Capstone (2016). Retrieved from Capstone Turbine Corporation: http://www.capstoneturbine.com/.

[10] Gas Research Institute and the National Renewable Energy Laboratory, Gas-Fired Distributed Energy Resource Technology Characterizations, U.S. Department of Energy, Oak Ridge, TN, 2003 Taken from: Energy and Environmental Analysis, Inc., estimates.

[11] P. Kiameh, Power Generation Handbook: Selection, Applications, Operation, Maintenance, first ed., McGraw-Hill Professional, New York, 2002.

[12] G. Masters, Renewable and Efficient Electric Power Systems, second ed., Wiley, Hoboken, NJ, 2013.

[13] R. Meyers, Encyclopedia of Sustainability Science and Technology, vol. 1, Springer, New York, 2012. pp. 17–46.

[14] A.-V. Boicea, G. Chicco, P. Mancarella, Optimal operation of a 30 kw natural gas microturbine cluster, U.P.B. Sci. Bull. 73 (1) (2011).

[15] G. Ming, M. Onofri, J. Ma, Dynamic emission characteristics of micro-turbine startup, in: Power and Energy Engineering Conference (APPEEC), IEEE, Asia-Pacific, 2011.

[16] V. Boicea, Essentials of Natural Gas Microturbines, Taylor & Francis Group, Boca Raton, FL, 2014.

[17] F. Weston, N. Seidman, C. James, Model Regulations for the Output of Specified Air Emissions from Smallerscale Electric Generation Resources, Regulatory Assistance Project (RAP), 2001.

[18] L. Charma, P. Mago, Micro-CHP Power Generation for Residential and Small Commercial Buildings, Nova Science, New York, 2009.

[19] S. Guda, C. Wang, M. Nehrir, A Simulink-based microturbine model for distributed generation studies, in: Power Symposium (pp. 269-274). the 37th Annual North AmericanNorth American: IEEE, 2005.

[20] W.I. Rowen, Simplified mathematical representations of heavy-duty gas turbines, J. Eng. Power Trans. ASME 105 (1983) 865–869.

[21] P. Krause, O. Wasynczuk, S. Sudhoff, S. Pekarek, Analysis of Electric Machinery and Drive Systems, 3rd ed., Wiley-IEEE Press, New Jersey, USA, 2013.

[22] A. Al-Hinai, A. Feliachi, Dynamic model of a microturbine used as a distributed generator, in: System Theory (pp. 209-213). Proceedings of the Thirty-Fourth Southeastern Symposium: IEEE, 2002.

[23] R. Noroozian, M. Abedi, G. Gharehpetian, S. Hosseini, Modelling and simulation of microturbine generation system for on-grid and off-grid operation modes, in: International Conference on Renewable Energies and Power Quality (ICREPQ'09). Valencia, 2009.

[24] A. Bertani, C. Bossi, F. Fornari, S. Spelta, F. Tivegna, A microturbine generation system for grid connected and islanding operation, in: Power Systems Conference and Exposition (pp. 360-365). IEEE, 2004.

[25] O. Fethi, L. Dessaint, K. Al-Haddad, Modeling and simulation of the electric part of a grid connected microturbine, in: Power Engineering Society General Meeting. 2, (pp. 2212-2219). Denver: IEE, 2004.

[26] T.A. Short, Electric Power Distribution Handbook, 2nd ed., CRC Press, Florida, USA, 2014.

FURTHER READING

[1] D. Gaonkar, Performance of microturbine generation system in grid connected and Islanding modes of operation, in: D.N. Gaonkar (Ed.), Distributed Generation, InTech, Vukovar, 2010, pp. 185–208.

[2] T. Giampaolo, Gas Turbine Handbook: Principles and Practices, 5th ed., Fairmont Press, Georgia, USA, 2013.

[3] B. Kolanowski, Guide to Microturbines, 1st ed., Fairmont Press, Georgia, USA, 2004.

[4] M.J. Moore, Micro-Turbine Generators, 1st ed., Wiley, New Jersey, USA, 2005.

[5] C. Soares, Microturbines: Applications for Distributed Energy Systems, 1st ed., Butterworth-Heinemann (Elsevier), UK, 2007.

[6] L.N. Hannet, A. Khan, Combustion turbine dynamic model validation from tests, IEEE Trans. Power Systems 8 (1993) 152–158.

[7] S. Grillo, S. Massucco, A. Morini, A. Pitto, F. Silvestro, Microturbine control modeling to investigate the effects of distributed generation in electric energy networks, IEEE Syst. J. 4 (3) (2010) 303–312.

[8] G. Li, G. Li, W. Yue, M. Zhou, K. Lo, Modeling and simulation of a microturbine generation system based on PSCAD/EMTDC, in: Critical Infrastructure (CRIS), IEEE, Beijing, 2010, pp. 1–6.

CHAPTER 5

Fuel Cells

Hamdi Abdi*, Ramtin Rasouli Nezhad†, Mohammad Salehimaleh‡
*Razi University, Kermanshah, Iran
†University of Western Ontario, London, ON, Canada
‡University of Mohaghegh Ardabili, Ardabil, Iran

5.1 INTRODUCTION

A fuel cell (FC) is an electrochemical device that works by reverse electrolysis operation, which produces electricity by the reaction of oxygen (air) and hydrogen (fuel). Therefore, a FC converts the chemical energy of a fuel directly into electrical energy. Because the output is DC, it needs inverters so as to convert DC output to AC, which will be explained precisely in Section 5.5.

All types of FCs consist of two electrodes (anode and cathode) and an electrolyte (depending on the model of FC, they are varied). Fuel (hydrogen-rich or other hydrocarbons) and oxidant (typically air) are delivered to the FC separately. The fuel and oxidant streams are separated by an electrode-electrolyte system. Fuel is fed to the anode (negative electrode) and an oxidant is fed to the cathode (positive electrode). Electrochemical oxidation and reduction reactions take place at the electrodes to produce electric current. The primary product of FC reactions is water and heat, which must be drained continuously from the FC. Therefore, outputs of FCs are absolutely clean from the environmental point of view.

A single FC is able to produce just less than one volt of electrical potential. To reach higher voltages, FCs are stacked on top of each other and connected in series as shown in Fig. 5.1. Cell stacks consist of repeating FC units, each comprised of an anode, cathode, electrolyte, and a bipolar separator plate. The number of cells in a stack depends on the desired power output and individual cell performance; stacks range in size from a few (<1 kW) to several hundred (250 + kW). Reactant gases—typically, desulphurized, reformed natural gas and air—flow over the electrode faces in channels through the bipolar separator plates. Because not all of the reactants are consumed in the oxidation process, about 20% of the hydrogen delivered to the FC stack is unused and is often "burned" downstream of the FC module.

Distributed Generation Systems
http://dx.doi.org/10.1016/B978-0-12-804208-3.00005-4

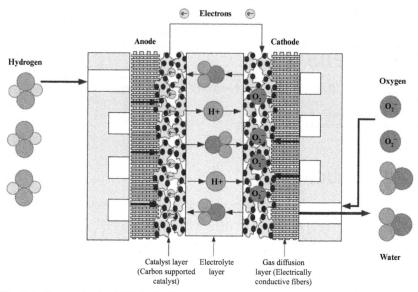

Fig. 5.1 The stack of a PEM fuel cell [1].

During this decade, power demand has been increasing dramatically, so governments and authorities have considered some alternative approaches to meet the power demand. In addition, climate changes like thawing glaciers and air pollution or even the ozone layer depletion make us use renewable energies that have no detrimental effects on Earth. Although Earth needs green energies, many kinds of renewable energies are inherently stochastic, so they associate to the other circumstances such as wind, sun, and so on. Thus, these energies are not accessible in every situation. The condition has been exacerbated, so that the power of fossil fuels cannot be neglected because they still account for the majority of generating power electricity.

Besides, with the last recession in the world, countries need cost-effective and environmentally friendly power generation methods that must be matched with conventional systems to work precisely and safely to compensate the shortage of load. FCs are the only chance make a compromise between fossil fuels and green energies. It should be noted that the FC is not stochastic and more reliable than the other renewable energies. Therefore, this energy is absolutely suitable for colossal uses because it can apply the power of hydrogen or other kind of hydrocarbons (depending on the features of different types of FCs), whereas the output is clean and reliable.

With the latest developments in FC technologies, both their power quality and flexibility have been improved considerably, whereas their price has plummeted.

In Section 5.2, FCs and their uses are introduced and the reasons and conditions of their practical applications are demonstrated. Voltage losses will be surveyed in Section 5.3 because it is the most important loss in FCs from an electrical point of view and helps readers to comprehend the perception of designing FCs. Furthermore, restructuring and the power market are discussed in Section 5.4 and their equations and models are shown while we assume to implement FCs as distributed generation (DG). Applying FCs as an uninterruptible power supply (UPS) or backup system will show how FCs can tackle myriad impedimentations to prevent the system from blackouts and decrease the rate of interruptions as they are very convenient and suitable in comparison with other kinds of batteries. FCs are appropriate for hybridizing with the other power resources such as fossil and renewable power and are also able to extract hydrogen or other hydrocarbons from them, which will be discussed in this section as hydrogen extraction methods. Furthermore, with combined heat and power (CHP) design, they acquire the highest percentage of efficiency, which is near to 90%. Eventually, in Section 5.5, power electronic devices, which are imperative in satisfying power quality standards for integrating DGs, will be discussed and conventional issues such as protection concerns and their related approaches also will be presented.

5.2 PRINCIPLES AND APPLICATIONS OF FCs

In this section, the historical development of FCs is presented. Then, different types of FCs are explained to demonstrate the uses of each one of them and also give comprehensive views to readers for helping them in choosing the best FC in different situations. Afterward, advantages and disadvantages of FCs are discussed, and finally the applications of FCs are presented.

5.2.1 Historical Development of FCs

Based on the Collen Spiegel [1] and Brian Cook [2] references, this topic has been written.

In 1801, Humphry Davy explained the initial concept of the FC. Christian Friedrich Schönbein in 1838 worked more on the FC conception. However, in 1839 Sir William Grove would discover this technology by accident during an electrolysis experiment [1]. When Sir William

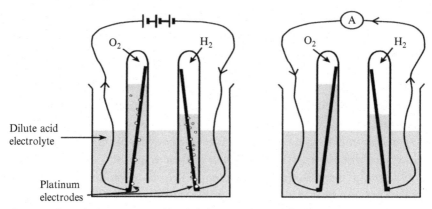

Fig. 5.2 Electrolysis operation (the left one) and Sir William experiment (the right one) [2]. *From B. Cook, Introduction to fuel cells and hydrogen technology, Eng. Sci. Educ. J. 11 (6) (2002) 205–216. Reproduced by permission of the Institution of Engineering & Technology.*

disconnected the battery from the electrolyzer and connected the two platinum electrodes together in a dilute sulphuric acid, he observed a current flowing in the opposite direction, consuming the gases of hydrogen and oxygen and generating 1 V. The Grove's gas battery is shown in the Fig. 5.2 [2].

To provide power for an electrolyzer as it is shown in Fig. 5.3, in 1842 Sir William produce a gas chain connecting a number of gas batteries in series so as to increase the voltage level of Grove's gas battery.

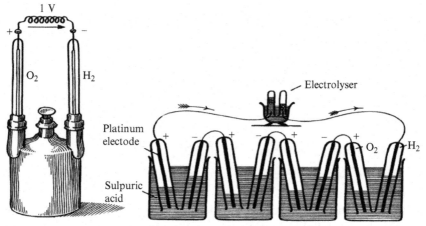

Fig. 5.3 Grove's gas battery (the left one) and Grove's gas chain (the right one) [2]. *From B. Cook, Introduction to fuel cells and hydrogen technology, Eng. Sci. Educ. J. 11 (6) (2002) 205–216. Reproduced by permission of the Institution of Engineering & Technology.*

In 1893, Friedrich Wilhelm Ostwald described the fundamental theories of FC operations. He could demonstrate the features of different components of FCs [1].

In 1889, Charles Langer and Ludwig Mond worked on FCs based on coal gas as a fuel. It was the first time that the term FC was used. Their FC reached 6 amp and 0.73 V per square foot while it had problems with liquid electrolytes [1]. During the same years, Charles R. Alder Wright and C. Thompson developed a FC, but they could not tackle chamber leaking problems, so these constraints caused them from reaching 1 V per cell. After them, Louis Paul Cailleteton and Louis Joseph Colardeau concluded that due to the high cost of FC components and the low price of coal, it was not cost-effective to use FCs. Although in those years, it was difficult to expand this technology and convert coal directly into power, now this is dramatically less ambiguous.

Over the early 1900s, Emil Bauruseda used solid electrolyte of clay and metal oxides to create high temperature devices.

In 1932, Francis Bacon from the University of Cambridge modified the research of Mond and Langer to attempt to produce an alkaline fuel cell (AFC). However, finally in 1959 he could present the first practical AFC, which used an alkaline electrolyte with porous nickel electrodes and generated 5 kW [2].

In the early 1960s, the first proton exchange membrane fuel cell (PEMFC) is invented by Thomas Grubb and Leonard Niedrach at General Electric for US military applications. Again, they used platinum, which made the FCs expensive. In the mid-1960s, their FC developed with NASA for use in the Gemini space program, which provided both electricity and drinking water for astronauts. So, NASA launched Apollo, which used a 1.5 KW AFC. In the early 1970s, scientists modified FC technologies and produced a 12 KW AFC to provide onboard power on all space shuttle flights [2].

In the 1970s, the world had two major problems including air pollution or climate change and an oil crisis, which led governments and authorities to use clean and efficient alternatives instead of fossil fuels. With the colossal progress in PEMFC technologies such as membranes or even hydrogen storage, many developed countries applied the FC to electric vehicles. Nowadays, many automakers produce fuel cell electric vehicles (FCEVs). Because of the price of oil, companies thought about using advanced technologies to increase the efficiency of power generators with zero harmful emissions. The phosphoric acid fuel cell (PAFC) was demonstrated in this era in many power stations such as a 1 MW unit in the United States. Then, because of the development of natural gas infrastructure such as internal reforming of natural gas to hydrogen, a molten carbonate fuel cell (MCFC) is applied in power stations.

In the 1980s, commercialization of the PAFC was on the agenda and a priority of many scientists. They predicted that with such rapid developments, by the end of the century, many power plants would change to PAFCs, but in reality it did not occur as they thought. However, with CHP technology, the PAFC was commercialized as the first commercial FC. Afterward, the US Navy worked on FCs in submarines to generate electricity, which had to be quiet, slight, and zero emission instead of polluted gas turbines or diesel generators that were not suitable for indoor use.

Over the 1990s era, commercialization of the solar oxide fuel cell (SOFC) and the PEMFC took place. Automakers used PEMFC for achieving a zero emission vehicle (ZEV). After that, ZEVs have had a strong role in the automotive industry. Ballard Company, with the cooperation of famous automakers, would commercialize the PEMFC for automotive applications. Furthermore, the PEMFC had been used in smaller power stations or even in small portable devices during this decade. Also, the SOFC became more advanced and many industries were content to apply this technology as it could work with more hydrocarbons as a fuel.

During the 2000s, climate change and global warming had a considerable effect on the growing green energies. FCs seemed convenient for reducing the amount of CO_2 emission, but they needed cost reduction and more subsidies to tackle the cost problems. With new developments in FC technologies and production, its price decreased. So, in many public transportation systems such as buses, they applied PEMFC. In this decade, both the PEMFC and direct methanol fuel cell (DMFC) have been used in auxiliary power units and equipment of infantry soldiers as they did not need heavy batteries any more while FCs had more durability and less weight and volume. One of the main uses of FCs is their backup role to prevent blackouts and critical interruptions.

Recently, uses of FCs in communication networks and also portable applications have been developing rapidly. During these years, with the increasing production of FCs, their price has plummeted. For example, Fig. 5.4 illustrates the price reduction of an 80 KW PEMFC based on high-volume manufacturing per year from 2002 to 2017.

As can be seen from the bar chart in Fig. 5.4, the initial estimate of 80 kW PEMFC based on 500,000 productions per year in 2002 was $275/kW. The

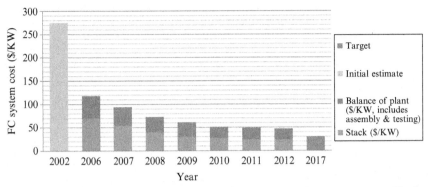

Fig. 5.4 Modeled cost of an 80 kW PEM fuel cell system based on a projection of high-volume manufacturing (500,000 units/year) [3].

price of the FC in 2006, which was the cost of stack and balance of plant, reduced to $94/kW. Reduction of both stack and balance of plant had plunged sharply to $47/kW in 2012. It is projected that by 2017 the price will decline to just $30/kW. Thus, in the future because of cutting edge technologies, more cost-effective FCs will be accessible.

Furthermore, economies of scale in the power market have played a significant role as manufactures can reduce the cost of production by increasing the number of units manufactured. Fortunately, there is now more willingness to use FCs, and this has a positive effect on them as shown for the 80 kW PEMFC in Fig. 5.5.

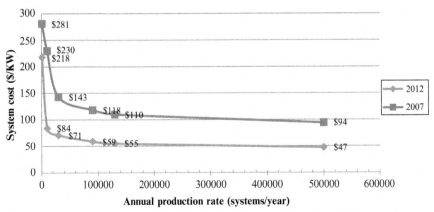

Fig. 5.5 Modeled cost of a 80 kW PEM fuel cell system based on different volumes of manufacturing in 2007 and 2012 [3].

The line chart in Fig. 5.5 illustrates the effect of manufacturing unit volume on the price of a 80 kW PEMFC in 2007 and 2017. As discussed before, with the development of renewable technologies the price decreases every year, so 2012 prices are lower than 2007 prices. From Fig. 5.5, it is obvious that if in 2012 we produced 1000 FC units per year, the price would be $218/kW, whereas with increasing the number of manufacturing, which was 10,000 units annually, the price would plummet to $84/kW. This downward trend would continue and reached to $47/kW in 2012 when manufacturing 500,000 per year. Therefore, economies of scale, which is one of the main factors of the power market, has been improving in the recent years.

Hence, FCs are now commercially available in wide areas from power stations to transportation and portable devices. However, this technology still needs more research and development.

5.2.2 Different Types of FCs

FCs are divided into five main categories as follows:
- PEMFC: proton exchange membrane fuel cell
- AFC: alkaline fuel cell
- PAFC: phosphoric acid fuel cell
- MCFC: molten carbonate fuel cell
- SOFC: solid oxide fuel cell

The main reason for different classifications of FC types is their catalyst differences. Their operating temperature is important, too, as in low-temperature FCs, all the fuel must be converted to be hydrogen-rich prior to entering the FC. In addition, the anode catalyst in low-temperature FCs (mainly platinum) is strongly poisoned by CO. However, these FCs can apply a reformer before injecting fuel to the FC unit to use the other hydrocarbons. In high-temperature FCs, CO and even CH_4, which is very common in industry, can be internally converted to hydrogen or even directly oxidized electrochemically.

Table 5.1 illustrates structural features and differences of five main types of FCs so as to deliver an overall conception about their traits, which can help readers to make a decision about one of these methods regarding which is more suitable with the load situations and consumer requirements. Information is extracted from [4].

Table 5.1 Summary of major differences of fuel cell types [4]

	PEMFC	AFC	PAFC	MCFC	SOFC
Electrolyte	Hydrated polymeric ion exchange membranes	Mobilized or immobilized potassium hydroxide in the asbestos matrix	Immobilized liquid phosphoric acid in SiC	Immobilized liquid molten carbonate in LiAlO2	Perovskites (Ceramics)
Electrodes	Carbon	Transition metals	Carbon	Nickel and nickel oxide	Perovskite and perovskite/metal cermet
Fuel	Pure H_2 (tolerates co_2)	Pure H_2	Pure H_2 (tolerates CO_2, 1% CO)	H_2, CO, CH_4, hydrocarbons (tolerates CO_2)	H_2, CO, CH_4, hydrocarbons (tolerates CO_2)
Efficiency (without CHP)	35%–45%	35%–60%	40%–45%	50%–60%	50%–60%
Catalyst	Platinum	Platinum	Platinum	Electrode material	Electrode material
Interconnect	Carbon or metal	Metal	Graphite	Stainless steel or nickel	Nickel, ceramic, or steel
Operating temperature	40–80°C	65–220°C	205°C	650°C	600–1000°C
Charge carrier	H^+	OH^-	H^+	$CO_3^=$	$O^=$

Continued

Table 5.1 Summary of major differences of fuel cell types [4]—cont'd

	PEMFC	AFC	PAFC	MCFC	SOFC
External reformer for hydrocarbon fuels	Yes	Yes	Yes	No, for some fuels	No, for some fuels and cell designs
External shift conversion of CO to hydrogen	Yes, plus purification to remove trace CO	Yes, plus purification to remove CO and CO_2	Yes	No	No
Prime cell components	Carbon-based	Carbon-based	Graphite-based	Stainless-based	Ceramic
Product water management	Evaporative	Evaporative	Evaporative	Gaseous product	Gaseous product
Product heat management	Process gas + liquid cooling medium	Process gas + electrolyte circulation	Process gas + liquid cooling medium or steam generation	Internal reforming + process gas	Internal reforming + process gas

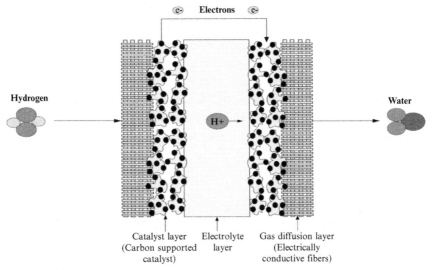

Fig. 5.6 A single PEMFC configuration [1].

5.2.2.1 Proton Exchange Membrane Fuel Cell

The PEMFC is relatively the most common type of FC, as it is applied in very large, broad areas from portable devices to DGs. One of the areas where the PEMFC has predominated is the automobile industry. Most of the vehicles that use FCs are working with a PEMFC (Fig. 5.6).

The following equations demonstrate anode, cathode, and equivalent reactions:

$$\text{Anode}: \ H_2(g) \rightarrow 2H^+(aq) + 2e^-$$

$$\text{Cathode}: \ \frac{1}{2}O_2(g) + 2H^+(aq) + 2e^- \rightarrow H_2O(1)$$

$$\text{Overall}: \ H_2(g) + \frac{1}{2}O_2(g) \rightarrow H_2O(1) + electric\ energy + waste\ water$$

Hydrogen is the main fuel of the PEMFC, but other fuels such as methanol, biomass, or the others can be injected with an external reformer that refines fuel and delivers pure hydrogen to the PEMFC. The electrolyte in this FC is a solid polymeric membrane placed between two platinum-catalyzed porous electrodes. PEMFCs typically operate at about 40–80°C, a temperature determined by both the thermal stability and the ionic conductivity characteristics of the polymeric membrane. To get sufficient ionic conductivity, the proton-conducting polymer electrolyte requires water.

Thus, temperatures are limited to <100°C. The low operating temperature allows the PEMFC to be brought up to steady state operation rapidly and makes it convenient for myriad applications.

Water management in the membrane of the PEMFC is critical for efficient performance. The FC must operate under conditions where the by-product water does not evaporate faster than it is produced because the membrane must be hydrated and needs a specific amount of water. Furthermore, electrocatalyst poisoning by carbon monoxide is another issue of PEMFC.

5.2.2.2 Alkaline Fuel Cell

The AFC is one of the first-developed types of FCs. In Apollo and other space missions, it has been applied by NASA and the US military since 1950. Historical development of FCs will be explained in more detail. Air and space and the military are the main areas that use AFCs. Using of this type of FC can produce both power electricity and drinking water for space crews and astronauts. Also, in comparison with gas turbine generators and batteries, the AFC has priorities for space uses as it needs no heavy fuel supply or numerous and heavy batteries (Fig. 5.7).

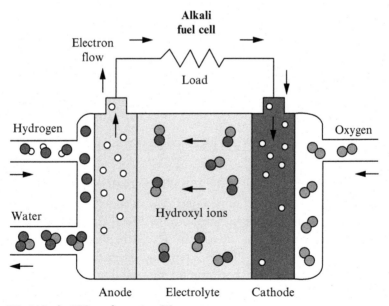

Fig. 5.7 A single AFC configuration [5].

Reactions that take place in anodes and cathodes are written as

$$\text{Anode}: H_2 + 2OH^- \rightarrow 2H_2O + 2e^-$$

$$\text{Cathode}: \frac{1}{2}O_2 + H_2O + 2e^- \rightarrow 2OH^-$$

$$\text{Overall}: H_2 + \frac{1}{2}O_2 \rightarrow H_2O + \text{electricity} + \text{heat}$$

The electrolyte in this FC is mobilized or immobilized potassium hydroxide in asbestos matrix. Operation temperature is between 65°C and 220°C.

The charge carrier in AFC is OH^- and like other low-temperature FCs, AFC works with hydrogen and oxygen. CO and CO_2 poison AFCs and reduce considerably the efficiency of an AFC, so it needs external reformer to provide pure hydrogen.

5.2.2.3 Phosphoric Acid Fuel Cell

The PAFC is the first type of FC that has been commercialized. It is commercially available in hotels, houses, hospitals, and some power stations in a range from 50 kW to 11 MW.

Efficiency of PAFC is ~35%–45%, which is higher than PEMFC, but lower than MCFC and SOFC. When it works with CHP, heat and power are applied simultaneously, so the efficiency grows dramatically and reaches to about 80%. Fig. 5.8 shows a configuration of PAFC.

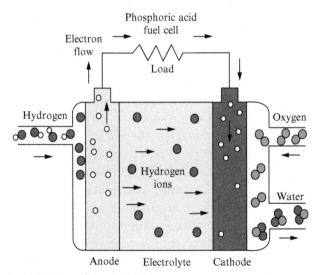

Fig. 5.8 A single PAFC configuration [5].

The next three equations present reactions of the PAFC:

$$\text{Anode}: H_2 \rightarrow 2H^+ + 2e^-$$

$$\text{Cathode}: \frac{1}{2}O_2 + 2H^+ + 2e^- \rightarrow H_2O$$

$$\text{Overall}: \frac{1}{2}O_2 + H_2 \rightarrow H_2O$$

The electrolyte in this FC immobilizes liquid phosphoric acid in SiC. The operating temperature is near to 220°C. In comparison with the other kinds of low-temperature FCs, the PAFC has more stability and tolerance, so it can work under higher temperature.

The PAFC has a lot of similarity with the PEMFC, among the other types of FCs. For example, electrodes and catalysts of both models are carbon and platinum, respectively. In addition, the charge carrier in a PAFC is H^+, exactly like a PEMFC.

On the other hand, while like other low-temperature FCs, the PAFC works with hydrogen and oxygen, it does not have a problem with CO_2 and can tolerate <1% of CO. However, it needs an external reformer to provide pure hydrogen. So, a PAFC is more flexible than a PEMFC or an AFC.

High cost is one of the main issues of the PAFC. It is expensive because a PAFC needs platinum and other material that must be tolerable against the concentrated acid.

5.2.2.4 Molten Carbonate Fuel Cell

The MCFC has a wide range of stationary scale and outdoor uses. Operation of this type of FC is different from other types. Due to high operating temperature, nickel and nickel oxide are applied in anode and cathode, respectively. So, it needs no expensive materials like a PAFC or a PEMFC.

The MCFC, with a 600–700°C operating temperature, is categorized as a high-temperature FC. Heat is a by-product of a MCFC that has myriad uses in industry. Also, with using CHP, the efficiency increases from under 60% to over 75%. Because of high operating temperature, heavy weight, and low start-up the major of uses for MCFCs is outdoor power stations, power plants, and marine applications and they are not very appropriate for domestic applications. In addition, high temperature can produce high-pressure steam, which is absolutely lucrative for turbines or even the chemical process. For instance, a MCFC because of its high temperature as opposed to low temperature FCs, has an internal reformer instead of an external reformer to use more hydrocarbons as fuel; a property that will be explained in the following equations.

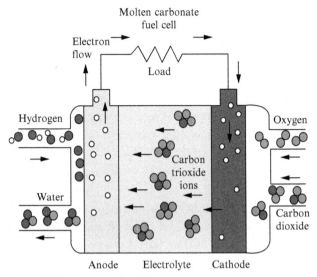

Fig. 5.9 A single MCFC configuration [5].

The configuration of a single MCFC is illustrated in Fig. 5.9, where H_2 is used as the fuel.

As explained before, a MCFC can work with some hydrocarbons such as CO or CH_4 as well as H_2. hydrogen and carbon monoxide can be injected directly to a MCFC while methane needs reformer.

$$\text{Anode}: H_2 + CO_3^= \rightarrow H_2O + CO_2 + 2e^-$$

$$\text{Cathode}: \frac{1}{2}O_2 + CO_2 + 2e^- \rightarrow CO_3^=$$

$$\text{Overall}: H_2 + \frac{1}{2}O_2 + CO_2(\text{cathode}) \rightarrow H_2O + CO_2(\text{anode})$$

From the above equations, it is obvious that the charge carrier is $CO_3^=$. The electrolyte of this FC is immobilized liquid molten carbonate in $LiAlO_2$. CO_2 is needed in the cathode reactions to stabilize the concentration of carbonate in the electrolyte.

With a reformer, MCFC can work with methane:

$$\text{First reaction}: \ CH_4 + 2H_2O \rightarrow 4H_2 + CO_2 \quad \text{Internal reforming}$$

Based on the abovementioned equation, in the first reaction, methane reacts with water steam and produces carbon monoxide and hydrogen.

Now, the generated hydrogen can be injected to the anode and react with $CO_3^=$ to produce electricity, water, and carbon dioxide:

$$\text{Second reaction}: \ 4H_2 + 4CO_3^= \rightarrow 4H_2O + 4CO_2 + 8e^-$$

Generated CO_2 from the second reaction can be injected to the cathode so as to extract $CO_3^=$ and heat:

$$\text{Third reaction}: 2O_2 + 4CO_2 + 8e^- \rightarrow 4CO_3^= + \text{Heat}$$

Although the MCFC can apply a wide range of hydrocarbons in comparison with the low-temperature FCs, its materials are endangered from corrosion because of high operating temperature. So, cell life is one of the big development barriers of the MCFC.

5.2.2.5 Solid Oxide Fuel Cell

Among other types of FCs, the SOFC is more similar to the MCFC. But, it has a higher operating temperature and can apply more hydrocarbons as fuel. From sub MW to 100 MW, SOFCs have been commercialized.

The SOFC has colossal uses, but it is commonly used in power plants and hybridized with a gas turbine. Due to the high operating temperature, $Co-ZrO_2$ or $Ni-ZrO_2$ cermet and Sr-doped $LaMnO_3$ are applied in anode and cathode, respectively.

The SOFC, with 600–1000°C operating temperature, has the highest operating temperature among all kinds of FCs. Heat as a by-product of SOFC has myriad usages in industry. Also, with using CHP, the efficiency increases from under 40% to almost 75% when the SOFC is hybridized with the gas turbine. A SOFC with a gas turbine has the highest rate of generating power of all types of FC (100 MW). Like other high-temperature FCs, the SOFC is convenient for power plants and DGs, but as opposed to the MCFC, recent developments in SOFC technologies lead SOFC to work in the range of about 600–800°C. So, now a SOFC can be applied in sub-MW applications too. In addition, SOFC (like MCFC) works with some hydrocarbons, has an internal reformer, and does not need an external reformer while in comparison with MCFC, more hydrocarbons can be injected to SOFC.

Configuration of a single SOFC is demonstrated in Fig. 5.10, where H_2 is used as the fuel.

The charge carrier in this type of FC is $O^=$. The electrolyte of this FC is nonporous metal oxide (ceramic), usually Y_2O_3-stabilized ZrO_2. Because the electrolyte is solid, there are varied shapes of a SOFC.

H_2, CO without reformer, and CH_4 with reformer are injected to the SOFC. The next equations show the reactions of SOFC when H_2 is the fuel:

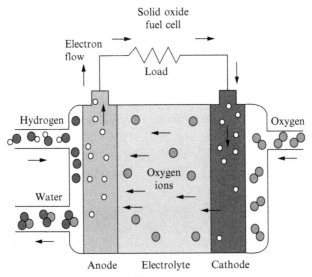

Fig. 5.10 A single SOFC configuration [5].

$$\text{Anode}: 2H_2 + 2O^{2-} \rightarrow 2H_2O + 4e^-$$
$$\text{Cathode}: O_2 + 4e^- \rightarrow 2O^{2-}$$
$$\text{Overall}: 2H_2 + O_2 \rightarrow H_2O$$

When carbon monoxide or methane is used in a SOFC, the following reactions take place, respectively:

$$\text{Anode}: CO + O^{2-} \rightarrow CO_2 + 2e^- \quad \text{when Carbon Monoxide use as fuel}$$
$$\text{Anode}: CH_4 + 4O^{2-} \rightarrow 2H_2O + CO_2 + 8e^- \quad \text{when Methane use as fuel}$$

5.2.3 Advantages and Disadvantages of FCs

FCs, like the other power generations, have both benefits and drawbacks. Although there are discrepancies between features of different types of FCs, we are striving to present a comprehensive explanation to provide information for readers about the traits of FCs.

5.2.3.1 Advantages of FCs

- Environmental pollution of FCs is near zero and in most cases, there is heat and water at the output, excluding electricity. Therefore, they are the best methods to tackle the air pollution and global warming disasters the will be discussed in more detail in the transportation discussion in this chapter.

- In impassable or remote areas where telecommunication stations are installed, sometimes it is very difficult to provide power for them via transmission lines. Also, diesel generators need monthly maintenance and cannot provide electricity for a long time, so they are not suitable for these places, whereas FCs with good features in these places can generate enough power for the telecommunication stations. In addition, the output of a FC is DC, so it can be injected to the telecommunication stations as they need a DC voltage, except in lightening usages.
- Their redundancy is considerable, as power capacity can be expanded by adding another FC unit while it does not need interruption of other units.
- There are a variety of fuel supplies and approaches, so as to extract fuels that are appropriate for FCs. For example, fuel from coal, oil, natural gas, an electrolyzer, and so on will be presented in different methods of extracting hydrogen to use in FC technologies.
- Efficiency of FCs is reasonable and when they are used as CHP can even reach to around 80% efficiency.
- Emissions of CO, CO_2, NOx or other greenhouse gases (GHGs) in the FCs are considerably lower than fossil fuel power plants.
- Hydrogen can be generated with the electrolyzer in any place, so it has a good decentralized potential to provide the fuel.
- FCs are one of the best UPSs and, as will be seen in the following topics, if FCs are used in networks, many blackouts will not occur.
- Operating time and endurance of FCs are much longer and better than conventional or even advanced batteries, which means that the operating time can be doubled just by doubling the amount of fuel without increasing the number of FC units.
- Installation and operation of FCs are cost-effective.
- Two of the main barriers of electricity are as follows. Electricity cannot be saved (conventional methods such as batteries or even reservoir storage do not have good efficiency) and cannot be transported except with transmission lines, which need expensive infrastructure. But, with FCs hydrogen can be used, which has been generated from different methods like electrolysis of water and also can be transported by fuel tanks. Thus, both of cost and transportation problems can be solved by hydrogen or other fuels, which can be injected into FCs.
- They can combine heat and power to increase efficiency, especially in high-temperature FCs.
- FCs can be hybridized with conventional power generators or even with solar and wind farms.

- Low-temperature FCs can be used in indoor places such as houses, hospitals, offices, air spaces, and so on because they are quiet, clean, and need no spacious spaces.
- With good design, they can respond to load variations or even transient loads.
- From 1 kW to 100 MW, depending on types of FCs, they can provide power for consumers and industries.
- Because FCs do not need mechanical and moving parts, noise pollution is very low in comparison with the other kinds of power generation. This feature makes them suitable for urban and indoor uses.
- The abovementioned advantage (no moving parts) makes them reliable from a maintenance point of view and reduces maintenance costs.

5.2.3.2 Disadvantages of FCs

- If a FC uses a fuel excluding hydrogen, the efficiency of the FC will be reduced because of electrolyte decomposition and catalyst poisoning. Sensitivity of different types of FCs is not equal. For instance, an AFC and a SOFC have a maximum and minimum sensitivity to the other fuels excluding hydrogen, respectively.
- Fuel reforming technology is expensive.
- FCs have noble materials to tolerate corrosion, poisoning, and so on, while they must have good conductivity traits. So, they are pricy and it is very difficult to find alternative materials for them.
- The response time of FCs in confronting transient loads is not appropriate. When we use a stand-alone FC, the situation deteriorates. But, by adding some devices like ultracapacitors, batteries, or even using a FC parallel with another line, the burden of transient load can be shifted, such as start-up current to the abovementioned devices. Thus, the design of a FC system becomes more important.
- Output of FCs is DC, so FC systems need inverters to convert DC to AC. But, inverters inject harmonics to the system. So, to overcome this problem, filters (usually low-pass filters) or power factor correction capacitors are applied. However, these power electronic devices make FC units more expensive.
- Endurance/reliability of higher-temperature units has not been developed.
- FCs need cutting edge and costly technologies to integrate into a network precisely and efficiently.
- Fuel storage of FCs is still critical as they need advanced storage.
- Infrastructure for FCs has not been developed.

5.2.4 FC Applications

One of the main advantages of FCs in comparison with other generators is that they are flexible and can be used in different environments. Some of them are suitable for indoor and others for outdoor applications, while they can provide power from a few watts up to more than 100 MW. In this part of the chapter, they are evaluated in three sections including portable devices, transportation, and DGs.

5.2.4.1 Portable Devices

Nowadays, the industry and society need portable devices more than ever because computers, cell phones, laptops, and other electronic devices are used regularly in our everyday lives. These applications must be clean, endurable, light, and small. Therefore, FCs have to be better than batteries in these items, while their prices do not have to be higher than batteries. It is obvious that FCs are cleaner, lighter, and smaller than diesel generators and other fossil-fueled generators for portable applications. In the bar charts in Fig. 5.11, FCs are compared with conventional batteries in terms of storage density by volume and weight:

Comparing Energy Density: Compressed hydrogen (3000 psi) vs. lithium-ion and lead-acid batteries

From these bar charts, it is concluded that the compressed hydrogen battery is 2.4 times smaller and lighter than the lithium-ion battery. In addition, for an equivalent amount of energy, the compressed hydrogen battery is seven times and well over three times lighter and smaller than the lead-acid battery, respectively.

After some charge and discharge, the charge capacity of a rechargeable batteries will be decreased while a FC does not have this problem and can provide power as the fuel is injected into it. All of these features make FCs the best choice for portable applications.

Another area that is important for portable devices is portable applications for the military. Communication equipment, GPS, night vision, and guns all use FCs. In addition to the abovementioned features of FCs, they are also quiet; a major benefit for military uses.

5.2.4.2 Transportation Industry

One of the main causes of global warming and air pollution is transportation pollutions. Numerous of vehicles emit GHGs into the atmosphere. The best practical green solution to tackle this problem is the applications of FCs in vehicles, as the output of the FCs are water and heat instead of the

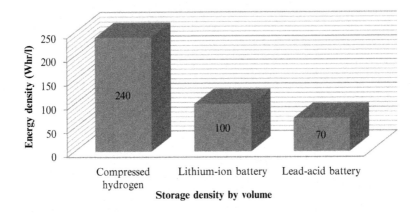

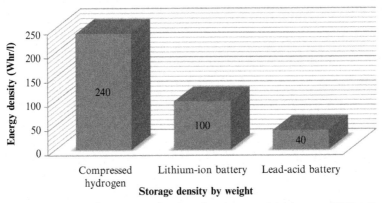

Fig. 5.11 Comparing the energy density of compressed hydrogen (3000 psi) with lithium-ion and lead-acid batteries [2]. *From B. Cook, Introduction to fuel cells and hydrogen technology, Eng. Sci. Educ. J. 11 (6) (2002) 205–216. Reproduced by permission of the Institution of Engineering & Technology.*

hydrocarbons that are exhausted by conventional cars. Nowadays, many automakers implement this technology in their products. For example Toyota, Nissan, Renault, DaimlerChrysler, Honda, Benz, Volvo, Ford, General Motors, and Volkswagen have been working on these projects in collaboration with the Ballard Company. They have been using gasoline, methanol, and pure hydrogen as the fuel. However, zero emission vehicles use pure hydrogen and many of these vehicles use an appropriate fuel processor to convert hydrocarbons to hydrogen. Since the 1970s, FCEVs have been working with PEMFCs, PAFCs, and AFCs but PEMFCs have played the main role in this industry. Their output is clean while the efficiency of these vehicles is greater than that of conventional internal combustion engines (ICEs).

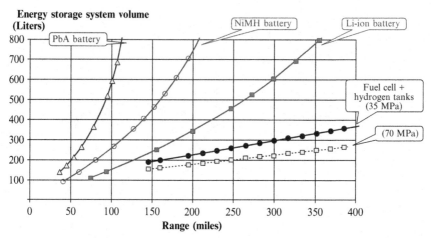

Fig. 5.12 Range of distance that a vehicle can go with batteries in comparison with a FC based on consumption in liters [6]. *From C.E. Thomas, Fuel cell and battery electric vehicles compared, Int. J. Hydrogen Energy 34 (15) (2009) 6005–6020.*

FCEVs have more durability than electric vehicles that use batteries. The line graph in Fig. 5.12 demonstrates the miles that can be driven in a vehicle with PbA, NiMH, or Li-ion batteries in comparison with a FC with hydrogen tanks (35 and 70 MPa) based on consumption in liters.

While a FC vehicle with 70 and 35 MPa could travel 400 miles with approximately 280 and 370 fuel liters, respectively, it can travel about 355, 210, and 120 miles with 800 fuel liters with the Li-ion, NiMH, and PbA batteries, respectively. Therefore, a FCEV needs a considerably lower storage tank, whereas it can travel a much longer distance.

5.2.4.3 Stationary Sector
With the rapid increase in power demand, conventional methods of generating power cannot meet the huge need of electricity appropriately. So, restructuring must be implemented to overcome the power shortages.

Stationary FCs can provide power to residential consumers; malls; critical loads in such specific industries as hospitals as UPSs, backups, or even primary generators; and also for helping power networks as DGs. Applying DGs have myriad benefits such as loss reduction, improvement of the voltage profile, increasing reliability, air pollution reduction, and so on, which will be discussed in more detail in Section 5.4. In this section, grid-connected and nongrid-connected applications of FCs are briefly introduced.

Grid-connected FCs provide primary or backup power for the system. In this case, FCs are more reliable than wind turbine generators (WTGs) or photovoltaic (PV) systems because, unlike WTG or PV, they are not stochastic and can generate power as long as we inject fuel to them. Usually, the PAFC, which is commercially available, and the PEMFC are more convenient for sub-MW use while the MCFC, DMFC, and SOFC are suitable for MW uses. When a FC is connected as a DG to the line, some standards must be met that are critical such as frequency and voltage of the FC which must be equal to the line. In addition, FCs usually cannot meet transient loads or start-up current. Therefore, we must consider all of these constraints in the FC design.

Nongrid-connected FCs such as stand-alone backup generators play a strong role in safety and reliability of systems. In comparison with conventional UPSs, FCs are lighter, cleaner, and smaller and they have a better durability and can respond to the interruption and load change quicker. Also, in remote areas that have no access to the power network, FC power plants can provide enough power as long as fuel is injected.

5.3 VOLTAGE LOSSES AND THEIR ROLES IN DESIGNING A FC

In Section 5.3, the initial designing of a FC is presented to explain the importance of the voltage of each cell and also to evaluate the methods of calculating the number of stacks that must be connected together in parallel and series to provide the desirable power. Then P-I and V-I polarization curves are introduced as they have a fundamental role in the characteristics of FCs.

The main body of this section explains different types of voltage losses such as ohmic, activation, and concentration losses to demonstrate them as the worst impediments in FCs, especially PEMFCs as the most common type of FC.

5.3.1 Initial Designing of a FC

The FC has many parameters that must be considered before designing it. The main parameter from an electrical point of view is voltage. When we want to design a FC system, first we have to think about voltage and current of each cell to set them appropriately in series and parallel together to reach a desirable power level. Voltage lost is the worst obstacle in our way to meet the required demand and it reduces efficiency considerably. If a manufacturer produces a FC that has ower voltage loss, then users can apply more efficient FCs, which need smaller area and less cell for the same power capacity. By this action, weight and size of the system are reduced while the efficiency is increased. The method of calculating the number of series and parallel FC stacks is explained by the next example. In this section we will study parameters that influence the voltage of a FC and at the end of this section we will reach the final equation of the voltage.

Example 5.1

Calculating the number of series and parallel stacks. Fig. 5.13 illustrates a FC system that is connected to the utility grid where it is needed that the FC units provide 480 kW of power. Our purpose in this example is evaluating the number of FC units and the number of FC stacks that are needed to be connected in series and parallel to compose each FC array to meet the desired power:

In this system, there is a boost converter with $V_{DC,output} = 480$ and 0.55 duty cycle (converter DC output voltage is determined by the inverter AC output voltage and the voltage drop of the inductor-capacitor (LC) filter) [8]. FCs need power electronic devices to interface to grid networks. Now we just want to show the initial designing of FCs. But a comprehensive explanation of power electronic devices that must be used in FCs and why we have to apply them will be discussed in more detail in the last section.

Assume that voltage and current of stacks are 27 V and 18.5 amp, respectively, and each FC unit can provide 48 kW of power while in this example. Note that we do not want to take many constraints into account to simplify equations.

FC units must generate 480 kW where the power of each one is 48 kW. So, the FC power plant comprises ten 48 kW units that are connected in parallel, which means that 10 boost converters are needed.

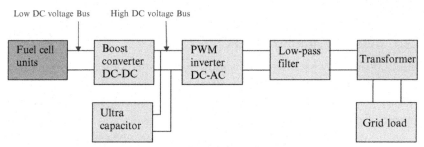

Fig. 5.13 Block diagram of a fuel cell distributed generation system [7].

To find out the number of stacks in series and parallel, first we have to calculate the voltage of each unit. As assumed before, the converter output voltage in this example is 480 V with 0.55 duty ratio.

$$\frac{V_{out}}{V_{in}} = \frac{1}{1-D} \Rightarrow (1-0.55) \times 480 \Rightarrow V_{out} = 216\,\text{V}$$

Therefore, the converter input voltage or the output of each FC unit is 216 V.

Now, the output current of each unit is evaluated by the following equation (ignoring losses):

$$I_{fc} = \frac{P_{fc}}{V_{fc}} = \frac{48000}{216} = 222.22\,\text{A}$$

The number of stacks per unit that must be connected in a series is extracted from the voltage of each unit as the sum of their stacks voltage in series demonstrates the voltage of the unit:

$$N_S = \frac{V_{fc}}{V_{st}} = \frac{216}{27} = 8$$

Calculating the number of parallel stacks per unit is done by both of the following equations. One of these methods is defined as the sum of the stacks current per unit in parallel, which shows the current of the unit. Another way is the use of power, which means the power of each unit can be obtained by multiplying the number of parallel and series stacks and the power of each stack:

$$N_P = \frac{I_{fc}}{i_{st}} = \frac{222.22}{18.5} \simeq 12$$

or

$$P_{st} = V_{st} \times i_{st} = 18.5 \times 27 = 0.5\,\text{kW} \Rightarrow N_P = \frac{P_{array}}{N_S \times P_{stack}} = \frac{48\,\text{kW}}{8 \times 0.5\,\text{kW}} = 12$$

In other words, each FC unit is comprised of 12 parallel branches with 8 series 0.5 kW stacks per branch.

The current can be obtained from the following equation:

$$i = \int jds \Rightarrow i = j \times A \tag{5.1}$$

where j is the current density (A/cm^2) and A (cm^2) is the FC area, so with increasing the FC area the current of the FC can be increased.

As will be explained in the following topics, V_{cell} is related to several parameters. The maximum theoretical potential of the cell is 1.229 V, which is called ideal voltage, but in reality V_{cell} is around 0.6–0.7 V. The cell power is equal to $P_{cell} = V_{cell} \times i_{cell} \Rightarrow$ so we have to evaluate the parameters that limit the value of voltage, because if we increase this value, the power will increase, too. However, we cannot increase the value of current and voltage as we wish based on P-I and V-I polarization curves.

5.3.2 P-J and V-I Polarization Curves

Operation and efficiency of FCs are associated with many factors and many of them can be comprehended by the V-I polarization curve. Fig. 5.14 shows output voltage based on output current of the FC. The ideal FC is defined as one where current can be provided as much as we desire while the voltage remains constant (does not drop). This voltage is called thermodynamic, open circuit voltage. But, because of limitations, the actual power is lower than the ideal voltage.

The i-v and power density curves of the PEMFC are plotted in Fig. 5.14, which indicates the dependence of the voltage and current to each other and their relationship with power density:

$$p = iv \qquad (5.2)$$

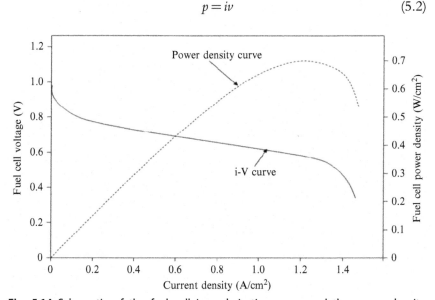

Fig. 5.14 Schematic of the fuel cell i-v polarization curve and the power density curve [9]. *From R. O'Hayre, Suk-Won Cha, Whitney Colella, Fritz B. Prinz, in: Fuel Cell Fundamentals, third ed. John Wiley and Sons, New York, 2008.*

The graph in Fig. 5.14 illustrates that power density rises to one point, but because of voltage drop, which takes place because of voltage losses, it goes down at the end of the curve. In other words, by increasing the current to some point, the voltage drops and power density rises, whereas after this point, the growth of the current caused both power and voltage drops and consequently the FC will collapse. As a result of limitations in increasing the current density, the power density of the FC is less than the expected value. Therefore, FCs are designed to operate under their maximum power density. It is very important to note that that design must not be done at higher than the maximum power density as the FC will be faced with power and voltage collapses. Thus, the amount of current generated by the FC is proportional to the consumption rate of the fuel and when the voltage is decreased, the power will be reduced, too. Therefore, keeping the voltage at the highest value is very important. However, keeping voltage and current constant at the high value is ideal and impossible in the real situation because of constraints and losses.

An ideal cell of a FC (e.g., a PEMFC) at 25°C can produce about 1.23 V. When the temperature of the FC (PEMFC is our aim in this sentence) reaches 80°C in operating time, the ideal voltage of FC will decrease from 1.23 to 1.18 V at 80°C. In fact, for all types of FCs, their actual voltage will not exceed 1 V per cell. Now, different types of voltage losses are discussed in the following section.

5.3.3 Different Types of Voltage Losses in FC

The cell voltage is calculated from

$$V_{cell} = V_{rev} - V_{irev} \tag{5.3}$$

where V_{rev} is the reversible cell potential or ideal voltage and V_{irev} is an irreversible cell voltage loss, which is defined as three major losses, including the activation potential (η_{act}), ohmic over potential (η_{ohmic}), and concentration over potential (η_{conc}) and can be obtained from the following equation

$$V_{irev} = \eta_{act} + \eta_{ohmic} + \eta_{conc} \tag{5.4}$$

In Fig. 5.15, three main voltage losses of a low-temperature FC (e.g., a PEMFC) are shown.

As can be seen from the voltage and current density graph of a typical low temperature FC, the theoretical electromotive force (EMF) or ideal voltage is about 1.2 V. Activation loss reduces voltage to about 0.8 V. The voltage is dropped to well under 0.5 V by the ohmic polarization. Concentration

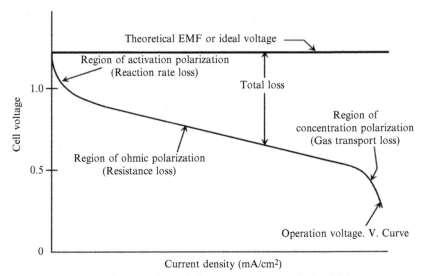

Fig. 5.15 Schematic of voltage losses of a low-temperature fuel cell [4].

losses cause voltage collapse at high current. Therefore, the most destructive loss is resistance loss.

The *j–v* curve is plotted for high-temperature FCs in Fig. 5.16.

Fig. 5.16 indicates that in high-temperature FCs such as SOFCs and MCFCs, losses that are related to activation have less importance and the concave part of the curve is barely visible. On the other hand, the losses

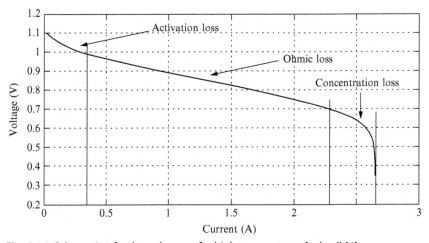

Fig. 5.16 Schematic of voltage losses of a high-temperature fuel cell [4].

associated with mass transport in the high-temperature FCs are more important and the convex part of the curve will be drawn more to the left in comparison with low-temperature FCs.

In the following topics, voltage losses will be formulated based on the low-temperature FC and especially the PEMFC.

5.3.3.1 Activation Losses

Activation loss occurs as a result of the energy needed to start the reaction. The better the catalyst, the less energy will be needed for activation. A limiting factor to achieve an available power density from the FC is the speed at which reaction can take place. For example, platinum can be a good catalyst to increase the reaction speed. Activation losses occur in both of the cathode and anode, but because the cathode reaction (oxygen reduction) is $\sim$100 times slower than the anode reaction, the cathode reaction has a stronger effect on limiting the power density.

Activation losses increase with the current density and decrease the voltage of a cell to <1 V based on the Tafel equation:

$$\eta_{act} = \frac{RT}{\alpha nF} Ln\frac{j}{j_0} \tag{5.5}$$

Also, we can solve activation problems with

$$\eta_{act} = \frac{RT}{\alpha 2F} Lnj - \frac{RT}{\alpha 2F} Lnj_0$$

where j and j_0 are the current density and the exchange current density of the reaction, respectively. Therefore, the amount of activation losses can be reduced by increasing the exchange current density. α is the transfer coefficient of an electron, which is between 0 and 1. F is the Faraday constant, which is 96,485 C/mol. n is the number of exchange protons per mole of reactant. T and R are temperature (kelvin) and resistance (Ω), respectively.

As mentioned before, the activation polarization includes the anode and cathode overpotentials:

$$\eta_{act} = \eta_{act-anode} + \eta_{act-cathode}$$

So

$$\eta_{act} = \frac{RT}{\alpha nF} Ln\left(\frac{j}{j_0}\right)_{anode} + \frac{RT}{\alpha nF} Ln\left(\frac{j}{j_0}\right)_{cathode} \tag{5.6}$$

Example 5.2

Calculating the activation losses. Assume that there is a FC that operates at 250 kW, current density of 0.9 A/cm², and exchange current density of 10^{-6} with resistance of 6 Ω. So, the activation losses can be calculated by Eq. (5.5):

$j = 0.9$ A/cm², $j_0 = 10^{-6}$ A/cm², $R = 6$ Ω, $F = 96{,}485$ c/mol, $\alpha = 0.5$, $T = 250$ K

From Eq. (5.5)

$$\eta_{act} = \frac{RT}{\alpha 2F} Ln\frac{j}{j_o}$$

$$\Rightarrow \eta_{act} = \frac{6 \times 250}{0.5 \times 2 \times 96485} Ln\frac{0.9}{10^{-6}}$$

$$\Rightarrow \eta_{act} = 0.178\,V$$

Another way to solve the activation losses problem is applying the Tafel plot (see Fig. 5.17).

Simplified version of Eq. (5.5) is

$$\eta_{act} = a + bLnj \tag{5.7}$$

where b is the Tafel slope, which is achieved from the slope of a plot of η_{act} as a function of ln (i) [4]. a and b are calculated as below:

$$a = -\frac{RT}{\alpha nF} Lnj_0, \quad b = \frac{RT}{\alpha nF}$$

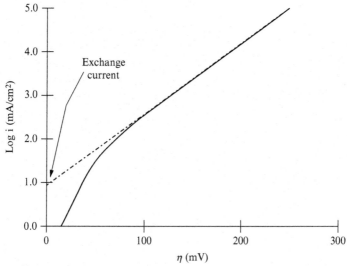

Fig. 5.17 Schematic of a Tafel plot [4].

As can be seen in the above mentioned example, the activation losses will be reduced by decreasing the amount of b. Thus, electro catalysts with lower Tafel slope are more attractive for us.

5.3.3.2 Ohmic Losses

Ohmic losses are due to ionic, electronic, and contact resistances, which occur in the electrodes and electrolyte, current collectors, contact resistance, and interconnects because every material has instinct resistance to charge flow. It indicates the losses of FC performance. Also, it depends on current density, temperature, and especially on materials that are used in cell and stack in comparison with activation and concentration losses, which depend on cathode and anode. Regarding to the cell type of FCs, the electronic, ionic, and contact resistances have different importance.

$$\eta_{ohmic} = iR_{ohmic} = i(R_{elec} + R_{ionic} + R_{contact})$$

Here the formulations of ohmic losses are evaluated for the PEMFC. In this type of FC, ionic losses are more important than other losses for studying ohmic losses, so the main topic that will be discussed in this part so as to reduce ohmic losses is increasing the ionic conductivity of the electrolyte. A simplified form of ohmic losses for the PEMFC is written as follows [1]:

$$\eta_{ohmic} = iR_{ohmic} = i(R_{elec} + R_{ionic}) \tag{5.8}$$

The conductivity of materials are calculated by the following equation:

$$\sigma = \frac{i}{R_{ohmic}} \tag{5.9}$$

To obtain the FCs resistance, we have to take into account the cross-sectional area of conductor $A(cm^2)$, its thickness and length L (cm), and conductivity $\sigma (\Omega^{-1} cm^{-1})$. Thus, R_{ohmic} can be computed by the following equation:

$$R = \frac{L_{cond}}{\sigma A_{cond}} \tag{5.10}$$

From Eq. (5.1) we have

$$i = \int jds \Rightarrow i = j \times A$$

Therefore, by combining Eqs. (5.1), (5.8) the ohmic losses can be calculated by the current density:

$$\eta_{ohmic} = j(ASR_{ohmic}) = j(A_{cell}R_{ohmic}) \tag{5.11}$$

Area specific resistance of the FC (ASR_{ohmic}) is calculated to make a comparison with different sizes of FCs and illustrates resistance rate with area. By combining Eqs. (5.11), (5.9):

$$\eta_{ohmic} = iR_{ohmic} = jA_{cel}\left(\frac{L_{thick}}{\sigma A_{fuelcell}}\right) = \frac{jL_{thick}}{\sigma} \tag{5.12}$$

From the abovementioned equation, it is concluded that one of the practical ways to reduce ohmic losses is using the thinner electrolytes in FCs because this action decreases the amount of L_{cond}. Another way is applying high-conductive electrodes and electrolyte material to increase conductivity.

Example 5.3

Calculating the ohmic losses. Assume that the PEMFC used in Example 5.2, has these specifications, too: the cross-sectional area of the conductor is 80 cm², with thickness of 40 μm and conductivity of 0.15 ohm^{-1} cm^{-1}. As assumed in Example 5.2, the current density of this FC is 0.9 A/cm². With an electrical resistance of 4 milliohms, the ohmic losses can be calculated as

$$A = 80 \, cm^2, \ j = 0.9 \, A/cm^2, \ L = 0.004 \, cm, \ R_{elec} = 0.004 \, \Omega,$$
$$\sigma = 0.15 \ \Omega^{-1} \ cm^{-1}$$

From Eq. (5.1),

$$i = \int jds \Rightarrow i = j \times A$$

So, the current is equal to $i = 0.9 \times 80 = 72 \, A$

The R_{elec} is given to us, so we must obtain the R_{ionic}. From Eq. (5.10),

$$R = \frac{L_{cond}}{\sigma A_{cond}} \Rightarrow \frac{0.004}{0.15 \times 80} = 3.33 \times 10^{-4}$$

Now, activation losses can be calculated by Eq. (5.8):

$$\eta_{ohmic} = iR_{ohmic} = i(R_{elec} + R_{ionic})$$
$$\Rightarrow \eta_{ohmic} = 72\left(0.004 + 3.33 \times 10^{-4}\right)$$
$$\Rightarrow \eta_{ohmic} = 0.312 \, V$$

From comparing Examples 5.2 and 5.3, it is obvious that the amount of ohmic losses is usually more than the activation losses. Now, the concentration losses are evaluated as follows.

5.3.3.3 Concentration Losses

These losses relate to the mass transport and they are a result of concentration reduction of hydrogen and oxygen gases in electrodes. New gases reactions should be available quickly in the catalyst side. By forming water at the cathode, especially at high current, the catalyst is blocked and degraded and the access to hydrogen will be limited. Therefore, the FC needs an appropriate supply and precision removal of reactants and by-products to generate current effectively. For instance, depletion and clogging of reactant and products will have a negative effect on the performance of a FC.

When the current density is low and the bulk reactant concentration is high, the concentration losses are not important, but under conditions of high current density and lower air and fuel concentration, mass transport losses will be critical.

The operating current density of a FC can be calculated by [4]

$$j = \frac{nFD(C_B - C_S)}{\sigma} \tag{5.13}$$

where C_B and C_S are the bulk and surface concentrations, respectively. D is the diffusion coefficient. N is the number of electrons that transferred per mole. In this equation, σ is defined as the thickness of the diffusion layer.

When the current density increases, consequently the reactant surface concentration plummets to zero. In this situation, limiting current density is formed. So, j_L is obtained as follows, if we put 0 for C_S in Eq. (5.13):

$$j_L = \frac{nFDC_B}{\sigma} \tag{5.14}$$

By combining Eqs. (5.13), (5.14)

$$-\frac{C_S}{C_B} = 1 - \frac{j}{j_L} \tag{5.15}$$

Based on the Nernst equation considering no current condition

$$E_{i=0} = E^0 + \frac{RT}{nf} \ln C_B \tag{5.16}$$

The Nernst equation when current is following is defined as:

$$E = E^0 + \frac{RT}{nf} \ln C_S \tag{5.17}$$

Therefore, concentration losses, which is the difference between Eqs. (5.16) and (5.17) is given as follows:

$$\Delta E = \eta_{conc} = \frac{RT}{nf} \ln \frac{C_S}{C_B} \qquad (5.18)$$

By substituting Eq. (5.15) in Eq. (5.18),

$$\eta_{conc} = -\frac{RT}{nf} \ln \left(1 - \frac{j}{j_L}\right) \qquad (5.19)$$

Because each electrode has these losses, the concentration losses are divided into cathode and anode losses. It is noticeable that limiting current densities of anode and cathode are different. However, when hydrogen is used as a fuel, we can neglect the anode loss because of high concentration and mass diffusivity. Therefore, in many cases, Eq. (5.19) is used instead of Eq. (5.20).

$$\eta_{conc} = \eta_{conc, anode} + \eta_{conc, cathode}$$

$$\eta_{conc} = -\frac{RT}{nF} \ln \left(1 - \frac{j}{j_{L, anode}}\right) - \frac{RT}{nF} \ln \left(1 - \frac{j}{j_{L, cathode}}\right) \qquad (5.20)$$

The minimum value of limiting the current density of anode and cathode is our reference limiting current density.

Precision design of flow channels of FCs caused the concentration losses reduction. In designing, a compromise between the pressure drop and water removal capability, which is one of the main problems of low-temperature FCs, must be taken into consideration to minimize mass transport losses.

There is a deviation between the practical value of concentration losses and its theoretical value, which is obtained from Eq. (5.20). Thus, a constant factor (B) is introduced to modify this equation, as follows:

$$B = \frac{RT}{nf} \Rightarrow \eta_{conc} = -B \ln \left(1 - \frac{j}{j_L}\right) \qquad (5.21)$$

By substituting Eq. (5.21) in Eq. (5.20),

$$\eta_{conc} = -B \ln \left(1 - \frac{j}{j_{L, anode}}\right) - B \ln \left(1 - \frac{j}{j_{L, cathode}}\right) \qquad (5.22)$$

Example 5.4

Calculating the concentration losses. Assume that the anode and cathode limiting current densities of our FC are 12 and 2 A/cm^2, respectively with the *B* factor equal to 0.06 V. Current density like previous examples is considered as 0.9 A/cm^2. So, concentration losses can be calculated as follows:

$$B = 0.06\,\text{V}, \quad j_{L,\,anode} = 12\,\text{A/cm}^2, \quad j_{L,\,cathode} = 12\,\text{A/cm}^2, \quad j = 0.9\,\text{A/cm}^2$$

With these parameters, Eq. (5.22) is used as follows:

$$\eta_{conc} = -B \ln\left(1 - \frac{j}{j_{L,\,anode}}\right) - B \ln\left(1 - \frac{j}{j_{L,\,cathode}}\right)$$

$$\Rightarrow \eta_{conc} = -0.06 \ln\left(1 - \frac{0.9}{12}\right) - 0.06 \ln\left(1 - \frac{0.9}{2}\right)$$

$$\Rightarrow \eta_{cocnc} = 0.04\,\text{V}$$

In this example, the maximum current density of the FC is limited by cathode as the amount of $j_{L,c}$ is lower than $j_{L,a}$. This means that our current density must be lower than 2 A/cm^2.

5.3.4 Cell Voltage

By now, the formulations of voltage losses are explained. So, now cell voltage can be obtained. From Eq. (5.3) we have

$$V_{cell} = V_{rev} - V_{irev}$$

If we describe V_{rev}, all of the parameters of the abovementioned equation will be known: V_{rev} or V_{OCV}, especially in a PEMFC defined as

$$V_{rev} = V^0 + \frac{RT}{2f} \ln \frac{P_{H_2} \times \sqrt{P_{O_2}}}{P_{H_2O} \times \sqrt{P_O}} \tag{5.23}$$

where V^0 or open cell voltage is about 1.229 V. P_O is standard pressure, which is usually equal to 1. P_{H_2}, P_{O_2}, and P_{H_2O} are hydrogen, oxygen, and water partial pressures, respectively.

Based on Eq. (5.4),

$$V_{irev} = \eta_{act} + \eta_{ohmic} + \eta_{conc}$$

By substituting Eqs. (5.6), (5.8), and (5.20) in V_{irev}, voltage losses will be defined as

$$V_{irev} = \frac{RT}{\alpha F} Ln\left(\frac{j}{j_o}\right)_{anode} + \frac{RT}{\alpha F} Ln\left(\frac{j}{j_o}\right)_{cathode}$$

$$+ \frac{RT}{nF} \ln\left(\frac{j_{L,a}}{j_{L,a}-j}\right) + \frac{RT}{nF} \ln\left(\frac{j_{L,c}}{j_{L,c}-j}\right) + iR_{ohmic}$$

(5.24)

Therefore, the cell voltage can be calculated by the following equation:

$$V_{cell} = V_{rev} - \frac{RT}{\alpha F} Ln\left(\frac{j}{j_o}\right)_{anode} - \frac{RT}{\alpha F} Ln\left(\frac{j}{j_o}\right)_{cathode} - \frac{RT}{nF} \ln\left(\frac{j_{L,a}}{j_{L,a}-j}\right)$$

$$- \frac{RT}{nF} \ln\left(\frac{j_{L,c}}{j_{L,c}-j}\right) - iR_{ohmic}$$

(5.25)

5.4 APPLYING FCs IN DG SYSTEMS

Nowadays, DGs are very common with an important role in power systems from generation to distribution. Our main purpose in this section is the description of FCs as one of the best choices of DGs to provide power for different applications. Restructuring and the power market is presented. Afterward, principle of operation and designing of DG systems are discussed. Demonstrating the effectiveness of FCs among the other kinds of DGs is the other topic that is studied. Therefore, readers can acquire information about comparison of them. Six main methods of producing hydrogen are introduced. Hybridizing the direct fuel cell with the gas turbine generator (DFC/T) as one the best ways of generating power by a fuel cell-gas turbine (FC/GT) system is evaluated. Finally, applying FCs as backup systems is discussed to show the effectiveness of the FC among the other types of backup systems.

5.4.1 Restructuring and the Power Market

There is no doubt that traditional methods of generating power are not good enough to satisfy the colossal needs for electricity as power demand rockets every year in comparison with the previous year. Furthermore, because in a conventional structure, there is a long distance between generators and end users, the losses are very large and a lot of power energy is wasted in transmission lines. Also, traditional methods need expensive devices and cables. For example, many substations are required to decrease or increase the power that is used while because of long distances, precise measurements and relays are necessary, too.

On the other hand, small-scale power generation units called DGs are typically located near the end users. So, many of the problems associated with traditional methods are solved by applying DGs. The need for restructuring and the competitive market is one of the reasons that convinced governments to support DGs to deregulate the power sector.

By reinforcing privatization and strengthening market competition, the power network structure will be optimized. For instance, reduction of power losses, optimization of voltage profile, minimization of voltage imbalance and other approaches for improving the performance and characteristics of power from generation to distribution are done by appropriate siting, sizing, and designing of DGs.

Fig. 5.18A shows the traditional structure of a power grid where generation, transmission, and distribution usually belong to the government and they sell electricity to consumers without rivals, while Fig. 5.18B illustrates the smart grid and restructuring [8].

As can be seen from Fig. 5.18, in the traditional structure of the power grid, end users do not have many options to choose their sources. Also, long-distance transmission lines, several transformers, substations, and other

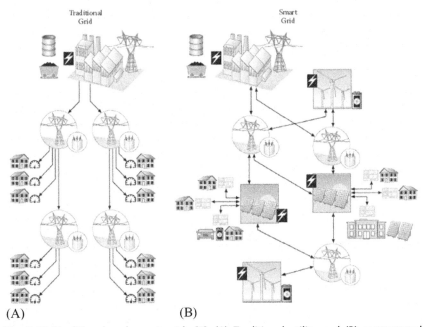

(A)   (B)

Fig. 5.18 Traditional and smart grids [8]. (A) Traditional utility and (B) restructured utility.

power system's equipment reduce the power quality while they increase expenses and losses.

Eliminating the monopoly of power companies and creation of a competitive environment of companies gives customers a chance to choose their electricity providers from different companies. With this new system configuration, a power market is created and competitions between companies becomes more important. Consumers want to buy power from companies that deliver more reliable and less pricy power with better regulations and power quality. Thus, every company must present better incentives and services so as to attract attention of customers. For example, they can offer more incentives to clients for changing their power consumption patterns, so as to avoid consuming power in peak demand periods. By this action, the need for some power plants that work just for generating power at peak times to compensate for power shortages would be solved.

Conventional power plants that operate based on fossil fuels such as coal and gas are one of the main causes of global warming and other environmental disasters. By deregulating and restructuring the power grid and using DGs near the customers, small-scale generators can provide electricity or even heat as a CHP for end users.

Besides, the necessity of developing new recourses and curtailing the emission of GHGs and fossil fuel reduction have led researchers to renewable energies such as wind, solar, FCs, and so on. However, FCs are more reliable than other resources such as PV and WTG because of their stochastic behavior. Furthermore, we can consider hydrogen, methane, or even other hydrocarbons based on the type of FCs as the input fuel. So, by using this technology, a compromise is reached between fossil fuels and green energies, which will be discussed in more detail in the following topics. It is worth noting that the efficiency of FCs is between 30% and 60%, which can be increased to 85% by using CHP. These efficiencies are higher than the efficiencies of other ways of generating power, especially renewable methods.

5.4.2 Principle of Operation and Designing of DG Systems

The basic issues that should be considered about DGs, especially FCs, include safety, operation and reliability, power quality, and accountability [10].

Safety. When a power outage in the network occurs in the presence of DGs, sometimes these generators are still energized. This phenomenon is called islanding and it lowers the safety for those who are working on the lines. Therefore, in this condition, based on the IEEE standard for detection of power islands, relays must detect and operate accurately to disconnect the grid-connected DGs from the lines in <2 s. But, we can use this problem in a good way by providing power as isolated DG for critical loads such as hospitals, factories, and so forth.

Therefore, if DGs are not disconnected properly from the network, one of the barriers of DGs will appear because it endangers the safety of those who work on the lines, but with appropriate measures they can provide the power of sensitive loads.

Operation and reliability. The most reliable renewable energy source between the FC, WTG, and PV is the FC as it does not depend on weather or geographic conditions. Unlike the WTG and PV, the FC can generate power continuously with high reliability. As we will explain in the future, as FCs are very reliable, we consider them as one of the best choices of backup generators.

Grid-connected DGs can regulate voltage by delivering or consuming the reactive power based on different circumstances. When the system is faced with a heavy load, DG must deliver both real and reactive power to the grid. The grid needs reactive power to boost its voltage. The FC is very good in this situation and can supply the amount of reactive power that the grid needs as will be indicated as an example in future topics, whereas WTGs, which usually apply induction generators, consume reactive power instead of injecting it and this action can deteriorate the situation and makes the system unreliable. In contrast, when a power system utility is under light load, our DG should receive the excessive reactive power from the grid [8]. We can set a FC for both of these needs while WTG and PV are not very reliable, because they do not have constant output power. Thus, sometimes they cannot provide the desired power as can FCs.

Power quality. Most renewable energies like the WTG, PV, and FC have power quality problems. For instance, some types of WTGs and all kinds of FCs and PVs generate low DC voltage, so we have to apply boost converters to increase and stabilize the DC output. Then, they must use inverters for obtaining AC voltage. Converters and inverters have some losses, so they decrease the efficiency of FCs. Nonlinear loads and power electronic devices inject harmonics to the grid. In this case, we just want to evaluate harmonics injected by power electronic devices such as inverters. Of course by designing the proper and precise filters, FCs can tackle this problem.

Integration issues such as how we can set the voltage and frequency of a FC equal to the grid and also power quality barriers and solving methods such as using a low-pass filter for filtering the harmonics and their equations will be described in the Section 5.5.

Accountability. For most end users it is not important from which methods they are provided power. They want a lower price with higher power quality of electricity. To overcome this obstacle, renewable energies need cutting edge technologies in their production process to increase reliability. As we illustrated in Fig. 5.5, development of economies of scale makes FCs cost-effective. While it is very critical that DGs continuously provide high-quality power to consumers, FCs just can do this duty appropriately in comparison with WTGs and PVs.

Predicting the time and quantity of delivering power and penalties for interruptions is the responsibility of organizations such as the Federal Energy Regulatory Commission (FERC). For example FERC sets penalties for DGs if their energies vary 1.5% from scheduled quantities. It is obvious that FCs can provide power based on previous scheduling.

5.4.3 Demonstrating the Effectiveness of FCs Among the Other Kinds of DGs

DGs are located near the loads and they are categorized as renewable and nonrenewable technologies. DGs can provide power as stand-alone, grid-connected, cogeneration, peak shaving, standby generators, and so on based on their capacity, size, amount of pollution, flexibility, and so forth. As we described in introducing the different types of FCs in Section 5.2, FCs can be used for all of these applications, while other DGs like PVs, WTGs, or even gas turbine generators, because of their restrictions, are suitable just for some applications. Now we want to separate sustainable and fossil energies and also make a brief comparison between them and FCs.

Nonrenewable DGs include [10]

- Reciprocating engine
- ICE including the natural gas, LPG, diesel, and gasoline ICE
- Combustion turbine
- Gas turbine
- Microturbine (MT)
- FC (extracting hydrogen from fossil fuels)

ICEs are convenient for standby, peak demand, and CHP applications in the commercial and industrial markets at capacities between 50 kW and 5 MW. MTs are small combustion turbines with outputs in the range of 10–500 kW, whereas reciprocating engines can provide power in the range of 30 kW to 6 MW. The highest amount of power is generated by the gas turbine generators, which is 1–100 MW. They are not stochastic and have good reliability. Although their efficiency is not good enough, they require low initial cost for construction, except in the FCs, in comparison with alternative methods. These technologies, excluding the FCs, are not quiet and because they use fossil fuels, they emit GHGs. But, it is impossible to ignore them as many of the current devices and systems have been designed and work based on fossil fuels.

On the other hand, renewable DGs include:
- Wind turbine generator (WTG) (onshore and offshore)
- Solar photovoltaic (PV)
- Solar thermal
- Low-head hydro
- Geothermal
- Biomass
- Nuclear
- Tidal, wave, ocean thermal energy conversion (OTEC)
- Fusion
- FCs (extracting hydrogen from sustainable methods)

The highest proportion of generating power from green energies belongs to the hydro DGs. They have good reliability and safety, whereas we need special places to construct them. While fusion is a very powerful method of generating power, this technology has not been yet developed as well as the other abovementioned methods and it will not be available commercially before at least 2050. Although, geothermal DGs are very convenient and have a reasonable efficiency for some sites, they cannot work every place as their placement is limited by geographic features. For OTEC technology, the thermal difference of water must be about 25°, so it cannot be used in many oceans.

WTGs and PVs are very common and many investors prefer to invest in these fields. Their efficiencies have increased while their initial costs have decreased considerably during this decade. But, they need a large space if we want to use them as large DGs. Also, in the placement of a WTG and a PV, some parameters such as enough wind and sunshine must be taken into account. Also, apart from placement problems, stochastic issues bother these technologies as their generations depending on weather conditions.

It is worth noting that all of the abovementioned methods have their specific advantages and they are lucrative in some applications and conditions. But, some traits such as high efficiency, reliability, quiet operation, low to zero emission, fuel flexibility, energy security, light weight, longevity, and so on make FCs one of the best choices of DGs. As it can be seen, FCs are classified as both renewable and nonrenewable DGs, so they are the best

Table 5.2 Electrical efficiency range of different types of distributed generation

DG types	Electrical efficiency range (%)
Reciprocating engine	30–42
Natural gas ICE	35 [11]
LPG ICE	35 [11]
Diesel ICE	44 [11]
Gasoline ICE	35 [11]
Micro turbine	20–30 [10]
Wind	<40 [10]
Solar PV	10–20 [10]
Biomass	10–20 [10]
PEMFC	40–65 [10]
AFC	40 [10]
PAFC	35–40 [10]
MCFC	50 [10]
SOFC	45–65 [10]

choice to make a comparison between fossil fuels and alternative energies by extracting hydrogen from both methods. The next topic will explain these approaches.

Table 5.2 compares the efficiency of different common DGs.

The efficiency will be increased by applying CHP technology. For example, in the FCs, the efficiency of MCFC and SOFC could exceed 80% in this mode.

Another mode is combining a FC with a gas turbine, which is very lucrative for us as the GT recuperates the energy in the FC exhaust stream thereby boosting the overall system efficiency [12]. But, FC/GT is only possible for the high-temperature FC technologies (such as SOFCs and MCFCs). The efficiency could reach 60%–80% in this mode. DFC/T technology will be evaluated more completely in the next topic.

Besides, by increasing the size of the system the efficiency rises. For instance, the efficiencies of a PEMFC with 10 kW and 1 MW capacities are 33% and just under 40%, respectively. From Table 5.2, it is obvious that the efficiency of FCs is greater than conventional DGs. Recently, by applying advanced technologies in the FCs, their efficiencies are increased while their costs are decreased significantly. Therefore, nowadays they are used in different applications more than ever before.

Fig. 5.19 shows the growing shipment of FC systems between 2008 and 2013. It can be concluded from this figure that stationary use of FCs has rocketed noticeably as they are proper for DGs. Now we have

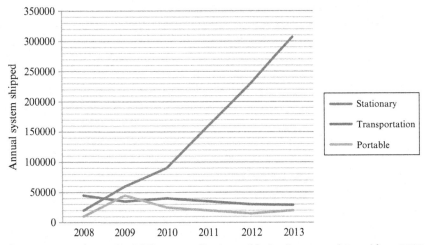

Fig. 5.19 Annual growth of different applications of fuel cell systems shipped from 2008 to 2013.

to know how hydrogen, which is the main fuel of FCs, can be produced. This study helps us to figure out how these energies can work together as hybrid systems.

5.4.4 Hydrogen Production Methods

Hydrogen is needed to generate electricity by FCs in many cases. Different methods are used to extract hydrogen from renewable and fossil energies, which are shown in Table 5.3.

The next sections discuss in some detail selected methods presented in Table 5.3.

5.4.4.1 Steam Methane Reforming Process for Producing Hydrogen from Natural Gas

Steam methane reforming (SMR) is the dominant technology used to produce hydrogen, as the highest proportion of extracting hydrogen is done by this method. Natural gas is very common in the industry and it is available in many places. Advanced infrastructure is already in place for extracting, transporting, and storage of natural gas. Furthermore, SMR can use other hydrocarbons such as gasoline and methanol in its process. Approximately 75% of the world's hydrogen production is produced by using methane and natural gas.

Table 5.3 Different production sources of hydrogen

Energy source	Energy type	Energy technology	Production method
Renewable	Power	Wind	Electrolysis
		Hydro	
		Waves	
		Sun	Photovoltaic
	Fuel	Biomass	Reforming
			Biological
	Power and fuel	Geothermal	Electrolysis
			Reforming
			Thermal
Fossil	Power	Nuclear	Electrolysis
			Nuclear thermo
	Fuel	Natural gas	Reforming
		Methanol	
		Oil	
		Coal	Gasification

If we apply natural gas, the efficiency of this technology will be about 70%. However, if we use some sources like methane, which contains sulfur or other impurities, the efficiency will be reduced because they need a pretreatment cleanup unit to purify the input fuel for the SMR process. Hydrogen-rich gas is the output of the SMR process, which it shown in Fig. 5.20.

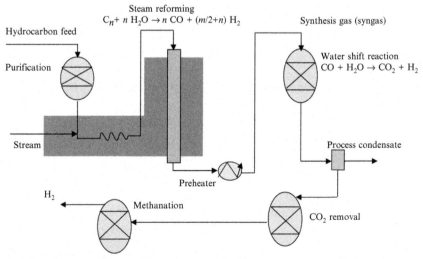

Fig. 5.20 Steam methane reforming process [10].

- First, hydrocarbons must be fed into a purification system to remove sulfur and other impurities from the fuel.
- Now, purified hydrocarbons react with steam at 850°C and 1.5–3 MPa to produce a syngas including CO, CO_2, and H_2O based on the following reaction:

$$C_nH_m + nH_2O \rightarrow nCO + (m/2 + n)H_2$$

So, if we use methane, the equation will be changed to

$$CH_4 + H_2O \rightarrow CO + 3H_2$$

- Then the synthesis gas is injected into the water shift reaction unit. The CO reacts with the water in this unit to produce more hydrogen:

$$CO + H_2O \rightarrow CO_2 + H_2$$

- Afterward, the process is continued by injecting the product into the CO_2 removal unit where liquid absorption system separates CO_2 from our product.
- Finally, to achieve higher purity hydrogen, the product is fed into a methanation process [10].

5.4.4.2 Gasification Process for Producing Hydrogen From Coal

Coal plants are widespread throughout the world, and 41% of all electric power is generated by these technologies. Pollution from coal plants is greater than from the other major methods of generating power and it is not very efficient. Therefore, it would be beneficial if this technology were replaced with a DG such as the FC.

Besides, hydrogen and other lucrative materials can be obtained by the gasification process. Then, we use the by-products in other industries as well as injecting hydrogen into FCs, especially in high-temperature FCs such as the MCFC and the SOFC. By means of this measure, clean, cost-effective energy with high efficiency will be available. The gasification process is illustrated in Fig. 5.21.

Initially, crushed coal is fed into the gasifier, where it is combined with steam at high pressure and temperature to convert it into gaseous components. The result of this reaction is

$$C_{(s)} + H_2O_{(g)} \rightarrow CO + H_2$$

A small, solid fraction like ash must be blown out by removal downstream. Also, there is some sulfur in coal, which reacts with hydrogen. The results of this reaction are H_2S and COS. Furthermore, nitrogen

Crushed coal

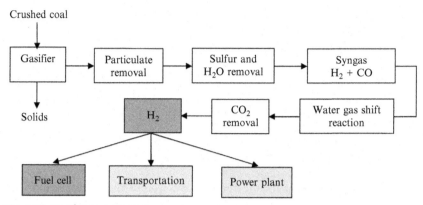

Fig. 5.21 Gasification process.

can react with hydrogen to produce ammonia. In the next step, all of these materials, which are valuable and can be used in other applications, must be removed from the product. The remaining product is called syngas.

Subsequently, just like the SMR process, synthesis gas should be fed into the water gas shift reaction unit where the CO reacts with water to produce more hydrogen.

$$CO + H_2O \rightarrow CO_2 + H_2$$

Then the process is completed by removing the CO_2 from the product. Finally, the rich hydrogen can be applied in power plants, transportation, and FCs. This process has a prominent role in producing hydrogen, as 18% of hydrogen products are extracted from coal.

5.4.4.3 Producing Hydrogen from Nuclear Energy
In contrast to the methods that use fossil fuels for producing hydrogen, nuclear power plants can provide hydrogen without emitting a large amount of GHGs. This technology is very reliable, whereas its safety is critical.

In this method, hydrogen can be produced by means of electrolysis or thermochemical process. For this action, the water temperature must be increased to about 700–1000°C. It is worth noting that if this method is applied to deliver the required heat for the SMR process, it will be more clean, efficient, and economical than before. Also, the need for natural gas in the hydrogen production process will be reduced [10].

5.4.4.4 Electrolysis Process for Producing Hydrogen from Water

The electrolysis process is defined as separating hydrogen and oxygen from water by electrical energy. Thus, in contrast to previous methods explained above, the electrolysis process needs electrical energy. Therefore, it would be reasonable to use this technology to produce hydrogen for the FC if there is extra power in some situations. For instance, when there is light demand, the extra power generated from power plants or DGs can be saved in the form of hydrogen. Then, when there will be a heavy load at peak times, the hydrogen can be injected to FCs so as to generate power electricity for compensating for energy shortages.

Because electrolysis is more expensive than gasification and SMR, this method normally is used for producing hydrogen by sustainable approaches such as wind, sun, geothermal, and hydro as demonstrated in Fig. 5.22.

5.4.4.5 Electrolysis Process for Producing Hydrogen by Solar Cell

Solar cells provide energy in two forms. First, they convert sunshine into electricity through a PV system. The second method is the heat generated by applying concentrating collectors [13]. However, the intermittent feature of this technology leads us to hybridize it with the FCs in two ways.

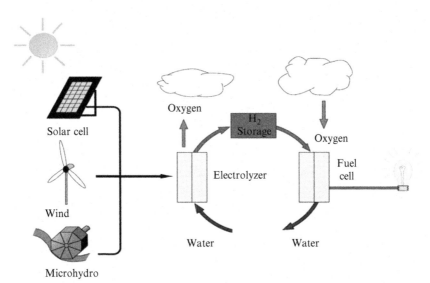

Fig. 5.22 Sustainable methods of producing hydrogen by electrolysis [2]. *From B. Cook, Introduction to fuel cells and hydrogen technology, Eng. Sci. Educ. J. 11 (6) (2002) 205–216. Reproduced by permission of the Institution of Engineering & Technology.*

Because solar panels cannot provide continuous power, FCs are used to provide base load power, whereas a PV provides additional energy, often coinciding with peak load. It is the conventional PV-FC system, which works as a hybrid system.

In real conditions, solar panels only generate power during the day while peak loads of electricity occur during the night, so it would be useful if the hydrogen needed were produced by the electrolysis process. In this method, when there is no need to provide a large amount of power, such as in the middle of the day, when electricity generation of solar cells is maximized, the solar cells deliver electricity to elctrolyzers to produce hydrogen. Afterward, when there is a need for power, the hydrogen is injected to the FCs.

5.4.4.6 Electrolysis Process for Producing Hydrogen by WTG

The efficiency of a WTG is noticeably higher than a PV, and it can generate power both day and night, although this technology is stochastic, too. Therefore, we can hybridize FCs with WTGs like PV-FC which was discussed earlier. The efficiency of a WTG with this action can increase considerably.

Like hydro energy, a WTG generates power first, after that this electricity is used for separating hydrogen and oxygen from water by the electrolysis process.

5.4.4.7 Other Methods of Generating Hydrogen, Especially by Hybridizing Renewable Energies

Geothermal, biomass, and hydro are other approaches to producing hydrogen. Also, it is suitable if we hybridize renewable energies such as the WTG, FC, and PV with each other. The reliability of power can be strengthened by this measure, because these energies complement each other. During the daytime, sunshine is available for the PV while during off-sunshine, the wind is usually accessible and vice versa, thus, helping to reach the goal of continuous power will be met precisely.

Another method is hybridizing the solar thermal and geothermal energies as hot water, which is formed from the geothermal process and can be heated again by the solar concentrating collectors to reach the desirable temperature. Then, enough hydrogen can be generated by high-temperature electrolyzers [13].

5.4.5 Hybridizing the FC With the Turbine

We have so far described how different resources can produce hydrogen for FCs. Now we want to evaluate how FCs can be hybridized with turbines. In other words, how FCs and turbines can complement each other. This technology is called direct fuel cell-turbine (DFC/T). The efficiency of DFC/T for 100 MW capacity reaches well over 75%. Natural gas and coal from conventional fossil resources and biogas as renewable resources can be fed into a reformer of FC/GT as a fuel. Biogas can be extracted from the solid digestion. We can hybridize FC with wastewater treatment facilities. It should be noted that if biogas is injected into the FC system, the efficiency will be higher than the same biogas system, while the pollution will plummet significantly. Also, CO, CH_4, H_2, or even carbonate and other hydrocarbons that are exhausted from natural gas and coal plants can be used in FC/GT technology with internal reforming. Therefore, the high temperature FCs (SOFC and MCFC) are suitable for this technology.

High-temperature FCs generate heat in addition to electric power. This generated heat must be released from the FC to set the temperature to the desired level. Then, the temperature is used for starting the gas turbine.

Fig. 5.23 shows a combination of a FC and gas turbine model that brings air into the FC with a compressor. After passing through the compressor the air is injected into the recuperative heat exchanger unit. The hot air and fuel are then fed to the cathode and anode of the FC, respectively. Output air and fuel of the FC are flowed into a combustor unit to be mixed with each other

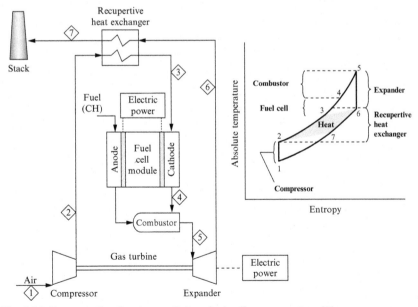

Fig. 5.23 Regenerative Brayton cycle in a fuel cell power system [4].

and burned. The output of the combustor is a flow with both high pressure and temperature. This flow goes into a turbine and after rotating the turbine blades the exhaust gases of the turbine are injected again to the heat exchanger. By this method, efficiency of the system improves because this system reuses the exhausts from the FC. Therefore, the total power, which was P_{FC}, is changed to $P_{FC} + P_{turbine}$.

In a sub-MW power application, the DFC300 is used, which consists of a single stack and has a 300 kW capacity. For generating power in the MW scale, the DFC1500 and DFC3000 are used, which consist of one four-stack and two four-stacks with capacities of 1.4 and 2.8 MW, respectively.

For example, in South Korea a 58.8 MW FC power plant has been constructed and works by applying 21 units of DFC3000. Covering 5 acres land, this power plant was constructed in a few months amd is the largest FC power plant in the world. The construction was very simple and repetitive because all 21 DFC3000 units there were identical. Therefore, this power plant did not need a complex design like conventional power plants. Each of the DFC3000 units has 2.8 MW capacity, so the total output of this power plant may be computed as follows:

$$2.8\,MW \times 21\,units = 58.8\,MW$$

In addition to electric power, for 70% of households required heat is provided by using DFCs as CHP units. Also, we can hybridize them with GTGs (gas turbine generators) in order to increase efficiency and capacity.

Example 5.5

Calculating the electrical capacity of a DFC when it changes to a DFC/T. Assume that the electrical efficiency of the abovementioned power plant is about 50%. Because our generation capacity is 58.8 MW, if this DFC power plant is changed to DFC/GT, based on DOE (Department of Energy) information, the efficiency must be increased to 75%. Therefore, the capacity of the new plant will be as given below:

Generation capacity of DFC with 50% efficiency = 58.8 MW

DFC/T efficiency is 75%

So, the generation capacity of DFC/T would be

$$\Rightarrow \frac{75}{50} \times 58.8\,MW = 88.2\,MW$$

Thus, high power can be achieved by precise and appropriate designing. However, the shaft speed control issues and working on the variable load conditions are complex matters that will be discussed in more detail in the next section for a SOFC/GT system.

5.4.6 Applying FCs as Backup Systems

FCs can provide backup power in the grid-connected mode when utility power is not available. They are usually designed like Fig. 5.24 to help critical loads such as hospitals, banks, and so on during interruptions or like Fig. 5.25, so they can supply a load in parallel with a utility.

In the configuration shown in Fig. 5.24, the utility is supplying both critical and noncritical loads, whereas FCs have been connected to the critical loads to help them when the utility is interrupted. Supercapacitors or even batteries are often connected to the FC units to provide the transient load and inrush current, as FCs cannot meet these loads. In steady state condition, the FCs just use some percentage of their capacities because they must be ready for substituting with the utility during the failure modes.

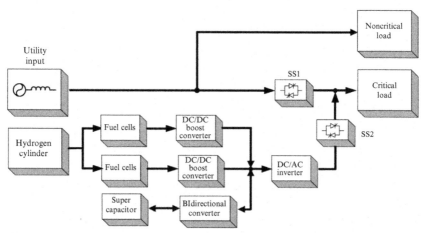

Fig. 5.24 Two fuel cell units for supplying backup power to a load connected to a local utility [4].

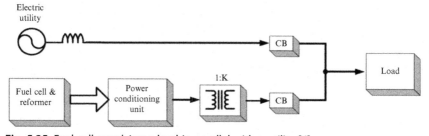

Fig. 5.25 Fuel cell supplying a load in parallel with a utility [4].

Fig. 5.25 illustrates another method by which FCs can help a utility: where the peak power and transient loads are provided by the utility and FCs are connected to the grid in parallel.

Integrating FCs to the power grid must be done precisely as we have to take many issues into consideration. For example, many conditions and rules such as synchronizing FCs with the utility must be met before connecting. All of these concerns and also power electronic devices like converters, inverters, and so on will be presented in Section 5.5.

5.4.6.1 The Biggest Blackouts in History

Considering reliable power sources as the backup for power plants and DGs that are susceptible to interruption are important because their failures will cause significant damage to the critical loads. In addition, these interruptions in some cases have cascading effects on grids and create blackouts. Backup systems must compensate critical shortages. Therefore, they play a prominent role in power systems.

Blackouts are caused by natural disasters, technical malfunctions, unintentional, and even intentional human errors. Table 5.4 shows some of the major blackouts in the world so as to demonstrate the constructing role of backup systems during the power outages:

5.4.6.2 Introducing FCs as the Best Choice for Backup Energy

The characteristics of FCs, which are explained in Section 5.2 convince us to consider them as valuable resources for backup systems. Now, the vital elements of a good backup system are described to make a comparison between FCs, batteries, and engine motors (EMs) as common sources in this field [14]:

- Range of power: Backup systems must be able to supply a wide range of power. Batteries are not so good in delivering high power while engine generators are not so appropriate for low loads. FCs can provide a wide range of power from 1 kW to 100 MW as described earlier.
- Capacity: Backups have to supply power for a long period of time. The battery discharges in a specific period of time, so it cannot provide power like a FC and engine generator, which can work as long as the fuel is injected into them.
- Safety: Although all of the three methods are dangerous to some extent, storage of a FC is more perilous than others and needs especial technology.

Table 5.4 The biggest blackouts in history

Place	Date(s)	People without power (in millions)	Reason	Restoration time
India	July 30–31, 2012	670	T[a]	2 days
India	January 2, 2001	226	T[a]	20 h
Bangladesh	November 1, 2014	150	T[a]	10 h
Pakistan	January 25, 2015	140	U[a]	20 h
Java and Bali, Indonesia	August 18, 2005	120	T[a]	24 h
Southern Brazil	March 11, 1999	97	N[a]	5 h
Turkey	March 31, 2015	70	T[a]	10 h
Brazil and Paraguay	November 10, 2009	67	N[a]	4 h
Italy	September 28, 2003	57	N[a]	18 h
Northeast United States and Canada	August 14–16, 2003	50	T[a]	2 days
Thailand	March 18, 1978	40	T[a]	9 h
Northeast United States and Northern Canada	November 9, 1965	30	T[a]	13 h
Germany, France, Italy, and Spain	November 4, 2006	15	U[a]	2 h
New York, United States	July 13–14, 1977	9	N[a]	24 h
Quebec, Canada	March 13, 1989	6	N[a]	12 h

[a]N, Natural disaster; T, Technical faults due to equipment and systems failures; U, Unintentional and intentional human mistakes or even terrorist actions.

- Efficiency: Based on the previous information, the greater efficiency belongs to FCs.
- Ability to respond to transient load: During start-up times or rapidly increasing demands, systems are faced with transient loads. A battery can respond to these conditions faster than an EM and a FC. However, the FCs can get help from batteries to compensate for their deficiency in rapid response.
- Starting time: Like responsiveness, batteries are faster than the other methods. Because FCs cannot compensate power in as short a time as conventional UPSs, in a steady state condition, the FCs use some

percentage of their capacities to be ready for substituting with the utility without delay.

- Reliability: The highest rate of reliability belongs to the FCs. They can work in many conditions and the lowest number of defects belongs to this type of backup system. In addition, like EM, the FC can provide required active and reactive power by a proper control system.
- Durability: High endurance is one of the main features of a good backup system. Batteries are the weakest, whereas FCs are the best choices that exist in this field.
- Cost: Although the initial cost of FCs is more than EMs and batteries, the cost of maintenance and replacing in FCs are much lower than the others. Therefore, they seem cost-effective. Also, based on Fig. 5.4, the initial cost of FCs decreases every year by applying cutting edge materials and technologies in manufacturing.
- Flexibility and scalability [14]: When there is not enough space and we need a small-scale and lightweight system, based on Fig. 5.11, the FCs are significantly smaller and lighter than batteries. It is obvious that in close spaces like buildings, EMs cannot be used.
- Outdoor and indoor usages: EMs are more convenient for use in outdoor places while batteries must be used indoors. But, FCs can be applied in both areas.
- Redundancy: The capacity of FCs and batteries can be increased simply and precisely by adding new numbers of them, whereas for EMs this is not easy and requires more money.
- Noise: EMs generate considerable noise while batteries and FCs do not have this problem.
- Pollution: EMs emit pollution so they cannot work in buildings and other enclosed spaces and have a negative effect on the environment by emitting GHGs. Because batteries and FCs work with an electrochemical process, it is obvious that they do not emit noticeable pollution.
- Maintenance: Because batteries need to charge and discharge, their lifetime is very short and need replacing; every 5–7 years they must be substituted by new batteries. EMs have moving parts, so they have a wear and tear problem and must be maintained at specific periods. Also, EMs are more commonly defective than the other methods, so they need more maintenance. FCs are the best choice, as they do not have moving parts, so they do not have wear and tear problems. Also, because they do not need to charge and discharge, in comparison with batteries, their lifetimes are much longer.

Therefore, it may be concluded from the abovementioned traits that FCs are more suitable for use as backup power systems than EMs and batteries. Also, some problems of FCs such as low response to transient loads can be solved by applying ultracapacitors or batteries with them.

5.4.7 Presenting an Example of a Grid-Connected FC to Demonstrate How it can Provide Desirable P and Q to the Power Network [7,15]

When a FC system injects power to the network as a DG in the grid-connected method, it must be able to manage supporting both real and reactive power. Now the FC system is applied using its voltage and the number of its series and parallel stacks that were calculated in Section 5.3. Based on Example 5.1, this system includes ten parallel 48 kW PEMFC units that generate 480 kW together. Each unit has 8 and 12 series and parallel 500 W stacks, respectively.

When the PEMFC DG system is connected to a power grid, a certain scheduled base load ($P_{Grid,sched}$), which is constant, is supplied by the utility grid and the rest of them, which are time varying, are provided by the FC system. X_{Grid}, X_{FC}, and X_{Load} are reactances of the utility, PEMFC, and load, respectively. Based on the load demand, the power grid must generate appropriate power. This amount of power is compared with the $P_{Grid,sched}$ and forms ΔP_{Grid}. Then, the load following the controller sets $\Delta P_{FC,ref}$ to make $P_{FC,ref}$ by adding the previous reference of the FC power ($P^0_{FC,ref}$) to it. Therefore, both the FC and grid supply load demands are as follows:

$$P_{Load} = P_{Grid} + P_{FC} \qquad (5.26)$$

P_{grid} was set to 100 kW (1 PU) to show how the FC can support the power grid when it needs more power [7].

This action shows that in the beginning, the load demand was 2 PU, so another 1 PU was delivered by the PEMFC. When the system needed 300 kW or 3 PU, because the FC could not respond rapidly to the load variation, first the utility supported 2 PU and after about 0.6 s, the FC injected 2 PU to the load while P_{Grid} decreased to its scheduled value, which was 1 PU. So, based on Eq. (5.26), 3 PU was delivered to the load. The reverse action took place in $t = 4.1$ s where the system at another time needed 200 kW, so first the grid reduced its generation and when the FC could reduce its value to 1 PU, P_{Grid} returned to its scheduled capacity [7].

Angle and amplitude of output voltage have a direct effect on the real and reactive power, respectively. Therefore, the reactive power can be controlled by the voltage amplitude ($|V|$) while the voltage angle (δ) controls the real power. Besides, the existence of the proper value of reactive power for preserving buses voltage at allowable margin and increasing the lines loading capabilities at these levels are obligatory.

First the load flow equations are introduced to calculate amplitude and angle of output voltage when there is a need for a certain P and Q in both heavy and light loads (For further study refer to [8].)

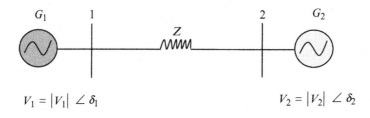

$$Z = |Z| \angle \theta_Z \Rightarrow |Z| \cos\theta_Z + j|Z| \sin\theta_Z \tag{5.27}$$

$$Z = R + jX \tag{5.28}$$

$$|Z| = \sqrt{R^2 + X^2} \tag{5.29}$$

$$\theta_Z = tg^{-1}\left(\frac{X}{R}\right) \tag{5.30}$$

Now P_{12} and Q_{12} are calculated as follows:

$$S_{12} = P_{12} + jQ_{12} \Rightarrow V_1 I_{12}{}^* \tag{5.31}$$

So, I_{12} is as

$$I_{12} = \frac{|V_1|\angle\delta_1 - |V_2|\angle\delta_2}{|Z|\angle\theta_Z} \tag{5.32}$$

By substituting Eq. (5.32) in Eq. (5.31),

$$S_{12} = \frac{|V_1|^2\angle\theta_Z}{|Z|} - \frac{|V_1||V_2|\angle\theta_Z + \delta_1 - \delta_2}{|Z|} \tag{5.33}$$

Therefore, the real and reactive power will be as follows:

$$P_{12} = \frac{|V_1|^2 \angle \theta_Z}{|Z|} - \frac{|V_1||V_2| \angle \theta_Z + \delta_1 - \delta_2}{|Z|} \qquad (5.34)$$

$$Q_{12} = \frac{|V_1|^2 \angle \theta_Z}{|Z|} - \frac{|V_1||V_2| \angle \theta_Z + \delta_1 - \delta_2}{|Z|} \qquad (5.35)$$

$$\text{if} \begin{cases} \delta_1 > \delta_2 \Rightarrow P_{12} > 0 \text{ generated real power,} \\ \qquad\quad \text{G1 is like a generator and G2 works as a motor} \\ \delta_1 < \delta_2 \Rightarrow P_{12} < 0 \text{ consumed real power,} \\ \qquad\quad \text{G1 is like a motor and G2 works as a generator} \end{cases} \qquad (5.36)$$

$$\text{If} \begin{cases} |V_1| > |V_2| \Rightarrow Q_{12} > 0 \quad \text{generated reactive power} \\ |V_1| < |V_2| \Rightarrow Q_{12} < 0 \quad \text{consumed reactive power} \end{cases} \qquad (5.37)$$

Fig. 5.26 shows the power flow between the PEMFC and the utility grid [7].

In this configuration, P and Q will be

$$P = \frac{|E||V_S|}{|Z|} \cos(\theta_z - \delta) - \frac{|E|^2}{|Z|} \cos(\theta_z)$$

$$Q = \frac{|E||V_S|}{|Z|} \sin(\theta_z - \delta) - \frac{|E|^2}{|Z|} \sin(\theta_z)$$

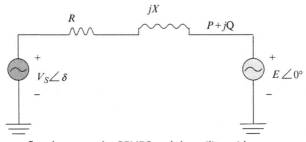

Fig. 5.26 Power flow between the PEMFC and the utility grid.

When the system needs a certain amount of real and reactive power, and the utility voltage is set before, the angle and amplitude of the FC output voltage are calculated as follows [8]:

$$|V_S| = \left[\frac{|Z^2|}{|E|^2} (P^2 + Q^2) + |E|^2 + 2P|Z|\cos(\theta_z) + 2Q|Z|\sin(\theta_z) \right]^{1/2}$$

(5.38)

$$\delta = \theta_z - \cos^{-1}\left(\frac{|Z|P}{|E||V_S|} + \frac{|E|}{|V_S|}\cos(\theta_z) \right)$$

(5.39)

Example 5.6

Calculating the angle and amplitude of the output voltage of the PEMFC, which is introduced in Example 5.1, under heavy loads to supply the desired real and reactive power to the utility grid.

Assume that the impedance between the PEMFC DG system and the utility in Fig. 5.26 is equal to $Z = 1 + j60$. Also, as can be seen from Fig. 5.13, the utility voltage is equal to 12.5 kV. Therefore, based on the Eqs. (5.29), (5.30), the amplitude and angle of the coupling and transmission line impedance will be

$$Z = 1 + j60$$

$$|Z| = \sqrt{R^2 + X^2} \Rightarrow \sqrt{1^2 + 60^2} \Rightarrow |Z| = 60.01$$

$$\theta_Z = tg^{-1}\left(\frac{X}{R}\right) \Rightarrow tg^{-1}\left(\frac{60}{1}\right) \Rightarrow \theta_Z = 89.045$$

Now the system response is evaluated to heavier loads. When the utility voltage is equal to 0.98 PU (12.25 kV), if we need 310 kW and 30 kVAr real and reactive power, respectively, the amplitude and angle of the PEMFC DG system output voltage would be calculated from Eqs. (5.38), (5.39):

$$E = 12,250\,V, \quad \delta_E = 0, \quad P = 310,000\,W, \quad Q = 30,000\,VAr$$

$$|V_{FC}| = \left[\frac{|Z^2|}{|E|^2}(P^2 + Q^2) + |E|^2 + 2P|Z|\cos(\theta_z) + 2Q|Z|\sin(\theta_z) \right]^{1/2}$$

$$\Rightarrow |V_{FC}| = \left[\frac{60.01^2}{12250^2}(400000^2 + 50000^2) + 12250^2 \right.$$

$$+ (2 \times 4000000 \times 60.01\cos(89.045))$$

$$\left. + (2 \times 50000 \times 60.01 \times \sin(89.045)) \right]^{1/2}$$

$$\Rightarrow |V_{FC}| = 12680 \text{ or } 1.014 \text{PU}$$

Based on the Eq. (5.37), as $|V_{FC}| > |E|$, so the PEMFC DG system injects 50 kVar reactive power to the grid.

The voltage angle can be obtained by Eq. (5.39):

$$\delta_{FC} = \theta_z - \cos^{-1}\left(\frac{|Z|P}{|E||V_{FC}|} + \frac{|E|}{|V_{FC}|}\cos(\theta_z)\right)$$

$$\Rightarrow \delta_{FC} = 89.045 - \cos^{-1}\left(\frac{60.01 \times 400000}{12250 \times 12680} + \frac{12250}{12680}\cos(89.045)\right)$$

$$\Rightarrow \delta_{FC} = 8.869°$$

Based on Eq. (5.37), as $\delta_S > \delta_E$, so the PEMFC DG system works as a generator and provides 400 kW real power to the utility.

It is worth noting that when the system is under light loading, Q can be negative, which means that the FC DG must consume reactive power while it provides real power to the power grid.

5.5 POWER ELECTRONIC INTERFACES, PROTECTION, AND POWER QUALITY CONCERNS

In the previous sections, the introduction of FCs and their varied applications were discussed. After that, the output voltage of the PEMFC and its voltage losses were evaluated to illustrate the electrical characteristic of FCs. Then, it was concluded that FCs are very suitable to use as DG or even a backup system in both the grid-connected and stand-alone methods. This section will discuss the power electronic interfaces and interconnection devices that must be used in FC systems so as to supply the proper power to end users and address their power quality and integration concerns.

5.5.1 Protection Concerns in Integrating FCs to the Grid

FCs as DG systems are integrated into the grid by some power electronic devices at the distributed or transmission lines based on their generation capacities. One of the main challenges to integration is the protection concerns that must be observed during integrating. There are some protection prerequisites such as fault clearing, reclosing, and avoiding unintentional islanding that must be done to ensure reliability during operation.

5.5.1.1 Change of Short Circuit Levels

Short circuit levels are defined by the equivalent system impedance at the fault points. Because systems are assumed to work without DG, by applying the FC DG system, equivalent network impedance can decrease, so fault levels will increase. Thus, in the fault condition, high fault currents can be exceeded the interrupting capacity of the circuit breakers, which were set before and as a consequence, circuit breakers cannot work properly. Current transformer (CT) saturation is another problem in this condition. The capacity of reclosers must be set as high as they can be to detect over/under currents [16].

Blindness of the protection phenomenon takes place when, as in Fig. 5.27, a fault occurs at a feeder that has one or more DGs.

In Fig. 5.27, the FC is placed between the fault point and the relay of the feeder. Therefore, the current that is detected by the relay is decreased due to the presence of the FC. The total of the network fault current will be surged, so the protection margin of the relay is restricted [17].

$$\text{Without FC} : I_s \approx I_r \approx I_f$$
$$\text{With FC} : I_r = I_s + I_{FC}$$

If the FC system is not interrupted immediately in this situation, which is shown in Fig. 5.27, the relay of the feeder cannot work appropriately and consequently the temperature of devices will increase while the power quality will decrease [17].

5.5.1.2 Reverse Power Flow

Usually a radial configuration is used in distributed systems, so the protection settings are designed based on unidirectional power flows. Therefore, the direction of the power flow can be reversed when a FC as a DG is integrated with the system. This phenomenon has a negative effect on the coordination

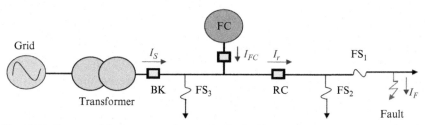

Fig. 5.27 Change of fault current due to the fuel cell interconnection [16]. *Used with permission from A. Rajapakse, D. Muthumuni, N. Perera, Grid integration of renewable energy systems, in: T.J. Hammons (Ed.), Renewable Energy, InTech, ISBN 978-953-7619-52-7, 2009. InTech © 2009.*

of protection relays [18]. When relays cannot work properly, hazardous events will happen for devices and also end users will receive low power quality.

5.5.1.3 False Trip of the Protection Relays

False trip of the protection relays happens when a fault occurs in the feeder adjacent to or near the substation, as shown in Fig. 5.28. The FC injects the fault current through the bus bar of the substation to the feeder at which the fault occurred.

$$I_r = I_{Utility} + I_{FC}$$

If this situation is not controlled properly, the current will be exceeded from the setting margin of the relay. Also, the relay will not detect the direction of the current, so it will send the wrong trip command [19]. Therefore, an unnecessary interruption will occur and reliability of the lines will plummet.

In Fig. 5.28, it is likely that the R_1 is interrupted incorrectly because of the upstream current fault.

5.5.1.4 Reclosing Concerns

Several types of relays are used in the distribution lines and also other protection devices such as breakers, reclosers, fuses, and so forth to protect the grid and increase the reliability. The most common faults at this level are transient ones, while just a few of them are permanent. Therefore, when a fault occurs, first reclosers interrupt the system for a very short time and then reconnect the system. In this condition, if the fault is removed, the recloser will let the system work. But, if the fault still exists, it will interrupt the

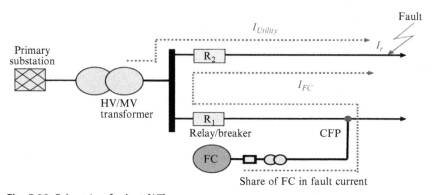

Fig. 5.28 False trip of relays [17].

system as long as the problem is not eliminated (the number of consecutive interruptions is adjustable). Assume that in the Fig. 5.28, a fault has taken place between R_1 and CFP (Connection point of the Feeder and DG). The relay interrupts the system and in this situation, if the FC is not interrupted from the system, we will encounter two problems. The first problem is that after solving the fault and reclosing, the relay detects the FC current as a permanent fault and interrupts the system again, incorrectly [17]. The second problem occurs because the FC supplies the system in an islanding mode, which is very dangerous for both the FC and the system, which is described in the following topic.

5.5.1.5 Unintentional Islanding

Unintentional islanding is defined as supplying the power grid by the DG independently when the DG should not inject power to the grid. In this situation there will be an undesired voltage and frequency in the island zone. In addition to the reduction of power quality, supplying the power system by two different frequencies is too dangerous (the system needs a check synchronizing relay to disconnect the DG). As mentioned in Section 5.2, islanding can be perilous for maintenance workers when they do not know that DG is still connected. Therefore, the independent DG must be disconnected immediately from the network.

Three main applied anti-islanding methods include the passive, active, and telecommunication methods. Under/over voltage, under/over frequency, rate of change of frequency (ROCOF), and voltage vector surge (VVS) relays are used for anti-islanding purposes [16].

On the other hand, by observing the standard rules, intentional islanding has considerable benefits. For instance, when a fault occurs and we want to supply a critical load, we can disconnect the DG thay has caused the fault from the system. Then this DG can inject power to the loads that are important for us as an islanded zone.

5.5.2 Power Electronic Interfaces

Fig. 5.29 illustrates the main power electronic interfaces which are usually applied in a FC system.

Initially, the reason for using the block diagram in Fig. 5.29 in FC systems is introduced concisely and then each one of its uses in the FC is described in more detail.

FCs, as in solar cell technology, generate DC voltage. It is a low DC voltage, so a step-up converter is applied to increase the DC voltage. On

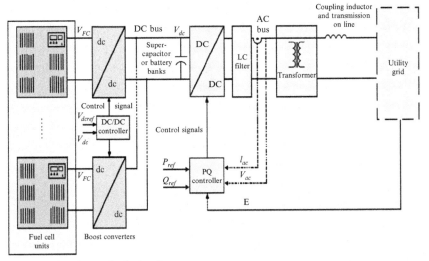

Fig. 5.29 Topology of a fuel cell system.

the high DC voltage bus, we usually get help from one or more ultracapacitors or battery banks (mentioned earlier) to stabilize the DC voltage and also to supply the transient load like starting current. Then, the high DC voltage goes through an inverter to present the AC voltage. As the inverter injects harmonics to the system, a precise filter should be designed, which is usually a low-pass filter, to eliminate harmful waves. Now a sinusoidal AC voltage and current must go through to a transformer to deliver desirable power to the grid or even to consumers.

It is worth noting that when a FC system is integrated into the grid, the amplitude, frequency, and phase of the output voltage of the FC and the grid must be exactly the same. A transformer can meet the desirable amplitude of the voltage where frequency has been matched with the grid by the inverter. In this situation, after the transformer, checking the synchronizing relay should be used to detect any fault.

5.5.2.1 Boost Converter

The output voltage of the FC is DC with a low magnitude, so the best way to step up the DC voltage is applying a boost converter. The circuit model of a boost converter is shown in Fig. 5.30.

Input voltage (V_{in}) is in series with the inductor (L), therefore it works as a current source. When the switch (S) is closed (on), the diode is reverse-biased,

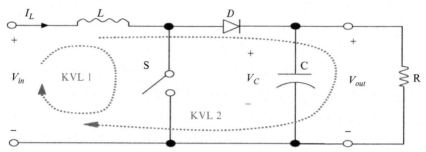

Fig. 5.30 Circuit model of a DC-DC boost converter.

so the inductor stores energy. In this situation, it is calculated that the voltage of the inductor from KVL 1 is

$$-V_{in} + V_L = 0 \Rightarrow V_L = V_{in}$$

In position two, when the switch is opened (off), the inductor is discharged and the diode conducts the current to the load. Therefore, both of source (V_{in}) and the inductor, which saved energy in the previous position, supply the load and as a consequence the output voltage will be higher than the input voltage. The voltage of the inductor in position two is obtained from the KVL 2 as

$$-V_{in} + V_L + V_{out} = 0 \Rightarrow V_L = V_{in} - V_{out}$$

The average output voltage level is varied by adjusting the time that the switch is closed and opened or the duty ratio [13]. The converter has some losses and its efficiency is between 80% and 98% based on different models. Therefore, in reality $P_{in} < P_{out}$.

Assuming an ideal converter without losses yields

$$P_{in} = P_{out} \Rightarrow V_{in}I_{in} = V_{out}I_{out} \tag{5.40}$$

So, in the ideal boost converter,

$$\frac{V_{in}}{V_{out}} = \frac{I_{out}}{I_{in}} \tag{5.41}$$

To calculate the duty cycle, the time integral of the inductor voltage over a period of time must be equal to zero (assume this converter works in the continuous conducting mode).

$$V_{in}t_{on} + (V_{in} - V_{out})t_{off} \tag{5.42}$$

$$\frac{V_{out}}{V_{in}} = \frac{I_{in}}{I_{out}} = \frac{T_s}{t_{off}} = \frac{1}{1-D} \qquad (5.43)$$

where T_s is the switching period. The duty cycle is <1. So, from the above equation it is obvious that V_{out} in the DC-DC boost converter is more than V_{in}.

The peak-to-peak output voltage ripple of the converter is equal to

$$\Delta V_{out} = \frac{V_{out} D T_S}{RC} \qquad (5.44)$$

Thus, the percentage of the output voltage ripple of the DC-DC converter based on the RC time constant (τ) is formulated as

$$\frac{\Delta V_{out}}{V_{out}} = \frac{D T_S}{RC} = D \frac{T_S}{\tau} \qquad (5.45)$$

The inductor and capacitor shown in the Fig. 5.30 reduce the ripple in the current and voltage, respectively. From Eq. (5.45), it can be concluded that the voltage ripple will be reduced by choosing a large capacitor or increasing the constant time (τ). As the input voltage of the inverter must be constant to get a better result, the capacitor keeps the converter output voltage constant.

Example 5.7

Calculating converter output voltage. Assume the input voltage of the converter at the low DC voltage bus, which is shown in Fig. 5.29, is 200 V. The duty cycle in this example is set to 0.6. The output voltage of the converter can be calculated from Eq. (5.43) as follows:

$$D = 0.6, \quad V_{in} = 200\,V$$

$$\frac{V_{out}}{V_{in}} = \frac{1}{1-D} \Rightarrow V_{out} = \frac{V_{in}}{1-D} \Rightarrow \frac{200}{1-0.6}$$

$$\Rightarrow V_{out} = 500\,V$$

5.5.2.2 UltraCapacitor and Battery

As mentioned in the previous sections, the FC cannot respond to transient loads like the starting current properly. Therefore, when a system encounters a load current transient, the battery or ultracapacitor supplies the difference between the steady state FC power and the transient load power, then the FC power is changed following a safe profile to restore the battery charge and supply the new steady state load power [20].

In Fig. 5.29 after the DC-DC boost converter an ultracapacitor unit has been connected in parallel with the system. A supercapacitor is an energy storage device that acts like a battery. The ultracapacitor works in a system that needs high power in a short time. Thus, it can be modeled by a capacitor in series with a resistor as shown in Fig. 5.31.

To reach the higher voltage, several ultracapacitors or batteries must be connected in series together to make a unit. Then, for example in the Fig. 5.29, this unit is connected in parallel with the high DC voltage bus. The Thevenin equivalent of a battery small signal model and its equations are presented in Fig. 5.32 [20]:

Because batteries are connected in series together, their voltages and resistances are added together. Each battery has specifications that are supplied by its manufacturer. If the voltage and resistance of each battery cell are considered equal to V_{BC} and R_{BC}, respectively, the following equations calculate the small signal parameters:

$$V_B = N_S \times V_{BC} \tag{5.46}$$

$$R_B = N_S \times R_{BC} \tag{5.47}$$

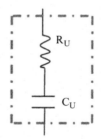

Fig. 5.31 Circuit model of an ultracapacitor unit.

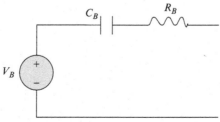

Fig. 5.32 Small signal model of a battery.

where N_S is the number of battery cells connected in series together. To obtain the capacitance in the small signal model, the maximum voltage must be used. Battery capacitance is written based on amperes per hour.

$$V_{\max, B} = \sqrt{2}\, V_B \tag{5.48}$$

$$V_C = V_{\max, B} - V_B \tag{5.49}$$

$$C_B = \frac{3600\, C_{AH}}{V_C} \tag{5.50}$$

Battery capacity (AH) is defined as a product of the current that is drawn from the battery while the battery is able to supply the load until its voltage is dropped to lower than a certain value for each cell. For instance, if a battery injects 6 amp for 10 h before its voltage drops to its critical margin, then its nominal capacity is 60 AH.

Example 5.8

Calculating the number of required batteries in a system. Assume in Fig. 5.29 that the system needs a battery bank with 96 kW power for helping the FC during transient loads. The input voltage of the inverter is 48 V. The specifications of the battery cell are assumed as

$$V_{BC} = 2V, \; R_{BC} = 0.04\Omega, \; I_{BC} = 2000A$$

The input voltage of the inverter is 48 V while the manufacturer mentioned that the voltage of each cell is 2 V. So, these cells should be in series to reach 48 V for the battery. From the Eq. (5.46),

$$V_B = N_S \times V_{BC} \Rightarrow N_S = \frac{V_B}{V_{BC}} \Rightarrow \frac{48}{2}$$

$\Rightarrow N_S = 24$, therefore 24 battery cells must be connected in series. Now the number of parallel cells can be achieved as

$$P = N_S \times N_P \times V_{BC} \times I_{BC} \Rightarrow N_P = \frac{P}{V_{BC} \times I_{BC} \times N_S} \Rightarrow \frac{96000}{2 \times 2000 \times 24}$$

$\Rightarrow N_S = 1$, so the number of all cells is equal to 24.

$$V_{\max, B} = \sqrt{2}\, V_B \Rightarrow \sqrt{2} \times 48 = 68V$$
$$R_B = N_S \times R_{BC} \Rightarrow 24 \times 0.04 = 0.96\Omega$$

5.5.2.3 Inverter

A FC generates DC voltage, so the voltage must be changed to AC except when it is used for DC applications such as in telecommunication stations, which need a DC voltage except in the lightening. The inverter converts the DC to AC power. Based on the application, importance, and cost of a project, there are different topologies for the inverter. Due to this fact, any of the following can be applied in the inverter: a bipolar junction transistor (BJT), a metal oxide semiconductor field effect transistor (MOSFET), a static induction transistor (SIT), a MOS-controlled thyristor (MCT), an insulated gate bipolar transistor (IGBT), or a gate turn-off thyristor (GTO).

Pulse-Width Modulated Voltage Source Inverter

A voltage source inverter (VSI), which is illustrated Fig. 5.33 and current source inverters (CSIs) with different waveforms such as a square wave, pulse width modulation (PWM), and so on are used in the industry. But, in the FC system usually a voltage-sourced PWM inverter is applied because this type can control both the frequency and amplitude of the AC output voltage. In inverters that are not PWM, any change in the output load directly affects the output voltage (when the load increases, the output voltage of the inverter decreases and vice versa) while output voltage of a PWM inverter can remain constant under a wide range of loads. For further information about other kinds of inverters, their topologies, and equations, please refer to [13], which is helpful in this regard.

As mentioned before, both frequency and output voltage magnitude can be adjusted to desirable ranges by applying the PWM technique. The modulation index is obtained as

$$V_{AC} = m V_{DC} \angle \delta \tag{5.51}$$

where m and δ are defined as the amplitude modulation index of the inverter and the firing angle of the inverter switches with respect to the angle of load bus voltage, respectively. The circuit model of a three-phase six-switch PWM VSI is shown in Fig. 5.34.

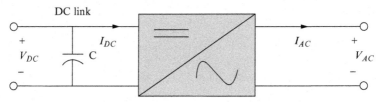

Fig. 5.33 Voltage source inverter.

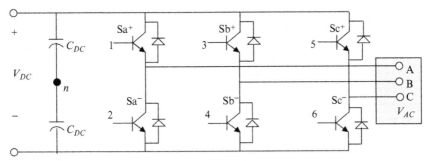

Fig. 5.34 Circuit model of a three-phase six-switch PWM VSI.

When the FC system is under a heavy load as in Example 5.6, if the hydrogen flow rate is constant, the FC output voltage will be decreased. But, based on Eq. (5.51), the amplitude of the output voltage can be regulated by adjusting modulation index. Also, as evaluated in the Example 5.6, drawing or injecting the desirable reactive power from the FC DG can be met by adjusting the output voltage amplitude.

Another parameter in Example 5.6 which is set for injecting enough real power to the grid, is the phase angle (δ) of the FC voltage with respect to the load voltage [21]. This action is done by adjusting the firing angle of the inverter. Therefore, the output voltage, real and reactive power, frequency, and phase angle of the system can be adjusted by the parameters of the inverter. Also, when a FC DG is connected to the grid, the amplitude (after applying a transformer), frequency, and phase angle of its voltage and the grid must be exactly equal, and these prerequisites are met simply by the inverter.

On the other hand, inverters inject harmonics to the system. So, it is vital to calculate these harmonics in the FC system.

Harmonics
Harmonics are defined as the sinusoidal voltages and currents whose frequencies are integral multiples of the power system's fundamental frequencies. The power quality and performance of the system are deteriorated by the harmonics. For example, overheating is one of the main harmful effects of them as the number of waves is increased by adding harmonics to the fundamental wave. Creating resonance with capacitor banks and filters is another disaster that takes place in the presence of harmonics. In this condition, the magnitudes of currents and voltages grow significantly [22].

Harmonic waves in the FCs are created by the switching circuits such as the inverters as well as harmonic current produced by the nonlinear loads connected to the system. Any distorted sinusoidal waveform can be represented by the Fourier series as follows:

$$f_{(t)} = \sum_{n=1}^{\infty} F_{nm} \sin(nw_o t + \theta_n) \qquad (5.52)$$

where F_m is the maximum of the F and is calculated as $F_m = \sqrt{2}F_{rms}$. Therefore, Eq. (5.52) is equal to

$$f_{(t)} = \sum_{n=1}^{\infty} \sqrt{2}F_{n\,rms} \sin(nw_o t + \theta_n) \qquad (5.53)$$

So, the distorted sinusoidal voltage and current are calculated as below:

$$v_{(t)} = \sum_{h=1}^{\infty} \sqrt{2}V_h \sin(hw_o t + \theta_h) \qquad (5.54)$$

$$i_{(t)} = \sum_{h=1}^{\infty} \sqrt{2}I_h \sin(hw_o t + \theta_h) \qquad (5.55)$$

where V_h and I_h are defined as the root mean square (RMS) values for the hth-order harmonic voltage and current, respectively [21]. Based on the Fourier series, it is obvious that each periodical wave for which its positive and negative parts are symmetrical, does not have even harmonics. Also, the frequency of the hth harmonic based on the fundamental harmonic is equal to

$$\omega_h = h \times \omega_0 \qquad (5.56)$$

Harmonics are combined with the main component of the voltage or current and causes the distortion waveform. Total harmonic distortion (*THD*) of voltage and current can be obtained as

$$THD_V = \frac{\sqrt{\sum_{h=2}^{\infty} V_h^2}}{V_1} \qquad (5.57)$$

In terms of acceptable power quality, the percentage of THD_V must be <5%.

$$THD_I = \frac{\sqrt{\sum_{h=2}^{\infty} I_h^2}}{I_1} \qquad (5.58)$$

Now the RMS of voltage and current are calculated based on their *THD* values:

$$V_{rms} = \sqrt{\sum_{h=1}^{\infty} V_h^2} \Rightarrow \sqrt{V_1^2 + \sum_{h=2}^{\infty} V_h^2} \Rightarrow V_1 \sqrt{1 + \frac{\sum_{h=2}^{\infty} V_h^2}{V_1^2}}$$

By substituting Eq. (5.57) in the above equation,

$$V_{rms} = V_1 \sqrt{1 + THD_V^2} \tag{5.59}$$

The RMS of current can be obtained like the voltage

$$I_{rms} = I_1 \sqrt{1 + THD_I^2} \tag{5.60}$$

Note that the harmonic problem is evaluated only in the steady state not in the transient mode.

Example 5.9

Calculating the RMS of voltage and current based on their *THD* values. Assume a capacitor bank with the nominal values below (capacitor banks are common in the FC DG systems):

$$Q_n = 1200\,KVAr\,(three\,phase), \quad V_n = 13.8\,KV(L-L), \quad X_c = 158.7\,K\Omega,$$
$$f = 60\,HZ$$

A voltage with fifth and seventh harmonics is injected with the following magnitudes. The case study has been taken from [23]:

$$V_1 = 100\%, \quad V_5 = 4\%, \quad V_7 = 3\%$$

The equivalent circuit of one phase of the capacitor is $V(t)$.

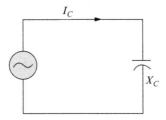

$$V(t) = V_1(t) + V_5(t) + V_7(t)$$

The question presented three-phase nominal voltage (L-L). To calculate the single-phase maximum voltage values, the voltage values should be multiplied by the square root of two, which is done for obtaining the maximum value, over the square root of three, which must be done for converting to the single phase:

$$V_{1\max} = 100\% \Rightarrow \frac{13800}{\sqrt{3}} \times \sqrt{2} \Rightarrow V_{1\max} = 11268\,\text{V}$$

$$V_{5\max} = 4\% \Rightarrow \frac{4}{100} \times 11268 \Rightarrow V_{5\max} = 451\,\text{V}$$

$$V_{7\max} = 3\% \Rightarrow \frac{3}{100} \times 11268 \Rightarrow V_{7\max} = 338\,\text{V}$$

Nominal current of the capacitor is calculated by the following equation:

$$I_{nC} = \frac{K\,var_{3ph}}{\sqrt{3}KV_{LLC}} \Rightarrow \frac{1200}{\sqrt{3} \times 13.8} \Rightarrow I_{nC} = 50.2\,\text{A}$$

The current harmonics can be formulated as

$$I_{C\,rms} = \frac{V_{rms}}{X_C} \Rightarrow I_C = jC\omega \times V_{rms}$$

So:

$$I_{C1} = jC\omega_1 \times V_1 = 100\%\,I_n = 50.2\,A$$

$$I_{C5} = jC\omega_5 \times V_5 \Rightarrow 5jC\omega_1 \times \frac{4}{100} V_1 = 20\%\,I_{C1} \Rightarrow I_{C5} = 20\%\,I_n = 10.05\,\text{A}$$

$$I_{C7} = jC\omega_7 \times V_7 \Rightarrow 7jC\omega_7 \times \frac{3}{100} V_1 = 21\%\,I_{C1} \Rightarrow I_{C7} = 21\%\,I_n = 10.54\,\text{A}$$

Percent THD_V and percent THD_i can be obtained using Eqs. (5.57), (5.58), respectively:

$$\%THD_V = \frac{\sqrt{\sum_{h=2}^{\infty} V_h^2}}{V_1} \times 100\% \Rightarrow \frac{\sqrt{\left(\frac{4}{100}\right)^2 + \left(\frac{3}{100}\right)^2}}{1}$$
$$\times 100\% \Rightarrow \%THD_V = 5\%$$

$$\%THD_I = \frac{\sqrt{\sum_{h=2}^{\infty} I_h^2}}{I_1} \times 100\% \Rightarrow \frac{\sqrt{\left(\frac{20}{100}\right)^2 + \left(\frac{21}{100}\right)^2}}{1}$$
$$\times 100\% \Rightarrow \%THD_I = 29\%$$

Finally, percent V_{rms} and percent I_{rms} based on their THD values are obtained by Eqs. (5.59), (5.60), respectively:

$$\%V_{rms} = \%V_1 \sqrt{1 + THD_V{}^2} \times 100\% \Rightarrow 1\sqrt{1 + \left(\frac{5}{100}\right)^2}$$

$$\times 100\% \Rightarrow \%V_{rms} = 100.1\%$$

$$\%I_{rms} = \%I_1 \sqrt{1 + THD_I{}^2} \times 100\% \Rightarrow 1\sqrt{1 + \left(\frac{29}{100}\right)^2}$$

$$\times 100\% \Rightarrow \%V_{rms} = 104.1\%$$

The values of harmonics in this question are acceptable. If they were not in the desirable ranges, one of the following solutions should be chosen: changing the size of capacitor, moving the location of the capacitor, eliminating the capacitor (which is not suggested in many cases), or putting the inductor in series with the capacitor bank, which will be discussed in the next topic.

Fig. 5.35 illustrates the harmonic sequences in the balanced condition.

Fig. 5.35A shows the phasors of the positive sequence harmonics, which include $h=1$, 7, 13, and so on. Fig. 5.35B presents the phasors of the positive sequence harmonics, which include $h=5$, 11, 17, and so on. In the zero sequence, as shown in Fig. 5.35C, there is no difference between the angles of phases, so their waveforms are identical. We call them triplen harmonics, which include $h=3$, 9, 12, and so on.

It is worth noting that the transformer connection groups have different effects on the harmonics. If the delta-connected is used in the primary windings of the transformer, the zero sequence (triplen) harmonics will not travel to the secondary side and get trapped in the delta loop. Therefore, the rest of the system will not have the triplen harmonics problems. That is why usually in the FC DG system, the delta configuration is used in the primary windings of the transformer.

Triplen harmonics are disastrous and they rank as the worse harmonics in the FC system. So, it is very important to eliminate or at least mitigate

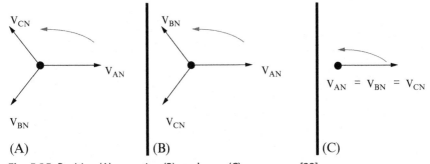

Fig. 5.35 Positive (A), negative (B), and zero (C) sequences [22].

them. Apart from the delta configuration, three-phase power electronic equipment such as three-phase converters and inverters do not generate triplen harmonics.

There are three main approaches to mitigate the harmonics [23]:
- Reducing harmonic current produced by the load such as nonlinear loads.
- Changing the system frequency response by filters, capacitors, and inductors.
- Applying filters.

5.5.2.4 Filters

As seen in the previous topic, nonlinear loads as well as converters and inverters inject harmonics and cause nonsinusoidal waveforms. In the FC system, both a converter and an inverter are used, so it is necessary to design an output filter, which is usually set after the inverter, to achieve better power quality.

Filters are categorized into two main types: passive and active filters. Passive filters consist of resistors, inductors, and capacitors, which are categorized into parallel and series filters. On the other hand, power electronic equipment such as inverters are applied in the active filters. Each model of passive and active filter has some advantages and disadvantages. Please refer to [22] and [13] for comprehensive information about them. Here, we just want to describe a low-pass filter concisely as it is the predominate technology in the FC system. The basic configuration of the low-pass filter is shown in Fig. 5.36.

L_f and C_f are the filter inductance and capacitance, respectively. The PWM VSI output of a FC is injected into a low-pass filter to provide a sinusoidal output voltage by eliminating certain harmonics. In real conditions, the inductor (L_f) has a low amount of resistance, so a small series resistor can be added to it. By changing the source frequency, in a certain value,

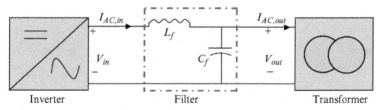

Fig. 5.36 Configuration of a LC filter.

the impedances of the inductor and capacitor are equal. This value is called the resonance frequency.

$$X_C = \frac{1}{\omega C} \Rightarrow \frac{1}{2\pi \times f \times C} \tag{5.61}$$

$$X_L = \omega L \Rightarrow 2\pi \times f \times L \tag{5.62}$$

$$X_L = X_C \Rightarrow \omega L = \frac{1}{\omega C} \Rightarrow 2\pi \times f \times L = \frac{1}{2\pi \times f \times C}$$

Therefore, the resonance frequency is

$$f_r = \frac{1}{2\pi \sqrt{L_f \times C_f}} \tag{5.63}$$

In the resonance frequency, the equal impedance is minimized while the current is maximized. This phenomenon causes a lot of damage.

Example 5.10

Calculating the current and equivalent impedance in different conditions. We want to calculate the current and equivalent impedance of the following circuit in the steady state and resonance mode with the following values:

$L = 10$ mH, $C = 60$ μF, $V_m = 170$ V, $f = 60$ Hz, $R = 0.1$ Ω

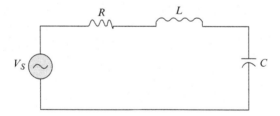

The RMS value of the source voltage is

$$V_{rms} = \frac{V_m}{\sqrt{2}} \Rightarrow \frac{170}{\sqrt{2}} \Rightarrow V_{rms} \approx 120\,\text{V}$$

Steady state mode:

The capacitor and inductor impedances are calculated from Eqs. (5.61), (5.62), respectively:

$$X_C = \frac{1}{\omega C} \Rightarrow \frac{1}{2\pi \times f \times C} \Rightarrow X_C = \frac{1}{2\pi \times 60 \times 60 \times 10^{-6}} \Rightarrow X_C = 44.2\Omega$$

$$X_L = \omega L \Rightarrow 2\pi \times f \times L \Rightarrow X_L = 2\pi \times 60 \times 10 \times 10^{-3} \Rightarrow X_L = 3.8\Omega$$

$$Z_{eq} = R + j(X_L - X_C) \Rightarrow \sqrt{R^2 + (X_L - X_C)^2} \Rightarrow \sqrt{0.1^2 + (3.8 - 44.2)^2}$$
$$\Rightarrow Z_{eq} \approx 40.4\Omega$$

$$I = \frac{V_{rms}}{Z_{eq}} \Rightarrow \frac{120}{40.4} \Rightarrow I \approx 3\,\text{A}$$

Resonance mode:

The resonance frequency is obtained from the Eq. (5.63) as

$$f_r = \frac{1}{2\pi\sqrt{L \times C}} \Rightarrow \frac{1}{2\pi\sqrt{10 \times 10^{-3} \times 60 \times 10^{-6}}} \Rightarrow f_r = 205\,\text{Hz}$$

When the frequency is equal to 205 Hz, the resonance mode takes place because in this frequency X_C is equal to X_L:

$$X_C = \frac{1}{\omega_r C} \Rightarrow \frac{1}{2\pi \times f_r \times C} \Rightarrow X_C = \frac{1}{2\pi \times 205 \times 60 \times 10^{-6}} \Rightarrow X_C \approx 13\Omega$$

$$X_L = \omega_r L \Rightarrow 2\pi \times f_r \times L \Rightarrow X_L = 2\pi \times 205 \times 10 \times 10^{-3} \Rightarrow X_L \approx 13\Omega$$

The new equivalent impedance will be

$$Z_{eq} = R + j(X_L - X_C) \Rightarrow \sqrt{R^2 + (X_L - X_C)^2} \Rightarrow \sqrt{0.1^2 + (13 - 13)^2}$$
$$\Rightarrow Z_{eq} = R = 0.1\Omega$$

The above equation demonstrates that the equivalent impedance is at the lowest value in the resonance mode. So, the current is maximized based on the following equation:

$$I = \frac{V_{rms}}{Z_{eq}} \Rightarrow \frac{120}{0.1} \Rightarrow I = 1200\,\text{A}$$

Now we need to describe why the LC filter shown in Fig. 5.36 is a low-pass filter. As can be seen, when the impedance of the inductor branch is lower than the capacitor branch, the current will go to the output of the filter, which is the input of the transformer. On the other hand, when the impedance of the capacitor branch is lower than the inductor branch, the current cannot go to the output of the filter because the capacitor branch acts as a short circuit.

Based on Eqs. (5.61), (5.62), the frequency has a direct and inverse relationship with the inductor and capacitor, respectively. Therefore, in the low frequency, X_C is high, whereas X_L has a small value. For instance, if the frequency is equal to zero,

$$X_C = \frac{1}{2\pi \times f \times C} \Rightarrow \frac{1}{2\pi \times 0 \times C} = \infty \quad X_L = 2\pi \times f \times L \Rightarrow 2\pi \times 0 \times L = 0$$

In this situation the current goes to the transformer as the high impedance of the capacitor. By increasing the frequency, X_C reduces while X_L increases. Thus, in the certain frequency (cutoff frequency) X_C and X_L will become equal. Up to the cutoff frequency, $X_C > X_L$ and the output current of the inverter goes to the transformer, but after this frequency, $X_C < X_L$ and the current is filtered and must return to the source instead of the transformer. As this LC filter lets low frequencies up to the cutoff frequency go to the output, whereas filters frequencies which are higher than the cutoff frequency do not, we call it a low-pass filter.

In the FC system, which must be connected to the grid, the filtered waveform is fed to the delta-wye step up transformer. Therefore, the transformer adjusts the voltage amplitude with the utility. Also, this transformer can correct power quality by eliminating the triple harmonics as well as provide isolation between loads and the FC. As mentioned earlier, the amplitude, phase, and frequency of the FC system voltage and the utility grid must be exactly equal before integrating the FC and the grid. By controlling the inverter parameters and the flow rates of the fuel and oxygen, we can reach the desirable values. By and large, the common topology of the grid-connected FC system can be drawn as in Fig. 5.37.

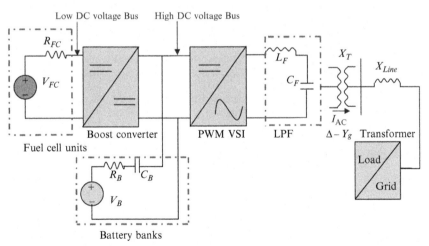

Fig. 5.37 Common topology of a grid-connected fuel cell.

REFERENCES

[1] C. Spiegel, PEM Fuel Cell Modeling and Simulation Using MATLAB, Academic press, USA, 2011.

[2] B. Cook, Introduction to fuel cells and hydrogen technology, Eng. Sci. Educ. J. 11 (6) (2002) 205–216.

[3] Available from: https://www.hydrogen.energy.gov/pdfs/12020_fuel_cell_system_cost_2012.pdf (Source: DOE).

[4] EG&G, Fuel Cell Handbook, seventh ed., EG&G technical services Inc., Albuquerque, 2004. National Energy Technology Laboratory.

[5] Smithsonian Institution. A Basic Overview of Fuel Cell Technology. Available from: http://fuelcells.si.edu/basics.htm.

[6] Available from: http://batteryuniversity.com/learn/article/will_the_fuel_cell_have_a_second_life.

[7] M.H. Nehrir, C. Wang, S.R. Shaw, Fuel cells: promising devices for distributed generation, IEEE Power Energ. Mag. 4 (1) (2006) 47–53.

[8] G.N. Sorebo, M.C. Echols, Smart Grid Security: An End-to-End View of Security in the New Electrical Grid, CRC Press, Boca Raton, FL, USA, ISBN 978-1-4398-5587-4, 2011.

[9] R. O'Hayre, Suk-Won Cha, Whitney Colella, Fritz B. Prinz, in: Fuel Cell Fundamentals, third ed., John Wiley and Sons, New York, 2008.

[10] M.H. Nehrir, C. Wang, Modeling and Control of Fuel Cells: Distributed Generation Applications, Vol. 41, John Wiley & Sons, New Jersey, USA, 2009.

[11] H. Fadali, Fuel Cell Distributed Generation, Power Conditioning, Control and Energy management, (2008). Electrical and Computer Engineering, Waterloo, Ontario, Canada.

[12] V. Tsourapas, J. Sun, A. Stefanopoulou, Incremental step reference governor for load conditioning of hybrid fuel cell and gas turbine power plants, IEEE Trans. Control Syst. Technol. 17 (4) (2009) 756–767.

[13] M.H. Rashid, Power Electronics Handbook: Devices, Circuits and Applications, Academic press, MA, USA, 2010.

[14] L. Birek, S. Molitorys, Hydrogen Fuel Cell Emergency Power System: Installation and Performance of Plug Power GenCore 5B48 Unit, 2010 MA, USA, 2010.

[15] C. Wang, M.H. Nehrir, H. Gao, Control of PEM fuel cell distributed generation systems, IEEE Trans. Energy Conver. 21 (2) (2006) 586–595.

[16] A. Rajapakse, D. Muthumuni, N. Perera, Grid integration of renewable energy systems, in: T.J. Hammons (Ed.), Renewable Energy, InTech, ISBN 978-953-7619-52-7, 2009.

[17] Available from: http://www.civilica.com/Paper-PSPC04-PSPC04_035.html.

[18] A. Von Jouanne, B. Banerjee, Assessment of voltage unbalance, IEEE Trans. Power Deliv. 16 (4) (2001) 782–790.

[19] J. Lopes, Integration of dispersed generation on distribution networks-impact studies, in: Power Engineering Society Winter Meeting, 2002. IEEE. Vol. 1. IEEE, 2002.

[20] V. Álvarez, et al., Design of a low power system based on fuel cells, Revista EIA (17) (2012) 85–103.

[21] M. Tanrioven, M.S. Alam, Modeling, control, and power quality evaluation of a PEM fuel cell-based power supply system for residential use, IEEE Trans. Ind. Appl. 42 (6) (2006) 1582–1589.

[22] T.L. Skvarenina (Ed.), The Power Electronics Handbook, CRC press, Akureyri, Iceland, 2001.

[23] R.C. Dugan, M.F. McGranaghan, H. Wayne Beaty, Electrical Power Systems Quality, McGraw-Hill, New York, NY, 1996.

FURTHER READING

[1] J. Wang, et al., Low cost fuel cell converter system for residential power generation, IEEE Trans. Power Electron. 19 (5) (2004) 1315–1322.

[2] J. Lee, et al., A 10-kW SOFC low-voltage battery hybrid power conditioning system for residential use, IEEE Trans. Energy Conver. 21 (2) (2006) 575–585.

[3] C.M. Colson, M.H. Nehrir, Evaluating the benefits of a hybrid solid oxide fuel cell combined heat and power plant for energy sustainability and emissions avoidance, IEEE Trans. Energy Conver. 26 (1) (2011) 140–148.

[4] E.M. Stewart, et al., Analysis of a distributed grid-connected fuel cell during fault conditions, IEEE Trans. Power Syst. 1 (25) (2010) 497–505.

[5] T. Niknam, et al., Multi-objective daily operation management of distribution network considering fuel cell power plants, IET Renew. Power Gen. 5 (5) (2011) 356–367.

[6] F. Peng, X.F. Wang, B.B. Mao, Discrete PLL control for single-phase fuel cell grid connected system, in: Control and Decision Conference (CCDC), 2011 Chinese. IEEE, 2011.

[7] A. Gebregergis, et al., Solid oxide fuel cell modeling, IEEE Trans. Ind. Electron. 56 (1) (2009) 139–148.

[8] Fuel Cell Energy–High Efficiency Direct Fuel Cell/Turbine ® Power Plant–US Department of Energy, Available from: www.fuelcelleergy.com.

[9] Fuel Cell Energy-Critical Components for Direct Fuel Cell/Turbine Ultra Efficiency system, Available from: www.fuelcellenergy.com.

[10] Fuel Cell Energy-Air Quality Benefits of Fuel Cell Distributed Generation-Air Quality Technology Symposium September 2011, Available from: www.fuelcellenergy.com.

[11] S.-Y. Choe, et al., Dynamic simulator for a PEM fuel cell system with a PWM DC/DC converter, IEEE Trans. Energy Conver. 23 (2) (2008) 669–680.

[12] W. Li, et al., Single-stage single-phase high-step-up ZVT boost converter for fuel-cell micro-grid system, IEEE Trans. Power Electron. 25 (12) (2010) 3057–3065.

[13] J. Doyon, M. Farooqus, M. Mau, The direct fuel cell stack engineering, J. Power Sourc. 118 (2003) 8–13.

[14] V.J. Gosbell, S. Perera, V. Smith. Voltage Unbalance. Technical Power Quality Center, University of Wollongong, Technical Note 6, 2002.

[15] A. Silvestri, A. Berizzi, S. Buonanno, Distributed generation planning using genetic algorithms, in: Electric Power Engineering, 1999. Power Tech Budapest 99. International Conference on. IEEE, 1999.

[16] T. Bäck, H.-P. Schwefel, An overview of evolutionary algorithms for parameter optimization, Evol. Comput. 1 (1) (1993) 1–23.

[17] Available from: http://www.freeenergyplanet.biz/fuel-cell-guide/heat-and-fuel-recovery-cycles.html.

[18] T.V. Nguyen, R.E. White, A water and heat management model for Proton-Exchange-Membrane fuel cells, J. Electrochem. Soc. 140 (8) (1993) 2178–2186.

[19] X.-M. Yu, X.-Y. Xiong, Y.-W. Wu, A PSO-based approach to optimal capacitor placement with harmonic distortion consideration, Elec. Power Syst. Res. 71 (1) (2004) 27–33.

[20] N. Hadjsaid, J.-F. Canard, F. Dumas, Dispersed generation impact on distribution networks, IEEE Comput. Appl. Power Mag. 12 (2) (1999) 22–28.

[21] H. Iyer, S. Ray, R. Ramakumar, Voltage profile improvement with distributed generation, in: Power Engineering Society General Meeting, 2005. IEEE, 2005.

[22] Available from: http://www.doitpoms.ac.uk/tlplib/fuel-cells/printall.php (Source: University of Cambridge).

[23] Available from: http://www.fuelcelltoday.com/applications/portable.

[24] Available from: http://texaselectricityalliance.wordpress.com.

[25] Available from: http://www.nrel.gov/.
[26] Available from: http://oilprice.com/Energy/Energy-General/Are-Hydrogen-Fuel-Cell-Vehicles-Dead-On-Arrival.html.
[27] Available from: www.miniHYDROGEN.com.
[28] Available from: http://sosteneslekule.blogspot.com/2015/12/future-of-fuel-cells-hydrogen-production.html.
[29] Available from: http://www.fuelcellenergy.com.
[30] Available from: http://www.fuelcelltoday.com/news-archive/2012/october/construction-of-worlds-largest-fuel-cell-power-plant-expected-to-commence-in-2012 (source:POSCO Energy).

CHAPTER 6

Design of Small Hydro Generation Systems

Morteza Nazari-Heris, Behnam Mohammadi-Ivatloo
University of Tabriz, Tabriz, Iran

6.1 INTRODUCTION

There are several fundamental procedures for generating electrical energy by the conversion of other forms of energy. One of the practical methods for generation of electrical energy is hydroelectric units in which the movement of water results in power production by driving a hydro turbine. Holding water behind large dams is a famous technique of hydroelectric units, which has the capability of generating large quantities of electrical energy. Large-scale hydroelectric generation systems are well-suited for systems with power generation greater than 1000 kW, which usually feeds into a large electricity network. Statistics show that hydroelectric units participate in 16% of the world's power production [1]. Hydro plants can be classified in different categories based on their power generation capacity: large, small, mini, and micro hydro generation units. Hydro installations with a capacity of more than 1000 kW are defined as large hydroelectric power plants. All hydroelectric power generation units with a generation between 500 and 1000 kW are called small hydro generation systems. Mini and micro hydro plants are defined as hydro units with a capacity of between 100 and 500 kW and less than 100 kW, respectively [2].

Small hydro power generation systems have been utilized in the past as a renewable energy source for electricity generation. The major advantages of these systems are elimination of the cost of transmission line construction and power loss, mitigation of environmental constraints, increment of network stability, and peak load reduction. Unlike thermal units, hydro generation units exert less influence on the environment and have no gas emissions [3]. Utilization of hydro power generation units is a key solution for reducing atmospheric pollution and decreasing our dependence on fossil fuels [4]. The fact that there are no requirements for the construction of huge

Distributed Generation Systems
http://dx.doi.org/10.1016/B978-0-12-804208-3.00006-6

dams and reservoirs are positive reasons in favor of setting up small hydro power generation systems in a local network that is operated as "run-of-river" systems.

The small hydro power systems (SHPs) are a suitable option for energy production in developing countries, especially in remote areas where the attainment of a power grid is unachievable. Any rural and urban water installations and canals, dams that have been created for irrigation purposes, and running rivers can be selected as a site for a small hydroelectric unit. The implementation of these systems in rural areas will be effective in developing the area economically, which has been reported in different studies in multiple areas of the world [5].

The chapter is organized as follows: Section 6.2 provides a general description of the generation basics of hydroelectrical energy systems, considering different categories of hydro power systems and components of a typical hydro power generation plant. Different technologies of hydro power generation systems with descriptions of the most common turbines is introduced in Section 6.3, which also analyzes the related formulation of the turbines from different points of view. Section 6.4 presents the design basics of a small hydro power generation plant, and provides the fundamental knowledge of designing a SHPs project. Economic analysis of SHPs is provided in Section 6.5, which includes analysis of the estimated cost of a SHPs project, the costs of turbines and a run-of-river hydro power generation system project. Finally, the chapter is concluded and summarized in Section 6.6.

6.2 GENERATION BASICS OF HYDROELECTRICAL ENERGY SYSTEMS

Hydro power generation plants are categorized as follows [6]:
- *Impoundment*: A large hydro power system in which river water is stored in a reservoir by utilization of a dam. Electricity generation is accomplished by using water stored in the reservoir.
- *Diversion*: A hydro power system that may not need the use of a dam, and utilizes a diversion facility on a portion of a river.
- *Run-of-river*: This hydro power system has few requirements or no impoundment, and utilizes the water flow for electricity generation.

Most SHPs are "run-of-river" with no or limited requirement to dam or store water. The required storage reservoir of this kind of SHPs is defined as pondage. The plants containing a pondage can be utilized for water flow regulation during the time period and especially for electrical energy production in on-peak hours. A typical run-of-river SHPs scheme can be seen in Fig. 6.1. Civil works and electromechanical equipment are two general

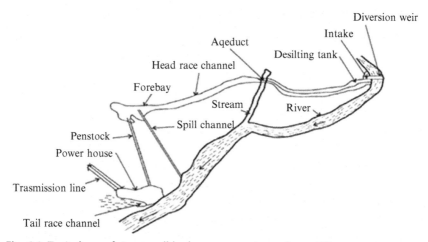

Fig. 6.1 Typical run-of-river small hydro power system scheme [3].

components of run-of-river SHPs. The civil works include a diversion weir and intake, desilting chamber, power channel including head race channel, forebay and spillway, penstock, powerhouse building, and tail race channel. Additionally, turbines with a governing system, and a generator with exci-tation system, switchgear, control, protection equipment, electrical and mechanical auxiliaries, and main transformer and switchyard equipment are enumerated as electromechanical equipment of run-of-river SHPs. Projects of run-of-river SHPs are classified in three types based on head. Low head (3–20 m), medium head (20–60 m), and high head (greater than 60 m) are the three main classifications of run-of river SHPs projects [7].

Components of a SHPs scheme are demonstrated in Fig. 6.2. Fundamen-tal components of SHPs include penstock, power house, tailrace, generating plant, and allied equipment [8].

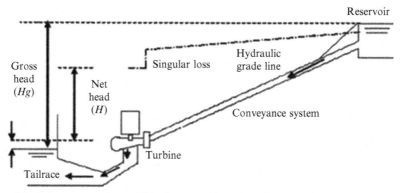

Fig. 6.2 Components of a small hydro power system scheme [3].

6.2.1 Basic Components of a Hydro Power Plant

In the following the basic components of a hydro power plant are defined:

Dam: A dam is utilized in most of the hydro power systems for holding back water, which is a large water reservoir. A dam is usually constructed across a river or a channel for utilization as water storage.

Penstock: A cavity or pipeline with a large diameter, in which conduction of water is done by opening the gates on the dams. In fact, the penstock plays the role of a conductor of water with high pressure from dams to the turbines.

Turbine: The blades of turbines are stroked by a high-pressure conduction of water, which turns the turbine. The turbine is attached to an electrical generator by a shaft. Different categories of turbines exist, which are divided into two general types: impulse turbines and reaction turbines. The different categories of turbines will be discussed in the next sections. Different types of vanes or buckets or blades are installed on a wheel, which is called the runner.

Tailrace: After working of high pressure water on the turbine, a channel carries water away from the turbine and the water re-enters to the river downstream, which is called tailrace. Moreover, the water surface in the tailrace is also referred to as tailrace.

Generator: By turning the blades of the turbine, the rotor inside the generator is turned. As a result, electric current is produced by rotating magnets inside the fixed-coil generator.

6.2.2 Hydroelectric Power Calculation

Conversion of water pressure to mechanical power is the basic operation of SHPs, which is then utilized for driving an electrical generator. Pressure head and volume flow rate are two essential components of generated power [9]. In general, the gross hydraulic power and the corresponding energy can be stated as follows [10]:

$$P_0 = \rho g Q H \tag{6.1}$$

$$E_0 = \rho g Q H \Delta t \tag{6.2}$$

In which P_0 (kW) and E_0 (kW h) over a time interval Δt (h) are the gross hydraulic power and the corresponding energy, respectively. Pressure head and volume flow rate are H (m) and Q (m^3/s), respectively. Also, ρ (kg/m^3) and g (m/s^2) are water density and acceleration due to gravity, respectively.

Discussion on turbine efficiency. Efficiency is considered as the ratio of the useful work performed in a process to the total energy expended, which can be stated as follows:

$$\eta = \frac{Power_{out}}{Power_{in}} \qquad (6.3)$$

where η is taken into account as efficiency, and $Power_{in}$ and $Power_{out}$ are referred to as the expanded power and useful power, respectively. Efficiencies of a hydro power system can be classified into two general categories: hydraulic efficiency and mechanical efficiency. Hydraulic efficiency is the obtained power by the runner of a turbine per power supplied at the inlet of a turbine. The probable power loss in a turbine is between the striking jet and the vane, which results in calling this type of efficiency as hydraulic efficiency. The equation of hydraulic efficiency η_h can be stated as

$$\eta_h = \frac{P_R}{P_W} \qquad (6.4)$$

where P_R and P_W are the respective elements utilized to demonstrate the runner power and the water power.

The ratio of available power at the shaft to the runner power is defined as mechanical efficiency. For calculation of mechanical efficiency η_m considering P_S as the shaft power, the following equation can be stated:

$$\eta_m = \frac{P_S}{P_R} \qquad (6.5)$$

Considering η as the total efficiency of the turbo-generator, the hydraulic power and the corresponding energy will be

$$\eta = \eta_h \eta_m \qquad (6.6)$$

$$P = \eta P_0 \qquad (6.7)$$

$$E = \eta E_0 \qquad (6.8)$$

Modern hydro turbines have an energy conversion efficiency of 90%, which is in a range of 60%–80% in micro-hydro systems.

Example 6.1

Suppose that the flow rate of a SHPs in a certain area with a pressure head of 100 m is 1 m^3/s. Calculate the hydraulic power with the consideration of total efficiency of 80% for turbo-generator and acceleration due to gravity of $g = 9.81$ m/s^2.

Solution

As $Q = 1\,\text{m}^3/\text{s}$, and $H = 100\,\text{m}$.

Then, from Eq. (6.1), $P_0 = \rho g Q H = 1000 \times 9.81 \times 1 \times 100 = 981000\,\text{W} = 981\,\text{kW}$.

Accordingly, considering Eq. (6.7), $P = \eta P_0 = 0.8 \times 981 = 784.8\,\text{kW}$.

Example 6.2

Consider an installation of a hydro power system in a site with a net head of 100 m. Hydraulic efficiency and mechanical efficiency are assumed to be $\eta_d = 95\%$ and $\eta_t = 85\%$, respectively. For a requested power of $P = 600\,\text{kW}$, determine the required pressure head in the site.

Solution

Since $\eta_h = 95\%$, $\eta_m = 85\%$, $H = 100\,\text{m}$, and $P = 600\,\text{kW}$.

Moreover, $\eta = \eta_h \times \eta_m = 0.95 \times 0.85 = 0.8075$.

Then, from Eq. (6.7), we have

$$P_{Total} = \eta \rho g Q H = 0.8075 \times 1000 \times 9.81 \times Q \times 100 = 600\,\text{kW}$$

Finally, $Q = \dfrac{600000}{0.8075 \times 1000 \times 9.81 \times 100} = 0.7574\,\text{m}^3/\text{s}$

Example 6.3

A hydro power system is considered for being installed at a site with a pressure head of 1.25 m³/s, which is available 10 h a day and has a net head of 100 m. Assume that it is expected to gain potential theoretical energy equal to 3 MWh. Is it possible to gain the expected potential energy? Additionally, calculate the hydraulic power taking into account hydraulic efficiency and mechanical efficiency of $\eta_d = 96\%$ and $\eta_t = 87\%$.

As $Q = 1.25\,\text{m}^3/\text{s}$, and $H = 35\,\text{m}$

Then, from Eqs. (6.2) and (6.8), we have

$$E = \eta E_0 = \eta \rho g Q H \Delta t$$

Moreover, $\eta = \eta_h * \eta_m = 0.96 * 0.87 = 0.8352$.

So that $E = 0.8352 \times 1000 \times 9.81 \times 1.25 \times 35 \times 10 = 3584.574\,\text{kW h}$.

So it is possible to attain the expected electrical energy. The hydraulic power is as follows:

$$P = 0.8352 \times 1000 \times 9.81 \times 1.25 \times 35 = 358.4574\,\text{kW}$$

6.3 DIFFERENT TECHNOLOGIES OF SMALL HYDRO GENERATION SYSTEMS

A turbine operation is the conversion of water pressure to mechanical energy. Two main kinds of hydro turbines exist: impulse and reaction turbines. The head and the flow rate of water at the site determine the option of the kind of turbine. Moreover, efficiency and cost are two other effective elements in turbine selection. Small hydro plants can be classified as based on turbine technologies as follows [11].

6.3.1 Impulse Turbines

Obtaining energy by the utilization of the momentum of flowing water, as opposed to the weight of the water, is the basic operation of impulse turbines. The kinetic energy of water is utilized for driving the runner by utilization of the impulse turbine, and is discharged to atmospheric pressure. An impulse turbine runner operates in air, which is moved by a jet of water. The application of impulse turbines is in sites with high head and low flow rate. Three types of impulse turbines are common for application for hydro power systems, including the Pelton, the cross-flow, and the Turgo.

6.3.1.1 Pelton Turbine

The Pelton turbine includes a wheel with a series of split buckets set around its rim. A high pressure jet of water is directed tangentially at the wheel, which hits each bucket and is split in half. Accordingly, turning of each half and deflecting back almost through 180° is attained. The application of Pelton turbines is in systems with large water heads, which are typically more than 250 m. A typical Pelton turbine is demonstrated in Fig. 6.3.

Calculations of Pelton turbines. In this subsection, calculations related to the Pelton turbine are provided. Consider velocity triangles for the jet striking the bucket as in Fig. 6.4. In this figure some parameters are demonstrated, which are defined in the following.

V_r is the relative velocity and is tangential to the blades. The absolute velocity and blade velocity are shown by V and u, respectively. V_w and V_f are respective identifiers of the velocity of whirl and the velocity of flow. V_{w1} and V_{w2} are the velocity component of flow at inlet and outlet along a tangential direction, respectively. The flow velocity at the inlet and outlet are shown by V_{f1} and V_{f2}, respectively. Moreover, α and β are utilized to show the angles between absolute velocities of the jet and vane at inlet and outlet, respectively. The inlet and outlet vane angles are demonstrated

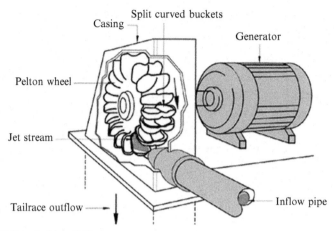

Fig. 6.3 Pelton turbine [12].

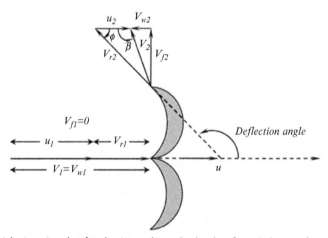

Fig. 6.4 Velocity triangles for the jet striking the bucket for a Pelton turbine.

by θ and φ, respectively. V_f and V_f are the respective radius of the wheel at the inlet and outlet of the vane. w is the tangential speed of the wheel.

u_1 which is the tangential speed of the vane at inlet, can be obtained as $u_1 = w \times R_1$.

Similarly, the tangential speed of the vane at the outlet, which is shown by u_2, is $u_2 = w \times R_2$.

Considering F_x as the rate of momentum change of the jet along the direction of vane motion, the related calculation formula can be stated as

$$F_x = \rho a V_1 [V_{w1} - (-V_{w2})] = \rho a V_1 [V_{w1} + V_{w2}] \quad (6.9)$$

The jet completed work on the vane per second is given by the product of force exerted on the vane and the distance moved by the vane in one second. The related equation is

$$\text{Work/Second} = F_x \, u = \rho a V_1 [V_{w1} + V_{w2}] u \qquad (6.10)$$

For obtaining the input to the turbine, the following equation can be utilized:

$$W_i = \frac{1}{2} \rho a V_1{}^3 \qquad (6.11)$$

The values N, P, and H are usually utilized for expressing performance or operating conditions for a turbine handling a particular fluid. Determination of the range of these operating parameters covered by a machine of a particular shape at high efficiency is an essential subject that should be taken into account. The type option of machine for a special application can be attained considering such information. Accordingly, utilizing these values, a starting point in the design of a particular machine, can be done. A parameter that is not based on the size of the machine (D-rotor or impeller diameter) can be useful, which will be the characteristic of all the machines of a homologous series. A parameter involving N, P, and H but not D is defined as a parameter called specific speed, which can be stated as

$$N_s = \frac{N \sqrt{P}}{H^{5/4}} \qquad (6.12)$$

Considering the turbine efficiency as the proportion of output per second to input per second or the proportion of work done per second to input per second, the efficiency can be calculated as

$$\eta = \frac{\rho a V_1 [V_{w1} + V_{w2}] u}{\frac{1}{2} \rho a V_1{}^3} = \frac{2u [V_{w1} + V_{w2}]}{V_1^2} \qquad (6.13)$$

Considering the inlet velocity triangle, we have

$$V_{w1} = V_1 \qquad (6.14)$$

Moreover, considering no frictional losses through the vane and no shock, we have

$$V_{r1} = V_{r2} = (V_1 - u_1) \qquad (6.15)$$

Considering the outlet velocity triangle, we have

$$V_{w2} = V_{r2} \cos\phi - u_2 = (V_1 - u) \cos\phi - u \qquad (6.16)$$

Accordingly, F_x can be calculated as

$$F_x = \rho a V_1 [V_1 + (V_1 - u)\cos\phi - u] = \rho a V_1 (V_1 - u)[1 + \cos\phi] \qquad (6.17)$$

The turbine efficiency can be obtained as

$$\eta = \frac{2u[V_1 + (V_1 - u)\cos\phi - u]}{V_1^2} = \frac{2u[V_1 - u + (V_1 - u)\cos\phi]}{V_1^2}$$

$$= \frac{2u}{V_1^2}(V_1 - u)[1 + \cos\phi] \qquad (6.18)$$

Example 6.4

Assume a Pelton turbine with a mean runner diameter of 1 m and a net head of 350 m. The runner is spinning at $N = 1000$ rpm. Considering the side clearance of $15°$ and discharge of 0.1 m^3/s, obtain the power available at the nozzle and calculate the hydraulic efficiency of the turbine.

Solution

As $Q = 0.1$ m^3/s, $H = 350$ m, $N = 1000$ rpm, $\phi = 15°$, and $D = 1$ m

Assuming $C_v = 0.98$

Then, the velocity of the jet can be provided as

$$V = C_v\sqrt{2gH} = 0.98\sqrt{2 \times 9.81 \times 350} = 81.21 \text{ m/s}$$

The velocity of the vane can be obtained as follows:

$$u = \frac{\pi DN}{60} = \frac{\pi \times 1 \times 1000}{60} = 52.36 \text{ m/s}$$

Power available at the nozzle can be obtained as the work done per second:

$$\text{Work/Second} = \rho g Q H = 1000 \times 9.81 \times 0.1 \times 350 = 343350 \text{ kW}$$

The turbine efficiency can be calculated as follows:

$$\eta = \frac{2u}{V_1^2}(V_1 - u)[1 + \cos\phi] = \frac{2 \times 52.36}{81.21^2}(81.21 - 52.36)(1 + \cos 15)$$

$$= 90.05\%$$

Example 6.5

Consider a Pelton turbine with a jet velocity of V and vane angle equal to ϕ. Obtain the relationship between bucket speed and jet velocity at the maximum efficiency of the turbine. Additionally, provide a calculation formula for the maximum efficiency of the Pelton turbine.

Solution

As $\eta = \dfrac{2u}{V_1^2}(V_1 - u)[1 + \cos\phi]$,

For obtaining the maximum efficiency, we have

$$\frac{d\eta}{du} = \frac{2d[(uV_1 - u^2)(1 + \cos\phi)]}{V_1^2 du} = \frac{(V_1 - 2u)(1 + \cos\phi)}{V_1^2} = 0$$

Then, $V_1 - 2u$

Or, $V_1 = 2u$

This means than when the jet velocity is the double of bucket speed, the maximum efficiency of the Pelton turbine is attained.

For providing a formula for maximum efficiency of the Pelton turbine, we have $\eta = \dfrac{2u}{V_1^2}(V_1 - u)[1 + \cos\phi]$, and $V_1 = 2u$

Then, $\eta_{max} = \dfrac{2u}{(2u)^2}(2u - u)[1 + \cos\phi] = \dfrac{1 + \cos\phi}{2}$

Considering the obtained formula for maximum efficiency of a Pelton turbine, it is obvious that the maximum efficiency of the turbine will be attained at the value of $\cos\phi$ equal to 1, and as a result ϕ should be $0°$. But it should be noted that this situation makes the jet complete deviation by $180°$, which forces the jet striking the bucket to strike the successive bucket on the back of it. So it acts such as a breaking jet, which should be avoided. Accordingly, for avoiding this condition, at least ϕ should be provided equal to $5°$.

Example 6.6

A Pelton turbine is considered for power production equal to 80 kW, being installed in a head of 100 m. The runner is spinning at 500 rpm, and the speed ratio and velocity coefficient are 0.48 and 0.98, respectively. Design this turbine for an overall efficiency equal to 72%.

Solution

$$p = \eta P_0 = \eta\rho gQH = 8000 \text{ W}$$

So that, $Q = \dfrac{P}{\eta\rho gH} = \dfrac{80000}{0.72 \times 1000 \times 9.81 \times 100} = 0.113 \,\text{m}^3/\text{s}$

So, the velocity of the jet can be calculated as follows:

$V = C_v\sqrt{2gH} = 0.98\sqrt{2 \times 9.81 \times 100} = 43.41 \text{ m/s}$

And velocity of the vane can be attained as

$u = \phi\sqrt{2gH} = 0.48\sqrt{2 \times 9.81 \times 100} = 21.26 \text{ m/s}$

Additionally, another equation exists for calculation of the velocity of the vane as

$$u = \frac{\pi DN}{60} = 21.26 \text{ m/s}$$

Hence, $D = \frac{60u}{\pi N} = \frac{60 \times 21.26}{\pi \times 500} = 0.81 \text{ m} = 81 \text{ cm}$

Jet diameter can be obtained as the following equation:

$$Q = V_1 \times \pi \times \left(\frac{d}{2}\right)^2$$

So that, $0.113 = 43.41 \times \pi \times \left(\frac{d}{2}\right)^2$

So that, we have $d = 2\sqrt{\dfrac{0.11}{43.41 \times \pi}} = 0.056 \text{ m} = 5.6 \text{ cm}$

The jet ratio is as follows:

$$\frac{D}{d} = \frac{81}{5.6} = 14.46$$

6.3.1.2 Cross-Flow Turbine

This type of turbine consists of a drum–like rotor, which includes a solid disk at each end. Moreover, the two disks are joined with gutter-shaped slats. A jet of water enters the top of the rotor through the curved blades. The turbine allows water to pass through the blades for a second time, which results in the emergence of the water on the far side of the rotor. Considering the shape of the blades, transferring the water momentum on each passage through the periphery of the rotor is done before falling away with an insignificant residual energy. A schematic of a cross-flow turbine is provided in Fig. 6.5. This kind of turbine is appropriate for installation in sites with a head between 5 and 200 m. Compared with other types of turbines, cross-flow turbines have low efficiency [13].

6.3.1.3 Turgo Turbine

A similar turbine to the Pelton turbine is the Turgo turbine, which differs in the shape of the buckets and the design of the turbine, which strikes the plane of the runner at a typical angle of 20, resulting in entrance of the water to the runner on one side and exit from the other side. Compared with the Pelton turbine, the Turgo turbine can provide equivalent power with a

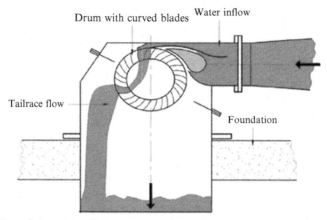

Fig. 6.5 Cross-flow turbine [12].

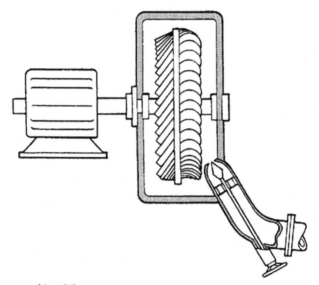

Fig. 6.6 Turgo turbine [9].

smaller diameter runner. Turgo turbines operate in systems with large water heads, which ensures high efficiency of the turbine. Fig. 6.6 shows a simple schematic of a Turgo turbine.

6.3.2 Reaction Turbines

Reaction turbines generate electrical energy by using the mutual action of pressure and moving water. When the rotor is completely filled in the water and is enclosed in a pressure casing, the operation of reaction turbines is attained. A draft tube is a diffuser that exists in all reaction turbines below the runner. The water discharges through the draft tube. Accordingly, the static pressure below the runner is reduced and the effective head increases. Application of reaction turbines is in sites with lower heads and higher flow rates. Two main kinds of reaction turbines are the Propeller and the Francis.

6.3.2.1 Propeller Turbine

Generally, this kind of turbine has an axial flow runner that has three to six blades based on the water head that the turbine is designed to be utilized. The application of propeller turbines is in systems with low water heads. Different types of propeller turbines exist, including bulb turbine, Kaplan, Straflo, and tube turbines.

Kaplan turbines are mostly being utilized in sites with low pressure head and large flow rates [3]. A tube-type propeller turbine, which is a typical Kaplan turbine, is shown in Fig. 6.7.

6.3.2.2 Francis Turbine

Francis turbines are the most well-known type of reaction turbines. This turbine has a radial flow runner or a mixed radial/axial flow runner. Radial water flow to the runner and axial emerge as a result of the runner spinning. Wicket gates and a draft tube are two other main elements of the Francis turbine. Systems with medium head size are appropriate for Francis turbine application. A layout of a Francis turbine is provied in Fig. 6.8.

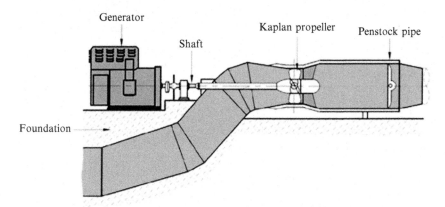

Fig. 6.7 Tube-type Propeller turbine (a typical Kaplan turbine) [12].

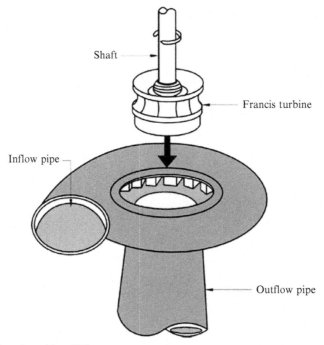

Fig. 6.8 Francis turbine [12].

Calculations of reaction turbines. This subsection provides calculations related to reaction turbines. Consider Fig. 6.9 as a layout of reaction turbines.

For obtaining the speed ratio and flow ratio, the following equations can be utilized:

$$R_s = \frac{u_1}{\sqrt{2gh}} \tag{6.19}$$

$$R_f = \frac{V_{f1}}{\sqrt{2gh}} \tag{6.20}$$

where R_s and R_f are speed ratio and flow ratio, respectively. Head and velocity of flow are shown by h and V_{f1}, respectively.

The head can be calculated as follows:

$$h = \frac{P_1}{\rho g} + \frac{V_1^2}{2g} \tag{6.21}$$

where the pressure at the inlet of the runner is shown by p_1.

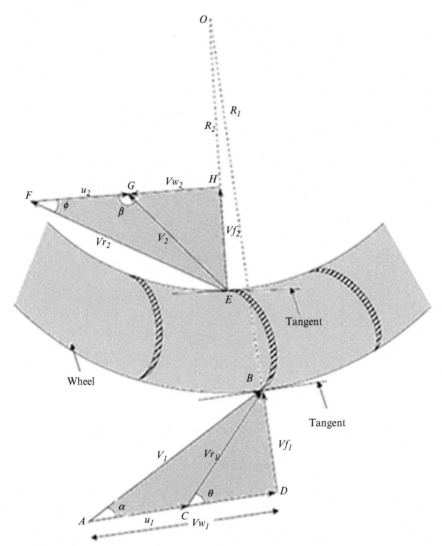

Fig. 6.9 Parameters of reaction turbines.

Discharge flowing can be stated as

$$Q = \pi D_1 B_1 V_{f1} = \pi D_2 B_2 V_{f2} \qquad (6.22)$$

where inlet and outlet diameters are shown by D_1 and D_2, respectively. B_1 and B_2 are respective elements utilized to illustrate the width of the runner at inlet and outlet. The inlet and outlet flow velocity are shown by V_{f1} and V_{f2}, respectively.

Considering t as the thickness of the vane and n as the number of vanes, discharge flowing can be obtained as

$$Q = (\pi D_1 - nt)B_1 V_{f1} = (\pi D_2 - nt)B_2 V_{f2} \qquad (6.23)$$

Work done per second on the runner can be obtained by utilization of

$$Work/Second = \rho a V_1 [V_{w1} u_1 + V_{w2} u_2] = \rho Q[V_{w1} u_1 + V_{w2} u_2] \qquad (6.24)$$

For obtaining efficiency, the following equation can be stated:

$$\eta = \frac{\rho Q[V_{w1} u_1 + V_{w2} u_2]}{\rho g Q h} = \frac{V_{w1} u_1 + V_{w2} u_2}{gh} \qquad (6.25)$$

Example 6.7

Consider a reaction turbine with an external diameter of 0.6 m. The width of the wheel and the velocity of flow at inlet are 140 mm and 2 m/s, respectively. Find the rate of flow passing through the turbine. Assuming the internal diameter of the turbine is equal to 0.2 m and considering a constant velocity of flow through the runner, calculate the width of the wheel at the outlet.

Solution

As $D_1 = 0.6$ m, $D_2 = 0.2$ m, $B_1 = 0.14$ m, and $V_{f1} = V_{f2} = 2$ m/s

For obtaining the discharge through the turbine, we have

$$Q = \pi D_1 B_1 V_{f1} = \pi \times 0.6 \times 0.14 \times 2 = 0.5278 \text{ m}^3/\text{s}$$

For solving the second part of the question we have

$$Q = \pi D_1 B_1 V_{f1} = \pi D_2 B_2 V_{f2}$$

Then, $Q = 0.5278 = \pi D_2 B_2 V_{f2} = \pi \times 0.2 \times B_2 \times 2$

Finally, we have $B_2 = \dfrac{0.5278}{\pi \times 0.2 \times 2} = 0.42$ m

Example 6.8

Consider a reaction turbine running at 600 rpm, which has an external diameter and a width of 600 mm and 200 mm, respectively. The absolute velocity of water at inlet is equal to 30 m/s and the guide vanes are at 25° to the wheel tangent. Obtain discharge through the turbine and inlet vane angle.

Solution

As $D_1 = 0.6$ m, $D_2 = 0.2$ m, $B_1 = 0.18$ m, $\alpha = 20°$, and $V_1 = 30$ m/s

The peripheral velocity is $u_1 = \dfrac{\pi D_1 N}{60} = \dfrac{\pi \times 0.6 \times 600}{60} = 18.85$ m/s

Considering Fig. 6.9, the discharge through the turbine can be obtained as the following:

$V_{f1} = V_f \sin\alpha = 30\sin 25 = 12.675$ m/s

$Q = \pi D_1 B_1 V_{f1} = \pi \times 0.6 \times 0.18 \times 12.675 = 4.3$ m^3/s

As seen in Fig. 6.9, $V_{w1} = V_f \cos\alpha = 30\cos 25 = 27.19$ m/s

$$\tan\theta = \frac{V_{f1}}{V_{w1} - u_1} = \frac{12.675}{27.19 - 18.85} = 1.52$$

Then, $\theta = \tan^{-1}(1.52) = 56.66°$

Example 6.9

There is a Kaplan turbine working in a head of 18 m and running at 150 rpm. Assume the flow ratio is equal to 4 m^3/s and the diameter is 4 m. Moreover, the boss and the inlet vane angle at the extreme edge of the runner are 10 m and 160°, respectively. The width of the wheel and the velocity of flow at inlet are 140 mm and 2 m/s, respectively. If the turbine has a radial discharge outlet, obtain the discharge, the hydraulic efficiency, the angle of the guide blade at the extreme edge of the runner, and the angle of the outlet vane at the extreme edge of the manner.

Solution

To better understand the question and related parameters, see Fig. 6.10.

As $H = 18$ m, $N = 150$ rpm, $\theta = 160°$, $D_0 = 4$ m, $D_{boss} = 10$ m, $V_{w2} = 0$ m/s

$$u_1 = u_2 = \frac{\pi D_0 N}{60} = \frac{\pi \times 4 \times 150}{60} = 31.42 \text{ m/s}$$

$V_{f1} = 0.43\sqrt{2gH} = 0.43\sqrt{2 \times 9.81 \times 18} = 8.08$ m/s

Considering Fig. 6.11, we have $\tan(180 - \theta) = \dfrac{V_{f1}}{u_1 - V_{w1}}$

Then, $\tan(180 - 163) = \tan(17) = 0.306 = \dfrac{V_{f1}}{u_1 - V_{w1}} = \dfrac{8.08}{31.42 - V_{w1}}$

As a result, $V_{w1} = 31.42 - 8.08/0.306 = 5.015$ m/s

For solving the second part of the question we have

$$\eta = \frac{V_{w1}u_1 \pm V_{w2}u_2}{gh} = \frac{V_{w1}u_1}{gh} = \frac{5.015 \times 31.42}{9.81 \times 18} = 89.24\%$$

For obtaining the angle of the guide blade at the extreme edge of the runner, we have

$$\tan(\alpha) = \frac{V_{f1}}{V_{w1}} = \frac{8.08}{5.015} = 1.611$$

So that $\alpha = \tan^{-1}(1.611) = 58.17°$

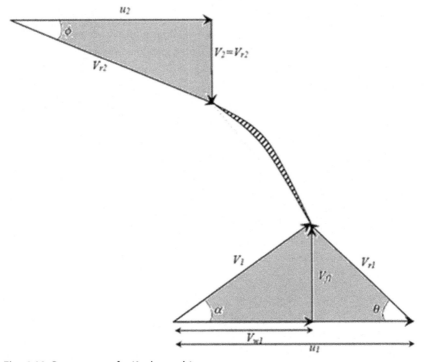

Fig. 6.10 Parameters of a Kaplan turbine.

Finally, the angle of the outlet vane at the extreme edge of the manner with the consideration of angles at Fig. 6.10 can be stated as follows:

$$\tan(\phi) = \frac{V_{f2}}{u_2} = \frac{8.08}{31.42} = 0.257$$

Accordingly, $\phi = \tan^{-1}(0.257) = 14.41°$

6.4 DESIGN OF SMALL HYDRO POWER SYSTEMS

In this section, the fundamental principles of design of SHPs are provided. Evaluation of the pressure head of a site and flow rate of the water are the primary steps in designing SHPs, which provides the power production potential of the considered system. Calculation of the pressure head and flow rate determines the turbine category, size of the system pipeline, size of the generator, and some other components of the system. The procedure of pressure head and flow rate of the water measurements are discussed below.

6.4.1 Head Measurement

The difference in the height of the intake of the pipeline and the turbine of a hydro system results in the existence of water pressure, which is defined as a head. An automatic calculation of height and distance can be obtained by the utilization of modern electronic digital levels, which provides the results in 4 seconds with an accuracy of 0.4 mm. Moreover, a global positioning system (GPS) can be used as an ideal option for positioning the field and rough mapping. After computing the gross head, for determination of the net head, losses along its path including losses of open channel, trash rack, inlet of the penstock, friction of the penstock, and gate or valve should be taken into account. Accordingly, the following equation can be stated for obtaining the net head H_N [14]:

$$H_N = H_G - (H_{Ch} + H_{Tr} + H_{En} + H_V + H_P) \qquad (6.26)$$

where H_{Ch} is loss of the open channel, and can be calculated as

$$H_{Ch} = \left(\frac{Q \times n_{Ch}}{A_{Ch} \times R_{Ch}^{2/3}} \right)^2 \times L_{Ch} \qquad (6.27)$$

where Q is the flow rate of water in a uniform open channel, n_{Ch} and A_{Ch} are the manning factor and the area of the open channel cross-sectional, respectively. R_{Ch} and L_{Ch} are the respective parameters of the hydraulic radius of the sectional area and length of the open channel.

Trash lack loss, which is demonstrated by H_{Tr}, can be obtained as follows:

$$H_{Tr} = K_{Tr} \times \left(\frac{t}{b} \right)^{4/3} \times \frac{V_{Ch}^2}{2g} \times \sin(\alpha) \qquad (6.28)$$

where K_{Tr} is a trash lack factor, and bar thickness and bar width of the trash rack screen are demonstrated by t and b, respectively. V_{Ch} and α are the respective parameters of velocity of open channel, and inclined angle with horizontal for trash rack.

For obtaining loss of an inlet of the penstock H_{En}, we have

$$H_{En} = K_{En} \times \frac{V_P^2}{2g} \qquad (6.29)$$

where K_{En} is a trash lack factor and velocity of penstock inlet is shown by V_P.

Calculation of valve loss H_V is

$$H_V = K_V \times \frac{V_P^2}{2g} \qquad (6.30)$$

where K_V is valve factor.

And friction of the penstock loss H_P can be stated as follows:

$$H_V = \frac{10.29 \times n_P^2 \times Q^2}{D_P^{5.33}} \qquad (6.31)$$

where n_P is the Manning factor of penstock, and diameter of the penstock is shown by D_P.

6.4.2 Flow Measurement

The second major part of required measurements for designing a hydro power plant is related to measuring the flow rate. Considering altering of flow rates through the different seasons, measurement of flow rates of water at various times of the year is necessary. Volume of water per second or minute is typically defined as flow rate of the water. Three concepts are involved in water flow measurement: container, float, and weir. For obtaining the flow rate of water, the container fill procedure is the most common method. For measuring the water flow by utilization of this method, a spot in the stream is identified, where all the water can be gained in a bucket. While this process is not possible, by building a temporary dam and forcing all of the water to flow through a single outlet, the flow rate of the water can be attained. Knowing the volume of the bucket or the temporary dam and using a stopwatch to measure the time taken to fill the container, flow rate can be determined [15].

6.4.3 Design of an Appropriate Turbine for Small Hydro Power Systems Projects

This subsection reviews hydro power turbines and their appropriate application, which is based on realistic parameters. Fig. 6.11 provides data for determining the application range of turbines. This figure shows the head and flow rate range of reviewed small hydro power turbines. As seen in this figure, Pelton and cross-flow turbines have the capability to operate in high head, which are in low flow rates.

Typical efficiency curves of reviewed hydro power turbines are shown in Fig. 6.12. One of the important factors in the comparison of hydro power systems is available efficiency. It should be noted that the Pelton, cross–flow, and Kaplan turbines attain high efficiencies in work points below designed flow. However, the application of Francis turbines at below half its normal flow results in high reduction of turbine efficiency. Additionally, the

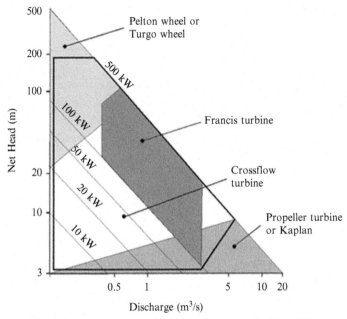

Fig. 6.11 Head and flow rate ranges of small hydro power turbines [3].

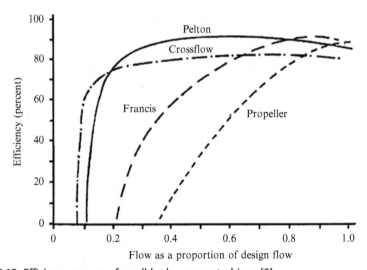

Fig. 6.12 Efficiency ranges of small hydro power turbines [9].

efficiency of fixed pitch propeller turbines decreases when running except at flow rates greater than 80% of full flow rate.

Example 6.10

Consider a SHPs project site with the head available equal to 230 m of water at a flow rate of 0.04 m³/s. Select and design a suitable turbine to generate power, assuming the required coefficients with justification and stating all the relevant parameters.

Solution

Considering an available head of 400 m and a flow rate equal to 0.2 m³/s, an appropriate turbine is to be designed. Economic analysis is the Pelton turbine.

Assuming 10% loss in the pipeline, the net head available at the nozzle is obtained as

$$H = 90\% \times 400 = 360\,\text{m}$$

Total efficiency of energy conversion is assumed to be 90%. Accordingly, the available power can be stated as follows:

$$P = \eta P_0 = \eta \rho g Q H = 0.9 \times 1000 \times 9.81 \times 0.2 \times 360 = 635688\,W = 635.7\,\text{kW}$$

Then, the velocity of the jet can be obtained as

$$V = C_v \sqrt{2gH} = 0.98\sqrt{2 \times 9.81 \times 360} = 82.362\,\text{m/s}$$

For calculation of jet diameter, the following equation can be written:

$$Q = V_1 \times \pi \times \left(\frac{d}{2}\right)^2$$

Accordingly, $0.2 = 82.362 \times \pi \times \left(\frac{d}{2}\right)^2$

So that we have $d = 2\sqrt{\dfrac{0.2}{82.362 \times \pi}} = 0.0556\,\text{m} = 5.56\,\text{cm}$

Assuming the speed ratio is equal to 0.46, the peripheral velocity is

$$u = 0.46 \times V_1 = 0.46 \times 82.362 = 38.01\,\text{m/s}$$

By consideration of possible rotor speeds of 3000 rpm, 1500 rpm, 1000 rpm, 750 rpm, 600 rpm, and so on, each of the speeds can be opted for finding the diameter, so that the peripheral velocity is u.

As obtained before, the peripheral velocity u is equal to 38.01 m/s

Then, for obtaining diameter based on rotor speed, we have

$$u = \frac{\pi DN}{60}$$

So that, for calculating diameter, the following equation will be utilized:

Table 6.1 Diameters based on different possible rotor speeds

N	D
3000	$D = \dfrac{725.94}{3000} = 0.24198$ m $= 24.2$ cm
1500	$D = \dfrac{725.94}{1500} = 0.48396$ m $= 48.4$ cm
1000	$D = \dfrac{725.94}{3000} = 0.72594$ m $= 72.6$ cm
750	$D = \dfrac{725.94}{750} = 0.96792$ m $= 96.8$ cm
600	$D = \dfrac{725.94}{600} = 1.21$ m $= 121$ cm

$$D = \frac{60u}{\pi N} = \frac{60 \times 38.01}{\pi N} = \frac{725.94}{N}$$

Accordingly, considering different possible rotor speeds, the diameter can be obtained as presented in Table 6.1.

The procedure is with respect to two cases of the speed, $N = 3000$ rpm and $N = 1500$ rpm. The specific speed can be calculated as $N_s = \dfrac{N\sqrt{P}}{H^{5/4}}$

For $N = 3000$ rpm, $N_s = \dfrac{3000\sqrt{635.7}}{360^{5/4}} = 48.24$

For $N = 1500$ rpm, $N_s = \dfrac{1500\sqrt{635.7}}{360^{5/4}} = 24.12$

Both obtained values are accepted as alternate designs.

For obtaining the jet ratio, we have

For $N = 3000$ rpm, $\dfrac{D}{d} = \dfrac{24.2}{5.56} = 4.35$

For $N = 1500$ rpm, $\dfrac{D}{d} = \dfrac{48.4}{5.56} = 8.70$

As jet diameter is 1.1 cm in both cases, the cup dimensions are the same in the two cases.

Length $L = 2.3 \times d = 2.3 \times 5.56 = 12.788$ cm

Breadth $B = 2.8 \times d = 2.8 \times 5.56 = 15.568$ cm

Depth $D = 0.6 \times d = 0.6 \times 5.56 = 3.336$ cm

6.5 ECONOMIC ANALYSIS OF SMALL HYDRO POWER PLANT PROJECTS

This section provides an analysis of the required investment cost for installing SHPs. There are two fundamental parameters of installation cost of hydro power plants, including existing infrastructures and installation capacity. Two kinds of costs have been taken into account for obtaining the project cost, including investment and annual costs. Civil costs, electromechanical equipment, power transmission-related costs, and some other indirect costs can be included as investment costs. Moreover, costs related to depreciation of equipment, maintenance, operation, and replacement costs represent annual costs. Explanation of each of the enumerated costs follows [16,17].

6.5.1 Investment Costs

First, direct investment costs are explained in the following:

Civil costs. Costs related to the construction and structure of a hydro plant, which consist of a dam, water transmission system, penstock, head pond, the fore bay, the power house, the structure of the tailrace, and other structures related to civil section costs.

Electromechanical equipment costs. This item of investment costs contains turbines, generators, governors, gates, control systems, and a power substation. From 30% to 40% of a SHPs budget is allocated to electromechanical equipment cost. For predicting this item of a SHPs cost (€/kW), the following equation can be utilized:

$$Cost = aP^{b-1}H^c \tag{6.32}$$

where P is power, and H is the net head. Moreover, a, b, and c are the coefficients that are based on the geographical, space, or time field of the utilization case.

Power transmission costs. For transferring generated electrical energy from a hydro power plant to a power transmission network, an allocation of cost to a power transmission line is required. Related location, kind of transmission system, capacity of SHPs, and length of transmission lines are the basic elements of power transmission costs.

The indirect costs of investment section included engineering and design, supervision and administration, and inflation costs. Type, size, and location of the plant are the basic elements of engineering and design costs. Supervision and administration costs consist of land cost, management, and inspection and supervision costs. Additionally, the inflation rate should be taken into account during the course of the project.

6.5.1 Annual Costs

Annual costs calculation aims to attain the net benefit of a project. As mentioned before, depreciation of equipment, maintenance, operation, and replacement costs are enumerated as annual costs.

Depreciation of equipment. Depreciation and other effective factors on equipment should be taken into account in the economic analysis.

Maintenance and operation costs. Costs related to personnel salaries, taxes, landscape, and consumable materials are included in maintenance and operation costs. The annual inflation factor has an impact on this category of cost.

Replacement cost. Replacement includes replacing and renovating the main parts of the SHPs such as generator windings, turbine runners, and other parts. Considering the nature of SHPs, the costs related to replacement and renovation at year 25 is assumed equal to the total value of the equipment when it was purchased.

Considering a realistic viewpoint about the total cost of an installation of SHPs, the turbine type can be taken into account, which has been discussed in [18]. Estimation of costs of propeller, Francis, and Pelton turbines (£/kW), which is based on European data, follows.

6.5.3 Costs for Kaplan Turbines

The cost of Kaplan turbines C_K (£/kW) based on turbine type and flow rate Q is divided into two different sections. Considering Q between 0.5 m^3/s and 5 m^3/s, the cost of a turbine can be stated as follows:

$$C_{K1} = 15,000 \times (Q \times H)^{0.68} \tag{6.33}$$

or

$$C_{K1} = 3500 \times (kW)^{0.68} \tag{6.34}$$

The cost of a Kaplan turbine by taking into account a flow rate between 5 m^3/s and 30 m^3/s can be calculated as

$$C_{K2} = 46,000 \times (Q \times H)^{0.35} \tag{6.35}$$

or

$$C_{K2} = 14,000 \times (kW)^{0.35} \tag{6.36}$$

Accordingly, the obtained cost of a Kaplan turbine is based on the installed turbine capacity.

6.5.4 Costs for Francis Turbines

The related cost of this kind of turbines C_F (£/kW) is divided into three different sections based on flow rate Q. The first formulation is stated as follows for flow rates between 0.5 m^3/s and 2.5 m^3/s:

$$C_{F1} = 142,000 \times \left(Q \times H^{0.5} \right)^{0.07}$$ (6.37)

or

$$C_{F1} = 122,000 \times \left(\text{kW}/H^{0.5} \right)^{0.07}$$ (6.38)

Moreover, considering a flow rate between 2.5 m^3/s and 10.0 m^3/s, the cost of Francis turbines can be obtained as

$$C_{F2} = 282,000 \times \left(Q \times H^{0.5} \right)^{0.11}$$ (6.39)

or

$$C_{F2} = 223,000 \times \left(\text{kW}/H^{0.5} \right)^{0.11}$$ (6.40)

Finally, the cost of Francis turbines for Q greater than 10.0 m^3/s, can be estimated as follows:

$$C_{F3} = 50,000 \times \left(Q \times H^{0.5} \right)^{0.52}$$ (6.41)

or

$$C_{F3} = 16,500 \times \left(\text{kW}/H^{0.5} \right)^{0.52}$$ (6.42)

6.5.5 Costs for Pelton Turbines

For obtaining the Pelton turbines cost C_p (£/kW), considering a flow rate Q, the following equations can be utilized:

$$C_{K1} = 8300 \times \left(Q \times H \right)^{0.54}$$ (6.43)

or

$$C_{K1} = 2600 \times \left(kW \right)^{0.54}$$ (6.44)

6.5.6 Cost Analysis for Run-of-River Small Hydro Power Systems Projects

This subsection aims to analyze the cost of a run-of-river SHPs project, which is assumed to be a typical layout of run-of-river project with a tubular turbine. This category of turbine has a common utilization in the low head range. Runner diameter determines the turbine size, and the layout of the power house is worked out based on runner diameter. The following equations can be utilized for obtaining runner diameter D [7]:

$$D = \frac{84.6 \times \theta_3 \times \sqrt{H}}{N} \tag{6.45}$$

$$\theta_3 = 0.0223 \sqrt[3]{N_s^2} \tag{6.46}$$

$$N_s = \frac{1.358 N \sqrt{P}}{\sqrt[4]{H^5}} \tag{6.47}$$

where the specific speed of the turbine and the rotational speed of the turbine in revolutions per minute are demonstrated by N_s and N, respectively. H and P are the respective elements utilized to show the rated net head in meters and the rated available power in kilowatts at full opening of the gate. As mentioned above, the cost of a hydro power project contains the costs of civil works and the costs related to electromechanical equipment. In [7] a methodology has been presented for generating the cost data of a run-of-river SHPs project. It should be noted that all costs are taken in Indian rupees (Rs.). In the following discussion, the obtained equations for costs of different equipment of civil works of a run-of-river project are provided.

The cost of a powerhouse per kilowatt C_{PH} (Rs.), which is the major component in low head hydro power projects, can be stated as

$$C_{PH} = 92615 \times P^{-0.2351} \times H^{-0.0585} \tag{6.48}$$

The cost of a diversion weir and intake per kilowatt C_{DW} (Rs.) can be obtained as follows:

$$C_{DW} = 12,415 \times P^{-0.2368} \times H^{-0.0597} \tag{6.49}$$

The cost of a power channel per kilowatt C_{PC} (Rs.) can be obtained as follows:

$$C_{PC} = 85,383 \times P^{-0.3811} \times H^{0.0307} \tag{6.50}$$

The cost of a desilting chamber per kilowatt C_{DC} (Rs.) can be calculated as

$$C_{DC} = 20,700 \times P^{-0.2385} \times H^{-0.0611} \tag{6.51}$$

The cost of a forebay and spillway per kilowatt C_F (Rs.) can be calculated as

$$C_{PS} = 7875 \times P^{-0.3806} \times H^{0.3804} \tag{6.52}$$

The cost of a penstock per kilowatt C_{PS} (Rs.) can be calculated as

$$C_{PS} = 7875 \times P^{-0.3806} \times H^{0.3804} \tag{6.53}$$

The cost of a tail race channel per kilowatt C_{TR} (Rs.) can be calculated as

$$C_{TR} = 28,164 \times P^{-0.376} \times H^{-0.624} \tag{6.54}$$

As a result, the cost of civil works of a run-of-river SHPs project per kilowatt can be stated as

$$C_{CW} = C_{PH} + C_{DW} + C_{PC} + C_{DC} + C_F + C_{PS} + C_{TR} \tag{6.55}$$

In addition, the costs related to electromechanical equipment should be taken into account. The runner diameter and capacity of the plant considering prevailing market rates are the fundamental parameters that have been considered for providing a calculation formula for the cost of electromechanical equipment, as follows.

The cost per kilowatt of turbines with governing system C_T (Rs.) can be calculated as

$$C_T = 63,346 \times P^{-0.1913} \times H^{-0.2171} \tag{6.56}$$

The cost per kilowatt of generator with excitation system C_G (Rs.) can be calculated as

$$C_G = 78,661 \times P^{-0.1855} \times H^{-0.2083} \tag{6.57}$$

The cost per kilowatt of electrical and mechanical auxiliary C_{EM} (Rs.) can be calculated as

$$C_{EM} = 40,860 \times P^{-0.1892} \times H^{-0.2118} \tag{6.58}$$

The cost per kilowatt of transformer and switchyard equipment C_{TS} (Rs.) can be calculated as

$$C_{TS} = 18,739 \times P^{-0.1803} \times H^{-0.2075} \tag{6.59}$$

So, the cost per kilowatt of electromechanical equipment C_{EE} (Rs.) can be stated as follows:

$$C_{EE} = C_T + C_G + C_{EM} + C_{TS} \tag{6.60}$$

Considering other miscellaneous costs beyond the cost of the civil works and electromechanical equipment, the total required cost for a run-of-river SHPs project can be estimated. Miscellaneous cost include the cost of establishment including designs, audit and account, indirect charges, tools and plants, communication expenses, preliminary expenses on report preparation, survey and investigations, and cost of land. Miscellaneous cost C_M is estimated as 13% of the total costs related to civil works and electromechanical equipment. Accordingly, the total cost per kilowatt of low head run-of-river SHPs project, which is the sum of costs related to civil works and electromechanical equipment and miscellaneous cost, can be stated as follows:

$$C_{Total} = C_{CW} + C_{EE} + C_M = 1.13 \times (C_{CW} + C_{EE}) \qquad (6.61)$$

Investment costs of small hydro power projects are a bit higher, especially power plants with capacities of less than 1000 kW. Increase of the head and power capacity of a hydro plant results in reduction of investment costs of SHPs. For plants with power capacities between 1 MW and 7 MW in the United Kingdom, the capital costs per kW are between USD 3400 and USD 4000. However, SHPs with power capacities less than 1 MW require a capital cost per kW between USD 3400 and USD 10000.

Consideration of economic evaluation and investigation is one of the key factors of industrial projects. This investigation aims to analyze whether that foundation of SHPs is economical or not. For achieving this goal, sale price of electricity should be obtained and a comparison of sale price and suggested price of energy by the state is done. The net present value can be provided by subtracting the total capital cost from the present value of revenue. For obtaining electricity purchase price λ, formulations have been reported in [19], in which the 15-year return (N) has been considered. In addition, interest rate r and annual increment in electricity price $\Delta\lambda$ are taken into account equal to 6% and 10%, respectively. For obtaining a formulation for λ, V_{NP} should be considered equal to zero. Accordingly, we have [19]

$$V_{NP} = V_{RP} - C_{tpp} \qquad (6.62)$$

where V_{NP} is the net present value. Present value of revenue and the total capital cost have been demonstrated by C_{tpp} and V_{RP}, respectively.

$$V_{RP} = C_{tpp} \qquad (6.63)$$

$$V_{RP} = \sum_{n=1}^{N} \left(E_{net} \times 10^6 \times \lambda(1 + \Delta\lambda)^n \right) \frac{1}{(1+r)^n} \qquad (6.64)$$

$$\sum_{n=1}^{N} \left(E_{net} \times 10^6 \times \lambda(1 + \Delta\lambda)^n \right) \frac{1}{(1 + r)^n} = C_{tpp} \qquad (6.65)$$

6.6 SUMMARY AND CONCLUSIONS

An essential element of a country's development is electrical energy, which can be provided by the conversion of other forms of energy, utilizing several procedures for generating electrical energy. Renewable energy sources are defined as a valuable solution for the achievement of clean and sustainable energy. Hydro power has a significant role in electrical energy production in the world, which is widely distributed. Considering reports of world electrical energy production, hydro power systems account for 16% of total world power production. Small hydro power generation systems are introduced as an important solution for electrification in rural areas. Removal of the cost of transmission line construction and power loss, mitigation of environmental constraints, and less environmental influence are the consequent advantages of small hydro plants. Considering this fact, knowledge of small hydro power generation plants and basic principles of designing a small hydro plant is needed. This chapter provided an overview of the design of small hydro generation systems.

First, classifications of hydro power systems were introduced and a brief explanation of SHPs, which were the focus of this chapter, were provided. Second, basics of hydro power generation and components of SHPs were defined. The available hydraulic power of a hydro system and the corresponding energy were formulated. Third, different technologies of SHPs and a classification of turbine technologies were provided, taking into account the related formulation of each category. Design of an appropriate turbine for SHPs project was then analyzed. Fourth, an economic analysis of SHPs projects considering investment costs and annual costs was discussed. Estimation of costs of different technologies of turbines and a run-of-river project were formulated and studied. Finally, a cost analysis of designing a SHPs project and the costs of turbines and run-of-river hydro power generation system project was provided. This chapter can be useful to undergraduate students and early-career engineers for understanding the basics of small hydro power generation plants and fundamental principles of designing SHPs.

REFERENCES

[1] G. Ardizzon, G. Cavazzini, G. Pavesi, A new generation of small hydro and pumped-hydro power plants: advances and future challenges, Renew. Sustain. Energy Rev. 31 (2014) 746–761.

[2] M. Mohibullah, A.M. Radzi, M.I.A. Hakim, Basic design aspects of micro hydro power plant and its potential development in Malaysia, in: Power and Energy Conference, 2004. PECon 2004. Proceedings National, 2004, , pp. 220–223.

[3] D.K. Okot, Review of small hydropower technology, Renew. Sustain. Energy Rev. 26 (2013) 515–520.

[4] A. Rasoulzadeh-akhijahani, B. Mohammadi-ivatloo, Short-term hydrothermal generation scheduling by a modified dynamic neighborhood learning based particle swarm optimization, Int. J. Electric. Power Energy Syst. 67 (2015) 350–367.

[5] S. Adhau, R. Moharil, P. Adhau, Mini-hydro power generation on existing irrigation projects: Case study of Indian sites, Renew. Sustain. Energy Rev. 16 (2012) 4785–4795.

[6] J.H.I. Ferreira, J.R. Camacho, J.A. Malagoli, S.C.G. Júnior, Assessment of the potential of small hydropower development in Brazil, Renew. Sustain. Energy Rev. 56 (2016) 380–387.

[7] S. Singal, R. Saini, C. Raghuvanshi, Analysis for cost estimation of low head run-of-river small hydropower schemes, Energy Sustain. Dev. 14 (2010) 117–126.

[8] H. Balat, A renewable perspective for sustainable energy development in Turkey: the case of small hydropower plants, Renew. Sustain. Energy Rev. 11 (2007) 2152–2165.

[9] O. Paish, Small hydro power: technology and current status, Renew. Sustain. Energy Rev. 6 (2002) 537–556.

[10] P. Purohit, Small hydro power projects under clean development mechanism in India: A preliminary assessment, Energy Policy 36 (2008) 2000–2015.

[11] O. Paish, Micro-hydropower: status and prospects, Proc. Inst. Mech. Eng. A: J. Power Energy 216 (2002) 31–40.

[12] I. Loots, M. Van Dijk, B. Barta, S. Van Vuuren, J. Bhagwan, A review of low head hydropower technologies and applications in a South African context, Renew. Sustain. Energy Rev. 50 (2015) 1254–1268.

[13] Y. Keawsuntia, Electricity generation from micro hydro turbine: a case study of cross-flow turbine, in: Utility Exhibition on Power and Energy Systems: Issues & Prospects for Asia (ICUE), 2011 International Conference and, 2011, , pp. 1–4.

[14] B.A. Nasir, Design of micro-hydro-electric power station, Int. J. Eng. Adv. Technol. 2 (2013) 39–47.

[15] D. New, Intro to hydropower, Part 2: Measuring head and flow, [online]; December (2004) *Home Power*, Available on www.homepower.com.

[16] F. Forouzbakhsh, S. Hosseini, M. Vakilian, An approach to the investment analysis of small and medium hydro-power plants, Energy Policy 35 (2007) 1013–1024.

[17] S. Hosseini, F. Forouzbakhsh, M. Rahimpoor, Determination of the optimal installation capacity of small hydro-power plants through the use of technical, economic and reliability indices, Energy Policy 33 (2005) 1948–1956.

[18] G.A. Aggidis, E. Luchinskaya, R. Rothschild, D. Howard, The costs of small-scale hydro power production: Impact on the development of existing potential, Renew. Energy 35 (2010) 2632–2638.

[19] A. Naghiloo, M. Abbaspour, B. Mohammadi-Ivatloo, K. Bakhtari, Modeling and design of a 25 MW osmotic power plant (PRO) on Bahmanshir River of Iran, Renew. Energy 78 (2015) 51–59.

CHAPTER 7

Energy Storage Systems

Hamdi Abdi*, Behnam Mohammadi-ivatloo†, Saeid Javadi*,
Amir Reza Khodaei*, Ehsan Dehnavi*
*Razi University, Kermanshah, Iran
†University of Tabriz, Tabriz, Iran

7.1 INTRODUCTION

Distributed generation (DG) is getting more attention in recent years. This is mainly due to the various advantages of DGs, such as an electrical energy loss reduction in the distribution system, reduction of voltage fluctuations, increasing reliability, power quality improvement, energy cost reduction, and ultimately increasing customer satisfaction. Despite all the benefits associated with DG in power systems, interconnecting these new technologies to the national energy systems leads to some crucial problems such as changing the protection setting, power system stability, and islanding phenomena.

DGs may include different forms of electrical energy generation; renewable resources, mainly wind and solar power plants, or nonrenewable resources (conventional methods). Employing most of the renewable energy resources, such as wind farms and photovoltaic (PV) systems as DGs leads to the main challenges: changeability and uncontrollability of output power. Indeed, these main features lead to additional fears in DGs application in a power system. Using an energy storage system (ESS) is proposed and is one of the most appropriate solutions in this area. This new category enables engineers to manage the power system optimally.

Generally, the ESS operation is categorized as follows:

- *The charging period*: This process is applicable using the network electrical energy, during the off-peak intervals when the electrical energy is available at lower prices,
- *The discharging period*: In times of peak the stored energy in an ESS is used. It should be mentioned that in this period the network electrical energy has a higher price and use of DGs is more economical. Accordingly, application of an ESS system is mainly explainable for reducing or even eliminating the uncertainties of renewable DG.

It should be mentioned that the most commonly used methods in ESSs are based on the DC type, so using these systems is widely more intertwined with power electronic devices to connect with the national power grids.

Generally, a variety of ESSs can be provided in terms of technology, location, capacity, demand, and costs of investment.

In this chapter, different operational status related to DGs prescience in power system consisting of interconnected or isolated mode is presented. Modeling storage system devices considering practical small- and large-scale ESSs based on different applications is described. Furthermore, the governing relations with each technology are explained in detail. Finally, some important points related to the economics of ESS operation will be discussed.

7.1.1 Major Interconnection Issues

Generally, DG-based power systems can operate in independent or stand-alone and grid-connected modes. In fact, in the first mode, the capacity of a DG unit is selected only based on the load requirements. But, in the latter case, this constraint is not determinant. Although the grid-connected operation mode usually is preferred due to the bilateral energy exchanges, the islanding condition is a main concern that should be considered.

Islanding means that one or several power plants (i.e., DGs), isolated from the national power grid, supply a part of the electrical network independently and following some faults in the main network. Operating in islanding condition is undesirable, as this mode may cause unwanted problems such as creating hazards for maintenance and repair staff, and equipment damage due to the instability in voltage and frequency.

Certain issues that arise due to the interconnection are discussed below.

7.1.2 Technical Concerns

- Stability: Interconnection of the DG to the grid affects the rotor angle, voltage, and frequency stability of the grid. Based on the type and size of the generators, DG either improves or worsens the stability of the system.
- Power quality: The power quality of the grid has recently become a problem with the increased use of power electronic devices. Most of the distributed generators are interfaced to the grid through power electronic circuits. Use of these power electronic interfaces increases the already existing power quality problem.
- Voltage fluctuations: Power injected by certain DG technologies, such as wind turbines and PV power plants, is fluctuating. This results in the local voltage fluctuation.

- There is a limit on the number of DG units that can be connected to the system. This limit depends on the size and type of the system. The supplied reactive power must be equal to the reactive power demand to maintain the voltage level of the system in the allowed region. Connecting more DGs may increase the reactive power supplied, thereby increasing the voltage level of the system considerably.
- DGs increase the short circuit current during a fault, which creates more challenging protection requirements. Therefore, to use DGs, improved protection devices need to be used, which adds to the cost of the system.
- Increasing the penetration level of DGs is difficult and it also takes a long time to locate any fault in the system. Moreover, the direction of current flow becomes unpredictable.
- Existing radial distribution system and control are designed to handle the power flow in just one direction. When DGs are connected, power flows in both directions. Therefore, existing systems need to be upgraded.

7.1.3 Economic Concerns

- Because of the unpredictable nature of the cost of fuel, it is difficult to properly plan the interconnection of DGs to the grid. This results in financial problems for customers.
- The network operator has to differentiate between the power from the grid and the power from the DGs.
- As mentioned earlier, the enhancement of already existing protection devices is necessary, which adds to the cost of the system.
- Needed improvements in the system increase the total cost of the system.

7.1.4 ESS Capacity and Size

Up to now different ESS technologies have been proposed. Some of them are suitable for medium-scale applications such as pumped storage, compressed air energy storage (CAES), flywheel, and superconducting magnetic energy storage (SMES). Mainly they have the capacity in MW ranges [1].

A general overview of ESS is described in Fig. 7.1 in which the systems are classified based on their applications.

Also, the characteristics related to the most important ESS technologies are indicated in Table 7.1 [2]. Furthermore, Fig. 7.2 presents some useful comparisons about ESS technologies by their power and energy densities.

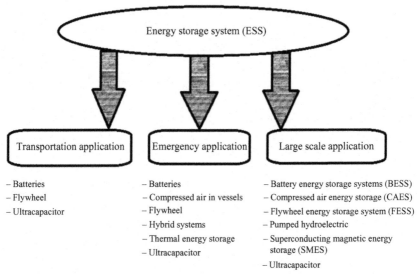

Fig. 7.1 Classification of the principal energy storage systems.

Table 7.1 Energy storage technologies

Technology	Power	Energy density	Response time	Efficiency
Pumped hydro	100 MW–2 GW	400 MWh–20 GWh	12 min	70–80%
CAES	110 MW–290 MW	1.16 GWh–3 GWh	12 min	90%
BESS	100 W–100 MW	1 kWh–200 MWh	Seconds	60–80%
Flywheels	5 kW–90 MW	5 kWh–200 kWh	12 min	80–95%
SMES	170 kW–100 MW	110 Wh–27 kWh	Milliseconds	95%
Super capacitors	<1 MW	1 Wh–1 kWh	Milliseconds	>95%

7.1.5 Power Electronic Interface

The electrical output power in all of the energy storage devices is in the form of DC power and it is required to be converted to AC using the power electronic devices and delivered to the power grid.

Generally, when the produced power is higher than the demand power, the extra power will be stored in the system (the system is charged).

Converting the AC power to DC power is inevitable in all of the electrical energy storage systems (EESSs). It means that when the power demand is more than the generated ones, the energy flows from the EESS to the

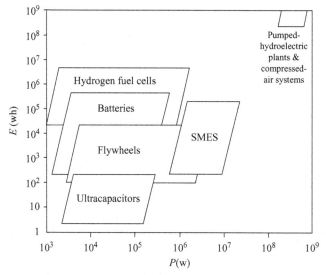

Fig. 7.2 Power and energy densities of different energy storage devices.

power grid. This means that the system enters the discharging mode. This interval requires converting DC power to AC.

Fig. 7.3 briefly shows the interconnection of an EES to the network through unidirectional and bidirectional converters. Because using two unidirectional converters is more costly than using a bidirectional converter, bidirectional converters are preferred in practical applications. As indicated

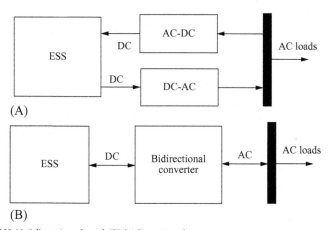

Fig. 7.3 (A) Unidirectional and (B) bidirectional converter.

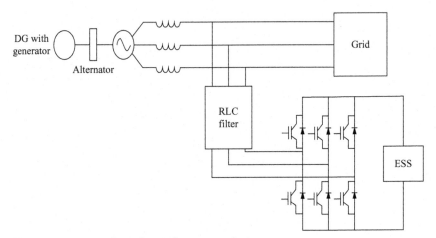

Fig. 7.4 Interconnection scheme for storage devices.

in Fig. 7.4, a bidirectional converter acts such as rectifier during the battery charging process [3]. This role will be changed to an inverter during the discharging process, in which the batteries are charged.

7.2 SMALL-SCALE ELECTRICAL ENERGY STORAGE SYSTEMS

As mentioned in the previous section, the output of most renewable energy resources such as wind and solar is unpredictable. In such cases, the energy storage devices can be used to augment DGs when it fails to fulfill the load requirements. In addition, the storage devices have also been identified in improving the impacts of DGs on the grid [4–6]. As it is necessary to understand the different types of storage devices, a brief overview of different storage systems is given. There are varieties of energy storage devices that have different techniques for storing energy.

Electrical energy can be saved in different forms; for example, mechanical, electrochemical, chemical, electromagnetic, thermal, and so on. This section provides an overview of the features of some small energy storage devices that are used in distribution applications. The small-scale EES systems described are batteries, super capacitors, electrochemical capacitors, as well as hydrogen storage.

7.2.1 Battery

Batteries are the most commonly used type of storage devices in distributed applications. Research has been carried out to study technical and economic performances of batteries [7,8]. Recently, the type of materials used in batteries has progressed a lot, which make them more feasible for use in a wide range of applications. Among the different types, lead acid batteries are the most commonly used ones. However, Li–ion batteries are gaining a lot of popularity mainly due to their high energy densities, energy capabilities, and lifetime compared to lead acid batteries. Though their maintenance costs are low compared to lead acid batteries, their capital cost is very high, which is the reason for their limited use. Batteries are the most efficient and economic devices for storing energy.

Regardless of batteries' application as EESSs in large scale, which has received much attention in recent years, many applications have been reported for batteries in recent years. However, batteries used in the new systems should have specific conditions. Some of these conditions are as follows [9]:

- Capability of reliably providing the required energy based on consumption patterns and different conditions.
- Economic necessity of batteries due to the continued application in the used period.

It should be noted that in addition to the abovementioned features, long durability and the ability to accurately measure the amount of stored energy are features of batteries that should be included.

In general, a battery is a source of the electric potential energy that after some internal chemical reactions, the chemical energy is converted to the electrical energy and this energy is accessible in its poles. The received energy in a battery's poles is called the electric propulsion and is measured by the volt unit. The positive pole is called the cathode and the negative one is called the anode.

Usually every battery is formed by one or some internal cells, which are parallel to increase the current. This combination is connected in series. Fig. 7.5 shows the operating principles of a battery. Every battery is formed from two half-cells, which are connected in series by an electric conductive called the electrolyte including the positive and negative ions. By connecting the battery to the electrical load, the negative ions are interned to the consumer through the conductor wire and after creating energy, they are moved to the positive ions and gradually neutralize the positive ions [10]. With the passing of time, more positive ions are neutralized and the battery's energy is decreased and its internal resistor is increased. In this state, after some specific time, which is a function of the battery's capacity, the battery is totally discharged.

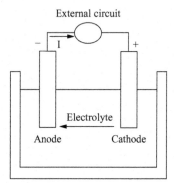

Fig. 7.5 Principle of operation of an electrochemical cell.

7.2.1.1 Classification of Cells

Batteries can be classified based on the following criteria:
- Recharge and use
- Material used for electrodes and electrolytes
- Power level

If a battery is used only once and cannot be charged to be used again, then it is called a primary battery. On the other hand, if the battery can be recharged and used for a number of times, it is called a secondary battery.

7.2.1.2 Self-Discharge Rate

Self-discharge of a cell is defined as the leakage in the current between the anode and the cathode, due to certain chemical reactions that take place within the cell. Even if the cell is not used, it can still get discharged to a certain level due to internal reactions. The rate at which the cell gets self-discharged depends on the cell chemistry as well as on the temperature at which it is being operated.

7.2.1.3 Discharge Rate

The rate of discharge of the cell also plays a vital role in its performance. Practical experiments have proven that the capacity of the cell decreases if the discharge time is more.

7.2.1.4 State of Charge

The state of charge (SOC) of a cell denotes the capacity that is currently available as a function of the rated capacity. The value of the SOC varies between 0% and 100%. If the SOC is 100%, then the cell is said to be fully charged, whereas a SOC of 0% indicates that the cell is completely discharged. In practical applications, the SOC is not allowed to go beyond 50% and therefore the cell is recharged when the SOC reaches 50%. Similarly, as a cell starts aging, the maximum SOC starts decreasing. This means that for an aged cell, a 100% SOC would be equivalent to a 75%–80% SOC of a new cell.

7.2.1.5 Important Consideration in the Batteries' Design

As mentioned before, the battery is a set of cells, which are connected in parallel or series to boost the current or voltage. However, the design of a set of (bank) battery is dependent to the system's operational conditions which will be explained below [11].

Bank voltage. The total voltage of the bank is met by connecting the cells in series. Fig. 7.6 shows the series connection of cells to obtain higher voltage levels. It can be seen that a 2X(V) battery bank is obtained by connecting two X(V) cells in series.

Bank capacity. Based on the consumer's needed capacity, by connecting the cells in parallel the output current should be chosen, so that the required capacity is provided (as illustrated in Fig. 7.7).

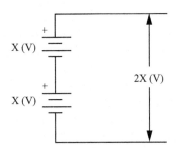

Fig. 7.6 Series connection of cells.

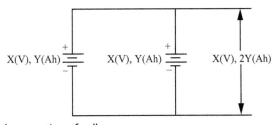

Fig. 7.7 Parallel connection of cells.

Example 7.1

Consider a 12 V and 18 Ah cell, if there are 30 of these cells including 6 parallel branches. There are five batteries in each branch. Calculate the output voltage and capacity?

Solution

Output voltage:

$$V_{out} = 5 \times 12 \text{ V} = 60 \text{ V}$$

Output capacity:

$$Ah_{out} = 6 \times 18 \text{ Ah} = 98 \text{A h}$$

Some of the common types of cells are listed below:
- Alkaline (zinc–manganese dioxide)
- Lead acid
- Nickel Cadmium (Ni-Cd)
- Lithium (lithium–copper oxide) Li–CuO
- Nickel-metal hydride (NiMH)
- Lithium (lithium–iron disulfide) LiFe S_2
- Lithium-ion (Li-Ion)
- Lithium-ion polymer
- Nickel oxyhydroxide (zinc–manganese dioxide/nickel oxyhydroxide)
- Zinc–Carbon
- Zinc–Chloride
- Lithium (lithium–manganese dioxide) LiMn O_2
- Mercury Oxide
- Zinc–Air
- Silver-oxide (Silver–Zinc)

7.2.2 Super Capacitor

Super capacitors are also electrochemical devices that have started to establish themselves as power sources capable of quickly recharging and supplying power. They have slowly started to replace traditional batteries in hybrid vehicles. In distributed applications, they have been used together with batteries to act as storage devices, but so far they have not been used independently. Similar to Li-ion, the capital cost of the super capacitors is very high, which is a major concern in their large-scale use. It has been suggested that if their capital cost is cut down, they would readily replace batteries and find a new market [12]. Super capacitors have very high power densities compared to batteries, with an almost unlimited number of charging and discharging cycles.

Super capacitors, also called ultracapacitors or electrical double-layer capacitors, are electrochemical storage devices that have gained popularity in recent years because of their large power densities. They are different from ordinary capacitors by the type of electrodes used. The carbon technology that is used in them creates a bigger surface area for a smaller separation of the electrodes. Though a super capacitor stores much less energy compared to a battery, it has almost unlimited life. Unlike batteries, super capacitors are not commercially available in large scale for DG applications, but will definitely be used in the near future due to their tremendous advantage compared to batteries. The most important drawback of the super capacitor that prevents its usage in large scale is its high cost. A number of super capacitor manufacturers are working on reducing the cost, attempting to make them available at lower initial costs in the market similar to batteries. Similarly, advancement continues to develop the technology to improve the energy density of the cells making them suitable for use in long-term applications. It is predicted that super capacitors will be used in the near future for storing energy from distributed systems and therefore it has been included in this discussion.

7.2.2.1 Cell Construction
Fig. 7.8 shows the basic construction of the super capacitor cell. Similar to the construction of the battery, it also has two electrodes immersed in an electrolyte. But, the difference lies in the material with which the electrodes are made as well as the material used as the electrolyte. The electrodes of the

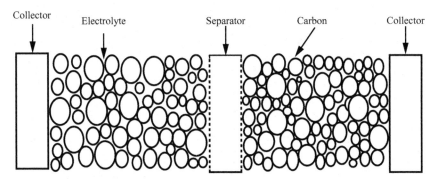

Fig. 7.8 An ultracapacitor cell construction.

super capacitor are generally made up of porous, high surface area particles. This increase in surface area results in an increased amount of ion absorption, which in turn results in higher power densities of the cells. The electrolyte is impregnated in between the anode and the cathode and is responsible for the movement of ions from one electrode to the other. The electrolytic solution is either potassium hydroxide or sulfuric acid. There is a thin membrane in between the electrodes called the separator. The main purpose of the separator is to prevent the electrodes from touching each other, but allow the flow of ions between them. Both the anode and the cathode are lined by current collectors, the purpose of which is to assure a proper interface between the electrodes and the external connections [13]. Based on the type of material that is used for each electrode, the super capacitor can be classified as either symmetrical or asymmetrical. In symmetrical capacitors, both the electrodes are made of the same material, which is usually carbon. In the asymmetrical type of capacitors, which are the newest types, the electrodes are made up of different materials. Carbon is used for one electrode and nickel hydroxide is used for the other. This is done mainly to make the charges on both sides of the capacitor unequal, thereby increasing the energy density and decreasing the leakage current [14].

7.2.2.2 Working Principle of the Cell

The storage mechanism in super capacitors is the transfer of energy between the electrodes and the electrolyte [15]. When a certain amount of voltage is applied, an electric field is created between the electrodes. As a result of the generated electric field, an absorption layer is formed on the activated carbon electrodes. The charging and discharging of the ions takes place in this absorption layer. The charged ions migrate toward the oppositely charged electrodes through this layer. The charge or discharge time depends on the physical structure of the cell and is usually a few microseconds. The energy stored in the cell is given by the following equations:

$$E = \frac{1}{2}CV^2 \qquad (7.1)$$

$$E\,(\text{Watt hour}) = E\,(\text{Joules}) / 3600\,(\text{s}) \qquad (7.2)$$

where E is the energy storage (Joules), C the capacitance (Farad), V the voltage (volts).

The amount of electrical energy that is stored in the capacitor depends on the three following factors [15]:

1. Surface area of the electrodes: the greater the surface area, the greater the amount of energy stored in the capacitor.
2. Distance between the electrodes: the amount of energy that can be stored varies directly with the distance between the electrodes.
3. Dielectric constant of the material used for the separation.

Most commercially available super capacitors have specific energy densities of about 5 W h/kg and specific power densities of about 20 kW/kg. The voltage level of super capacitors is about 2.5 V to 2.7 V per cell. The internal resistances of super capacitors are very low compared to batteries, thereby resulting in higher power densities than batteries.

Example 7.2

The energy density of a super capacitor is 5 W h/kg and its weight is 50 kg, the total voltage is 100 V. Specify the number of capacitors in series and the number of parallel branches. The voltage and capacity of each cell is 2.5 V and 200 mF, respectively.

Solution
Energy storage:

$$\text{Energy storage} = \text{Energy densities} \times \text{Weight} = 5 \text{ W h/kg} \times 50 \text{ kg}$$
$$= 250 \text{ W h}$$

$$E\,(\text{Watt hour}) = E\,(\text{Joules})\,/\,3600\,(\text{s}) \Rightarrow E\,(\text{Joules}) = 250 \text{ W h} \times 3600\,(\text{s})$$
$$= 900 \text{ kJ}$$

Total capacitance:

$$E = \frac{1}{2} \times C_t V^2 \Rightarrow 900 \text{ kJ} = \frac{1}{2} \times C_t \times 1000^2 \Rightarrow C_t = 1.8 \text{ F}$$

Number of series capacitors:

$$N_{C_s} = \frac{100 \text{ V}}{5 \text{ V}} = 20$$

Capacity of series capacitors:

$$C_s = \frac{200 \text{ mF}}{20} = 10 \text{ mF}$$

Number of parallel capacitors:

$$C_t = C_p \times N_{C_p} \Rightarrow N_{C_p} = \frac{C_t}{C_p} = \frac{1.8 \text{ F}}{10 \text{ mF}} = 180$$

Example 7.3

The area section of the pages of a capacitor cell 20×20 cm and the distance between the pages is 10 cm and a dielectric between the plates of the air. If there are 10 tuberculosis series in each branch and five parallel branches,

(A) Calculate the total capacity?

(B) What is the amount of energy stored at voltage of 400 volts?

Solution

(A) Area section:

$$A = 20 \times 10^{-2} \times 20 \times 10^{-2} = 0.04\,\mathrm{m}^2$$

Total capacity:

$$C = \varepsilon \frac{A}{d} = 8 \times 10^{-12} \frac{4}{0.1} = 32\mathrm{nF}$$

Stored energy:

$$E = \frac{1}{2}CV^2 = \frac{1}{2} \times 32 \times 10^{-12} \times 20,000^2 = 6.4\,\mathrm{mJ}$$

7.2.2.3 Cell Characteristics

Charging and Discharging

Super capacitors can be charged and discharged by either the constant current method or the constant power method.

- *Constant current charging.* Super capacitors are charged by a constant current input from the converter. The chosen converter can be either a buck or a boost converter. In most of the applications, a buck converter is preferred because of its continuous output charge current.
- *Constant power charging,* In this method, the current from the source is drawn at a constant voltage and used for charging the super capacitors. This method transfers all the available charge from the source to the super capacitors quickly. When compared to a constant current charging method, the time taken for charging in a constant power method is much higher.

7.2.3 Electrochemical Capacitor

Super capacitors, also known as electrochemical capacitors, store the using ionic absorption (two-layer electrochemical capacitors) or fast reactions.

These capacitors can be a suitable alternative or supplement to batteries in electrical energy storage applications when the electrical energy is required at high powers. Generally, electrochemical capacitors are more similar to batteries in configuration and applications [16].

With recent advances in electric charge storage structures, as well nanostructural materials significant improvements in performance are achieved.

Ions decomposition in smaller pores of dissolved ions using carbon electrodes micro-nano pores leads to manufacturing high-capacity capacitors for electrochemical double-layer capacitors. Thus, designing new devices with high-energy capacity by using different electrodes is possible.

The combination of pseudocapacitive nanomaterials contains acids, nitrides, and polymers with the latest generation of nano-lithium electrodes made advanced microelectrochemical energy volume capacitors, which makes it possible to build flexible and compatible devices.

The main parts of an electrochemical capacitor are described in the following discussion.

7.2.3.1 The Main Capacitor

Electrical energy is stored in the capacitor. Different types of capacitors include the electrolytic capacitor, the static capacitor, and the electrochemical capacitor.

A capacitor is made of two parallel plates and the space between them is filled with a dielectric such as air, paper, and so on.

The stored energy in an ideal electrochemical capacitor is presented as

$$E = \frac{1}{2}CV^2 \tag{7.3}$$

where C is the capacity in farads, and V is the voltage in volts.

Example 7.4

In an electrochemical capacitor with a voltage of 20 kV and 2 farad capacitor capacity, calculate the energy that is stored in it.

Solution

$$E = \frac{1}{2}CV^2 = \frac{1}{2} \times 2 \times 20,000^2 = 400 \text{ MJ}$$

Example 7.5

Calculate the amount of capacitance to voltage 400 V at 0.5 mA for 10 s.

Solution

Current:

$$I = qt \Rightarrow q = \frac{I}{t} = \frac{5 \times 10^{-3}}{10} = 500 \ \mu c$$

Capacitance:

$$C = \frac{q}{V} = \frac{500 \ \mu c}{400} = 1.25 \ \mu F$$

7.2.3.2 Electrochemical Capacitor Characteristics

An electrochemical capacitor is a component that includes two electrodes, a separator, an electrolyte, two current collectors, and an enclosure containing these devices [17]. The storing energy process is not same as the batteries. The electrolytes that are used in the electrochemical capacitors can be potassium hydroxide or sulfuric acid. In a double-layer capacitor, two capacitors are placed in series with the electrolyte. Fig. 7.9 shows this sample structure.

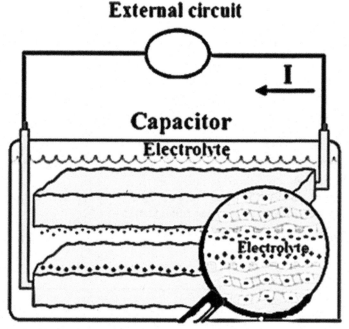

Fig. 7.9 A double-layer capacitor.

7.2.4 Hydrogen Storage

Hydrogen is one of a simplest and most abundant natural elements that can be a viable alternative to fossil fuels. A product resulting from the combustion of hydrogen is water vapor that is dispersed in the air and helps in softening the environment.

7.2.4.1 Fuel Cell

A fuel cell (FC) is a device in which hydrogen or natural gas (as input) enters into an electrochemical reaction with oxygen, and as long as this process continues, electrical energy will be generated in direct current (DC). FC efficiency in a favorable condition is almost double of that of an internal combustion engine (ICE), which is justification for increasing the application of these devices.

The key system components for providing DC power in a FC is described in Fig. 7.10.

7.2.4.2 Application of Hydrogen Cells

Hydrogen is the main element in hydrogen-powered fuel. Using hydrogen cells is desirable to prevent environmental pollution. This device only needs oxygen and hydrogen to operate, and the fuel is cheap and readily available. Hydrogen fuel cells are not yet widely released because their production costs are higher than those of fossil fuels. Perhaps the most important reason for their higher cost is that a precious metal such as platinum is used as the catalyst.

7.2.4.3 Hydrogen Fuel Cell Performance

The working mechanism of a hydrogen-powered FC is described as follows. A hydrogen molecule, that is, H_2, releases two electrons in the presence of the platinum and they are delivered to the metal (anode). Accordingly, the hydrogen molecule turns into protons, as in Eq. (7.4).

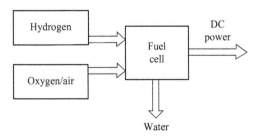

Fig. 7.10 The key system components for providing DC power.

$$H_2 \rightarrow 2H^+ + 2e \qquad (7.4)$$

Finally, when the protons are located in the vicinity of the metal, they are capable of getting two electrons (cathode), colliding with an oxygen atom, and consequently a water molecule is formed.

$$2H^+ + \frac{1}{2}O_2 + 2e \rightarrow H_2O \qquad (7.5)$$

The release of two electrons from the cathode and their receipt by the anode results in establishing a direct electrical connection between two poles.

This is a simplified discharging process in a hydrogen cell and the charging process is to the contrary. Generally, these two processes relay on the difference in power demand and generation [18].

Example 7.6

In a hydrogen container it is assumed that 12 molecules of hydrogen at the anode reacted.

(A) The number of free electrons at the anode

(B) The number of water molecules generated at the cathode

Solution

(A)

$$\text{Number of electrons} = 12 \times 2 = 24e$$

(B)

$$\text{Number of } H_2O \text{ molecules} = 12$$

The major applications of hydrogen FCs are

- Energy storage with frequently repeated charging and discharging modes
- Power generation
- Electric vehicles

7.3 LARGE-SCALE ELECTRICAL ENERGY STORAGE SYSTEMS

7.3.1 Flywheel

Flywheels have been used for a long time and still have many applications. The most common application of this technology is its use in the internal combustion engines. A flywheel is a simple form of mechanically storing energy. The flywheel is a heavy rotating disk that is used to store angular momentum. This equipment resists changes in rotational speed and when an irregular torque is applied to it, it tries to stabilize the rotation axis. In some small motors, flywheels are used to store energy for a long time and then release it in a shorter time, and they cause the power to increase in a short period. Their structures are usually made of cast iron and their size is different based on the motor's type. The size of a flywheel varies based on the cylinders and engine type and size. In general, with increasing the number of cylinders, the size of flywheels can be smaller.

Flywheels are similar to mechanical batteries, which work based on the rotation of a particular mass around an axis and stores energy as kinetic energy. This energy is used to accelerate a motor by an internal motor and the electrical energy is produced by this motor in the generator mode. Theoretically, with increasing the flywheel speed, the amount of stored energy is increased. However, it can be used in speed reduction.

7.3.1.1 Energy Storage Capacity

Flywheels store energy in the form of the spinning object's angular momentum. This spinning object is actually a rotor that stores the energy in the form of kinetic energy. The amount of stored kinetic energy in a spinning object depending on its mass and rotational velocity [19].

$$E = \frac{1}{2}I\omega^2 \tag{7.6}$$

where E is the kinetic energy, I is the moment of inertia (with units of mass distance), and ω is the rotational velocity (with units of radians/time). The moment of inertia depends on the geometry and mass of the spinning object. The stored kinetic energy is calculated using Eq. (7.7):

$$E = \frac{1}{4}Mr^2\omega^2 = \frac{1}{4}Mv^2 \tag{7.7}$$

where M is the mass of the disc, r is its radius, and v is the linear velocity of the outer rim of the cylinder (approximated by $r\omega$). It is concluded from Eq. (7.7) that to improve the energy capacity of a flywheel, increasing the rim speed is more effective than increasing the mass of the rotor. In practice, the design of the flywheel is limited by the strength of the rotor material to withstand the stresses caused by rotation.

Example 7.7

Calculate the stored energy, linear speed, and moment of inertia in a flywheel with the following characteristics:

  Radius of the disc: 0.5 m
  Mass of the disc: 7 kg
  Rotational velocity: 10 radians/time

Solution

Linear speed:

$$v \left(\frac{m}{s}\right) = r \ (m) \times \omega \ (radians/time) = 10 \times 0.5 = 5 \ \frac{m}{s}$$

Stored energy:

$$E = \frac{1}{4}Mv^2 = \frac{1}{4} \times 7 \times 5^2 = 43.7 \ J$$

Moment of inertia

$$I = Mr = 7 \times 10 = 70$$

7.3.1.2 Principles of Energy Conversion

Flywheels' structure is based on storing energy in the form of kinetic energy and delivering it to consumers in the form of electrical energy. So, studying this energy conversion, which is done by the electromechanical effects in the machine, is very significant. Regardless of this, it should be noted that in the charge mode, to maximize the stored energy, the speed and acceleration of the machine should be maximized and these quantities are reduced during the discharge period for the use of energy. Power electronic devices and the clutch system are two main methods in this field. The first one is technologically advanced, but its cost is high, and the second one causes more losses in spite of its simplicity.

7.3.1.3 Energy Losses

The stored energy in a flywheel is proportional to the flywheel's mass and the square of the rotational speed. Flywheels are usually made of a ring made of titanium and a cylinder of composite material, which are able to rotate up to tens of thousands of revolutions per minute. To reduce friction in high speed (mainly because of the rotor contact to its surroundings) and consequently reducing energy losses, this equipment should not have any contact with any object. So, flywheels are usually installed on magnetic bearings in such a way that the contact between the fixed and moving parts can be ignored. Accordingly, the stored energy in the magnetic bearing system is low and does not cause the depletion of the flywheel.

7.3.1.4 Flywheel Subsystems

The main subsystems of a flywheel are as follows [20].
- *Rotor*: stores the energy in the kinetic energy form.

- *Bearings*: hold the rotor.
- *Motor-Generator*: mainly converts the stored kinetic energy into the electrical energy appropriate for the demands, or vice versa.
- *Conversion part or power electronic device*: its main duty is to provide the appropriate voltage and current requested by the demand based on the various conversion equipment such as converters, rectifiers, and so on.
- *Control and measurements*: monitor the system operation conditions based on using the miscellaneous control and instrumentation devices to ensure that the flywheels are operated in permissible operating constraints.
- *Framework or housing*: includes all of the abovementioned devices and protects the flywheel against all of the hazards and failures.

7.3.1.5 Motors and Generators
In the motor mode, electrical energy is converted into kinetic energy, or a flywheel will be charged. This state is called the charging period. Also, in the discharge period, called the generator mode, the mechanical energy will be transformed into electrical energy. A motor generator system may include a motor and a generator, separately, or an aggregated system capable of operating in both cases, according to the current direction [21]. Fig. 7.11 shows the flywheel structure.

7.3.1.6 Features and Limitations
This section reviews some of the main specifications and limitations related to flywheels.

(A) *Maximum power*. Calculating the energy stored in the flywheel was previously described. As is clear, the stored energy is a function of the rotation speed and mass. However, the maximum power is dependent on the motor-generator specification, as well as the power electronic devices. As a result, the stored energy in a flywheel is completely independent of the power.

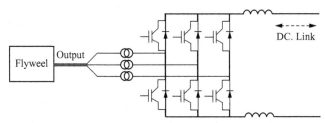

Fig. 7.11 Flywheel structure.

Flywheels can deliver high power in a relatively short period of time. There are flywheels that can provide power capacity from 100 to 2000 kWac in a short time period, between 5 and 50 seconds.

(B) *Energy and efficiency*: As mentioned before, the stored energy in a flywheel is completely dependent on the speed and mass. Accordingly, the efficiency is a function of loss amounts and friction.

(C) *Charge and recharge times*: The charging time in flywheels is relatively low, but they need a relatively longer time for the discharging process. Both of these times are in the range of seconds.

7.3.1.7 Advantages and Disadvantages of Flywheels

The main advantages are as follows:
- Compactness and low volume
- High performance
- Low maintenance
- Lack of sound and noise

The main disadvantages are:
- Safety considerations
- High cost of raw materials
- High cost of magnetic bearings

7.3.2 Superconducting Magnetic Energy Storage

SMES devices are used to improve power system stability and reduce the energy resource changeability. The electrical energy stored in the magnetic field is generated by DC. This device has the capability to be charged or discharged such as the battery, frequently, without any changes in its magnetic field properties. The superconducting coils have a very high efficiency of about 95% during the storage process. Furthermore, they are used to smooth the load curve and increase the general load factor referred to the system.

Different disturbances such as sudden load changes, switching, and transmission line outages leads the power system to an imbalance status and instability conditions. In this case, the kinetic energy of shaft synchronous generators is forced to increase the required energy. As a result, the control loops will be activated to maintain power system stability. This process leads to sudden changes in various parameters of the system, such as frequency and transmission line flows. Accordingly, occurring cascading problems are very probable. If a suitable amount of electrical energy stored in the system is available, the abovementioned problems will be reduced or even eliminated by the quick energy exchange with the power system.

SMES has a fast dynamic response. This is mainly due to converting the electrical energy to magnetic energy or vice versa. SMES units are made mostly in two high-capacity scale; that is, about 1800 MJ to balance the load profile and low capacity (several MJ) to increase the damping and improve system stability. Superconducting coils are connected and charged using power system converters by controlling the DC voltage through changing the trigger angle in a wide range of positive and negative voltage values.

The input signals can be selected as grid voltage changes, frequency deviation, or changing at synchronous machine speed. The output signal is the absorbed power. The most important feature of SMES is the independence between stored energy and consumption, which is followed by multiple advantages such as optimal operation, dynamic performance improvement, and pollution reduction.

The main part of the SMES system is a superconducting coil, which is necessary to maintain in the superconducting state, by preserving its temperature at a very low degree. This process can be practical using a cooling coil. Some good examples are replacing the coil in a vacuum chamber or liquid helium, to reduce the electrical resistance to a value more close to zero. There are no moving parts in SMES, so the system lifetime is very high, and it is a low-maintenance system. Furthermore, the response time is very fast, about a few milliseconds.

Appropriate examples for SMES applicable in large-scale systems can be found in [22,23]. It should be noted that because of the rapid discharge capabilities of these systems they are mainly proposed in power systems for pulsed-power and system stability applications.

7.3.2.1 Coil and Superconductor

The main part of SMES is the superconducting coil, which its stored energy is expressed as Eq. (7.8):

$$E = \frac{1}{2} L i^2 \tag{7.8}$$

where L is the inductance of the coil, I is the current, and E denotes the stored energy.

Also, the volume of energy density is as follows:

$$\mu_D = \frac{B^2}{2\mu_0} \tag{7.9}$$

Example 7.8

Consider a SMES system with a charge current of 2A, current discharge of 1.25A, primary current of 0 A, and inductance of 4 henry. If it is charged for 2 min and discharged in 1.5 min, calculate the superconducting power after this process.

Solution

Power:

$$P_{\text{Finale}} = P_{\text{Charging}} - P_{\text{discharging}} = E_{\text{Charging}}$$
$$\times t_{\text{Charging}} - E_{\text{Discharging}} \times t_{\text{Discharging}}$$
$$= \frac{1}{2} \times 4\left(2^2 \times 2 - 1.25^2 \times 1.5\right) = 11.3 \text{ Watt}$$

7.3.2.2 Energy Conversion System

SMES is completely different from the other storage systems. During the charging and discharging process the electrical energy does not convert into any other forms of energy and it remains only in the electrical energy form. Based on the differences between produced and demand powers the charging and discharging processes are performed and the electric current direction will be changed proportionally.

7.3.2.3 Control System

There is a control link between the generated power and the demand, as well as the stored energy, which completely controls the power flow. As a most important note, the power disagreement should be supplied by the SMES. This process determines the system status, charging or discharging mode.

7.3.3 Pumped Storage Power Systems

In a pumped storage system, which is called a pumped hydroelectric storage (PHS) power plant, there are two reservoirs with different heights, containing water. When the power demand is less than the power generation, the additional electric energy supplies an electric machine in the motor mode. Accordingly, the stored water is transferred from a downstream reservoir to an upstream one. Consequently, the water is stored in the upstream reservoir. Accordingly, when the power demand is greater than the generated power this process is done vice versa; the water flows from the upstream reservoir to the downstream one, and consequently electric energy is produced and delivered to the power system. Fig. 7.12 illustrates a typical pumped storage system.

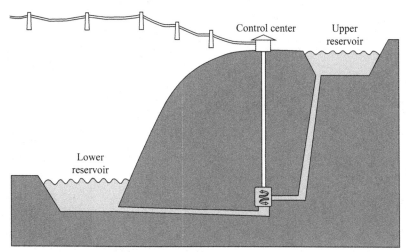

Fig. 7.12 A sample pumped storage system.

The largest installed PHS capacity is in Japan [24], which is followed by China [25] and the United States [26]. Table 7.2 shows the installed PHS capacities in world [27].

The main components of this system are a concluding motor-generator, water channel, and upstream and downstream reservoirs.

This system is a low-cost operation, even though it has a long lifetime and high reliability. Construction of this system in a large capacity is possible. Furthermore, the corresponding response speed is appropriately high. In such systems using the sea, a lake, or a river as a downstream reservoir is possible. The system efficiency is about 70%–80%, based on unit capacity, diameter of water transmission channels, hydro turbine capacities used, and also the level difference between upstream and downstream reservoirs.

7.3.3.1 Kinds of Pumped Storage Power Systems

There are two kinds of pumped storage systems as:
1. Pure or off-stream PHS, which rely entirely on water that was previously pumped into an upper reservoir as the source of energy (closed-loop systems).
2. Combined, hybrid, or pump-back PHS, which use both pumped water and natural stream flow water to generate power.

7.3.3.2 Pumped Storage Power Systems Efficiency

The round-trip efficiency (electricity generated divided by the electricity used to pump water) of facilities with older designs may be lower than 60%, while a state-of-the-art PHS system may achieve over 80% efficiency.

Table 7.2 Installed PHS capacities

Country	Total installed PHS capacity (MW)
Japan	27,438
China	21,545
USA	20,858
Italy	7071
Spain	6889
Germany	6388
France	5894
Austria	4808
South Korea	4700
United Kingdom	2828
Switzerland	2687
Taiwan	2608
Australia	2542
Poland	1745
Portugal	1592
South Africa	1580
Thailand	1391
Belgium	1307
Czech Republic	1145
Luxembourg	1096
Bulgaria	1052
Iran	1040
Slovakia	1017

7.3.3.3 New Technology of Pumped Storage Power Systems

- Most existing PHS facilities are equipped with a fixed-speed pump turbine. While those fixed-speed PHS facilities may provide economical bulk electricity storage, they can only provide frequency regulation during its generating mode.
- New variable speed technology allows PHS facilities to regulate frequency at both pumping and generating modes. Japan has pioneered the variable-speed PHS technology and has successfully operated such systems at the Okawachi PHS station for over 20 years in generating mode, but not in pumping mode.
- Seawater PHS: The Okinawa seawater PHS station, which commenced operation in 1999, is the world's first seawater PHS.
- The Okinawa PHS station uses the open sea as the lower reservoir together with a constructed upper reservoir at 150 m.

- Compressed Air PHS: This innovative design could potentially free PHS from geographic requirements and make it feasible at almost any location with flexible and scalable capacity.
- Undersea PHS: It uses electricity to pump water out of the tank to store energy, and generates electricity when sea water is filling into the tank through the generator.

7.3.4 Compressed Air Energy Storage System

CAES is one of the energy storage technologies with high efficiency and low emission. This technology is based on compressing air in off-peak times and storing it in a suitable tank. This period is called the charging process. At peak hours, when the power is requested by consumers, the compressed air enters the combustor of a gas turbine and combined with the fuel. Consequently, the electrical energy is generated using a gas turbine. This period is called the discharge interval.

This system is more similar to a gas turbine power plant. It should be mentioned that the fuel consumption in this case is less than 40% of the fuel needed in a conventional gas turbine power plant with the same amount of electrical energy produced.

Natural underground cavities, caves, and dummy cavities are good examples of what can be used as storage tanks in the CAES system [28].

To store compressed air at large-scale capacity, water tanks and depleted gas fields usually are more economical. This is a low-loss system and energy can be stored for a long time. The main disadvantage of this system is its requirement to store the compressed air in a bulk tank. Usually, building a new, large reservoir for energy storage is not cost-effective. Fig. 7.13 shows the energy storage schematic using compressed air systems.

Multiplicity in the energy conversion process is one of the main disadvantages of using CAES. This is mainly due to storing the mechanical energy as compressed air, and then converting it into electrical energy. As a result, the total efficiency of the system will be decreased. In addition, construction and development of CAES on a large scale are still two nontechnical concerns that should be taken into account. These two items are the need for special geographical features and heavy investment.

However, in small-scale CAES systems, the abovementioned disadvantages and limitations are overcomed as follows:

(1) The energy storage process is carried out in small-scale high-pressure tanks (made of steel or carbon fiber). Therefore, implementing this system is possible at any location, even in the small area.

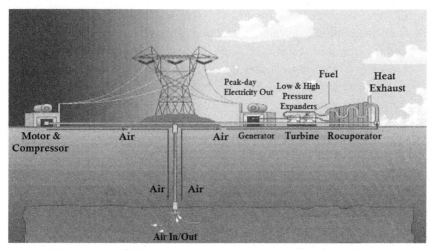

Fig. 7.13 The process of energy storage using a compressed air system.

(2) Although the initial investment required is still a major parameter in the compressed air storage tanks, reducing the storage size will reduce the amount of investment needed drastically.

(3) The parallel performance of these systems with renewable energy resources (such as solar small power plants), in various operation modes concluding connected and isolated leads to a lot of diverse and complementary benefits.

(4) Very fast advances and developments in turbines and engines working directly with compressed air lead to many benefits, such as implementing various strategies to decrease the environmental effects, costs, and so on.

7.4 ECONOMICS OF USING ENERGY STORAGE SYSTEMS WITH DISTRIBUTED GENERATORS

7.4.1 Introduction

This section deals with the economic analysis of distributed generators and storage devices. Different DG technologies like the diesel generator, wind turbine, biomass, and storage devices like batteries and super capacitors are used in the analysis. Each DG is used with each of the storage devices and the cost of each combination is analyzed.

Most of the previous work related to the economic analysis of DGs has used HOMER to obtain the optimal combination of the DGs and battery.

It should be noted that for analyzing the energy storage device, the battery data is replaced by the energy storage device due to the lack of a built-in model in HOMER.

7.4.2 Hybrid Optimization for Electric Renewables

Hybrid optimization for electric renewables (HOMER) is optimization software that has been developed by the National Renewable Energy Laboratory (NREL) for evaluating both grid-connected as well as the off-grid hybrid systems [29]. It suggests a list of cost-effective systems under the conditions specified by the user. It has a very user friendly graphical user interface (GUI), by which the necessary components are added or removed from the system. The major components in HOMER include power sources (including FCs, PVs, biomass generators, and hydro and wind turbines), loads, and batteries. Once the components are selected and assembled, the size, cost, and other data of each can be set through the component windows. The left part of Fig. 7.14 shows a system with built-in HOMER and the toolbar on the right is where the components are added or removed.

After the simulation is run, two types of results are presented; namely, the sensitivity analysis and optimization. Sensitivity analysis is used to determine the effects of sensitive variables (such as the wind speed, diesel price, etc.) on

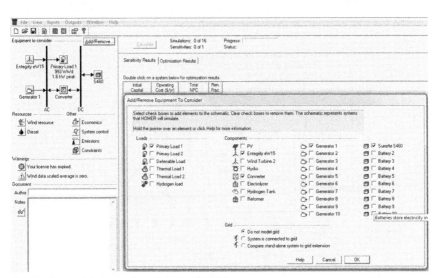

Fig. 7.14 Graphical user interface in HOMER.

the total cost of the system. The optimal result suggests a list of various sizes and combinations of the components and their total cost. Apart from the technical and cost issues, HOMER also deals with certain environmental constraints such as the amount of CO and CO_2 emission by the system.

In summary, HOMER can be used for three different purposes, as follows:
- To analyze the cost of a hybrid system before installation.
- To find out the cost-effective sizes and combinations of the necessary components in the system.
- To analyze the cost of an already built or existing system.

7.5 EXAMPLE OF NETWORK TECHNICAL ISSUES

As previously stated, the storage devices improve network parameters such as stability, power quality, reduce losses, reduce costs, and so forth. In this section, for proofing this allegation, an 8-bus system with or without some storage items listed in the previous section has been evaluated in MATLAB/ SIMULINK. Fig. 7.15 shows a one-line diagram of the 8-bus test system.

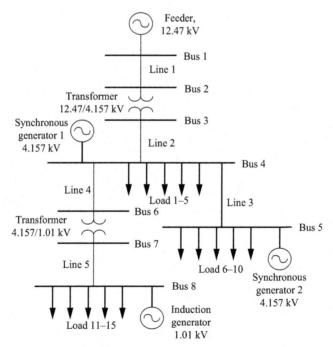

Fig. 7.15 One-line diagram of the 8-bus test system.

7.5.1 Description of the Test System

The system consists of a main source, which is connected to bus 1 at a voltage level of 12.47 kV and three distributed generators connected to buses 4, 5, and 8. The DGs connected to buses 4 and 5 are represented by synchronous generators and the DG that is connected to bus 8 is represented by an induction generator. Storage devices were connected to the same buses as DG interfaced with power conditioning devices. Buses 4 and 5 are at a voltage level of 4.157 kV and bus 8 is at lower voltage level of 1.01 kV. Three sets of loads are connected to each of the buses 4, 5, and 8. Each set of loads consists of induction motors and RL loads.

Transient disturbances such as faults, load switching, and load shedding might result in the change of system states and may lead to system instability. System disturbance indicators generally used for transient stability are rotor angle, rotor speed, terminal voltage, and frequency. The type and size of the DGs or storage present in the system can influence the magnitude and frequency of these oscillations.

7.5.2 Transient Stability Indicators

The transient stability of a system can be assessed by means of certain indicators. In this chapter, four different indicators have been chosen to analyze the stability of the test system. The four chosen transient stability indicators are as follows [30].

(1) *Rotor speed deviation:* When a fault is applied to the system the rotor angle starts increasing, and once the fault is cleared the rotor angle starts to decrease and settles back to a constant value. The stability of the system is assessed based on the maximum amount of deviation in the rotor speed during a fault.

(2) *Oscillation duration:* This is the time taken by the oscillations to reach a new equilibrium after the clearance of the fault

(3) *Rotor angle:* When a fault is applied to the system the rotor speed is distorted, and once the fault is cleared the rotor speed settles back to a constant value. The stability of the system is assessed based on the amount of change in the rotor speed during a fault.

(4) *Terminal voltage:* When a fault occurs in the system voltage drops during the fault and as soon as the fault is cleared voltage goes back to its previous value. In this case the transient stability is expressed with voltage drop during the fault. The variation in the terminal voltage of the DG due to different fault conditions is monitored.

7.5.3 Results

In Fig. 7.15, energy storage devices with distributed generators in buses 4, 5, and 8 can be installed. As stated in Subsection 7.4.2, the ESS connected to the buses has been mentioned by bidirectional converter. The 8-bus system with and without an ESS takes into account a fault at the time of 1.6 s and it can be cleared at the time of 1.8 s is simulated in MATLAB. Figs. 7.16, 7.17, and 7.18 show the rotor angle, terminal voltage, and rotor speed with and

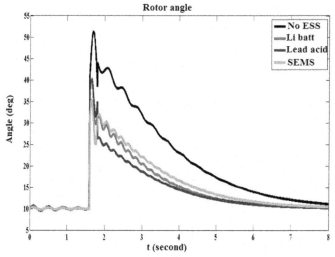

Fig. 7.16 Rotor angle with and without ESS with fault in 1.6 s and clearing at 1.8 s.

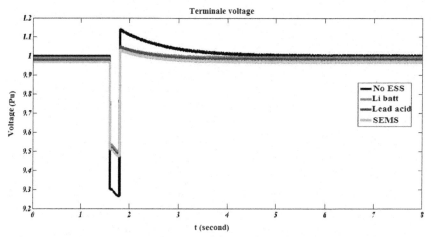

Fig. 7.17 Terminal voltage with and without ESS with fault in 1.6 s and clearing at 1.8 s.

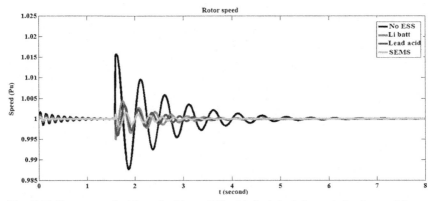

Fig. 7.18 Rotor speed with and without ESS with fault in 1.6 s and clearing at 1.8 s.

without storage devices, respectively. As is clear from the figures, an ESS reduces terminal voltage deviation, rotor speed deviation, rotor angle deviation, and oscillation duration that means making the system more stable. Note that the ESS used here is lithium–ion batteries, superconductors, and battery acid.

7.6 ANOTHER EXAMPLE OF NETWORK TECHNICAL ISSUES

The purpose of the economic analysis is the comparison of the total cost of a system with and without energy storage devices. HOMER is the software that has been used for the cost analysis of the system with distributed generators and energy storage devices.

The main components of the system are the distributed generators, electric grid, energy storage devices, converter, and the loads. The capital, operation, maintenance, fuel and other costs contribute to the total cost of the system.

7.6.1 Results

Fig. 7.19 shows diesel generators with load and a network that is connected to an AC bus. Fig. 7.20 shows diesel generators with load and a network with an ESS that is connected to an AC bus. Note that the cost analysis is done within a year for each network in HOMER software and the total cost in a year is achieved.

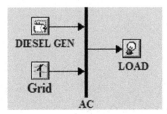

Fig. 7.19 Diesel system without storage device.

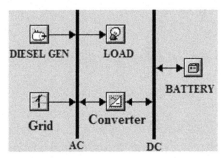

Fig. 7.20 Diesel system with storage device.

Finally, based on the obtained results of economic analysis, it is known that energy storage devices decrease total cost for some DGs.

7.7 CONCLUSIONS

With an increased focus on the DG concept, the analysis of impacts of DGs on the network becomes critical. Among the many issues that persist in interconnecting the distributed generators to the grid, the major issue is the technical impacts. The energy storage device is an important component of the DG. The energy storage devices are critical components, especially in systems where the renewable energy resources such as wind and solar energy are used. This chapter focused on describing the EES. For this purpose, the first models of energy storage device was explained. Then, the technical and economical considerations in the utilization of energy storage systems with DGs was discussed.

REFERENCES

[1] H. Zhao, et al., Review of energy storage system for wind power integration support, Appl. Energy 137 (2015) 545–553.

[2] B. Zakeri, S. Syri, Electrical energy storage systems: a comparative life cycle cost analysis, Renew. Sustain. Energy Rev. 42 (2015) 569–596.

[3] M. Nick, R. Cherkaoui, M. Paolone, Optimal allocation of dispersed energy storage systems in active distribution networks for energy balance and grid support, IEEE Trans. Power Syst. 29 (5) (2014) 2300–2310.

[4] A. Castillo, D.F. Gayme, Grid-scale energy storage applications in renewable energy integration: a survey, Energ. Conver. Manage. 87 (2014) 885–894.

[5] A. Date, et al., Progress of thermoelectric power generation systems: prospect for small to medium scale power generation, Renew. Sustain. Energy Rev. 33 (2014) 371–381.

[6] O. Erdinc, Economic impacts of small-scale own generating and storage units, and electric vehicles under different demand response strategies for smart households, Appl. Energy 126 (2014) 142–150.

[7] B. Huskinson, et al., A metal-free organic-inorganic aqueous flow battery, Nature 505 (7482) (2014) 195–198.

[8] Z. Wang, et al., Enhancing lithium–sulphur battery performance by strongly binding the discharge products on amino-functionalized reduced graphene oxide, Nature Commun. 5 (2014) 1–8.

[9] T. Dragičević, et al., Supervisory control of an adaptive-droop regulated DC microgrid with battery management capability, IEEE Trans. Power Electron. 29 (2) (2014) 695–706.

[10] T. Kousksou, et al., Energy storage: applications and challenges, Solar Energy Mater. Solar Cells 120 (2014) 59–80.

[11] O. Ellabban, H. Abu-Rub, F. Blaabjerg, Renewable energy resources: current status, future prospects and their enabling technology, Renew. Sustain. Energy Rev. 39 (2014) 748–764.

[12] U.B. Nasini, et al., Phosphorous and nitrogen dual heteroatom doped mesoporous carbon synthesized via microwave method for supercapacitor application, J. Power Sources 250 (2014) 257–265.

[13] X. Xia, et al., A new type of porous graphite foams and their integrated composites with oxide/polymer core/shell nanowires for supercapacitors: structural design, fabrication, and full supercapacitor demonstrations, Nano Lett. 14 (3) (2014) 1651–1658.

[14] L. Wang, et al., Three-dimensional Ni (OH) 2 nanoflakes/graphene/nickel foam electrode with high rate capability for supercapacitor applications, Int. J. Hydrogen Energy 39 (15) (2014) 7876–7884.

[15] G. Wang, et al., Solid-state supercapacitor based on activated carbon cloths exhibits excellent rate capability, Adv. Mater. 26 (17) (2014) 2676–2682.

[16] V. Martynyuk, M. Ortigueira, Fractional model of an electrochemical capacitor, Signal Process. 107 (2015) 355–360.

[17] C.J. Raj, et al., Electrochemical capacitor behavior of copper sulfide (CuS) nanoplatelets, J. Alloys Compd. 586 (2014) 191–196.

[18] S. Dutta, A review on production, storage of hydrogen and its utilization as an energy resource, J. Ind. Eng. Chem. 20 (4) (2014) 1148–1156.

[19] A. Dhand, K. Pullen, Review of battery electric vehicle propulsion systems incorporating flywheel energy storage, Int. J. Automot. Technol. 16 (3) (2015) 487–500.

[20] M.A.M. Ramli, A. Hiendro, S. Twaha, Economic analysis of PV/diesel hybrid system with flywheel energy storage, Renew. Energy 78 (2015) 398–405.

[21] R. Sebastián, R. Peña-Alzola, Control and simulation of a flywheel energy storage for a wind diesel power system, Int. J. Electric. Power Energy Syst. 64 (2015) 1049–1056.

[22] K. Zhang, et al., Optimal control of state-of-charge of superconducting magnetic energy storage for wind power system, IET Renew. Power Gen. 8 (1) (2014) 58–66.

[23] J.Y. Zhang, et al., Electric energy exchange and applications of superconducting magnet in an SMES device, IEEE Trans. Appl. Supercond. 24 (3) (2014) 1–4.

[24] T. Ma, et al., Pumped storage-based standalone photovoltaic power generation system: modeling and techno-economic optimization, Appl. Energy 137 (2015) 649–659.

[25] H.S. de Boer, et al., The application of power-to-gas, pumped hydro storage and compressed air energy storage in an electricity system at different wind power penetration levels, Energy 72 (2014) 360–370.

[26] J.I. Pérez-Díaz, J.I. Sarasúa, J.R. Wilhelmi, Contribution of a hydraulic short-circuit pumped-storage power plant to the load–frequency regulation of an isolated power system, Int. J. Electric. Power Energy Syst. 62 (2014) 199–211.

[27] Federal Energy Regulatory Commission, http://www.ferc.gov/industries/hydropower/gen-info/licensing/pump-storage.asp (accessed October 21, 2014).

[28] H. Safaei, D.W. Keith, Compressed air energy storage with waste heat export: an Alberta case study, Energ. Conver. Manage. 78 (2014) 114–124.

[29] H. Shahinzadeh, et al., Optimal planning of an off-grid electricity generation with renewable energy resources using the HOMER software, Int. J. Power Electron. Drive Syst. 6 (1) (2015) 137.

[30] P. Kundur, in: N.J. Balu, M.G. Lauby (Eds.), Power System Stability and Control, McGraw-Hill, New York, 1994.

CHAPTER 8

Market Design Issues of Distributed Generation

**Jeremy Lin*, Fernando H. Magnago[†,‡], Elham Foruzan[§],
Ricardo Albarracín-Sánchez[¶]**
*PJM Interconnection, Audubon, PA, United States
[†]Nexant Inc., Phoenix, AZ, United States
[‡]Universidad Nacional de Rio Cuarto, Rio Cuarto, Argentina
[§]University of Nebraska-Lincoln, Lincoln, NE, United States
[¶]Universidad Politécnica de Madrid (UPM), Madrid, Spain

In this chapter, we provide a brief description of various types of distributed generation (DG) and technical regulations related to interconnecting DG into the distribution grid. We also provide a review of the current status of compensation schemes available to DGs. Then, we will present the obstacles and challenges that prevent DGs from participating on a level playing field with other conventional generators in the setting of both the traditional utility and the current electricity markets, which are succinctly introduced. We will also provide potential solutions that can be used to overcome these challenges. For example, one of the challenges is that most of the DGs are located at the distribution system while the wholesale markets operate at the transmission level. One of the potential solutions that we will propose includes a market mechanism to properly account for the energy output from DGs rather than as behind-the-meter generators (out of market).

8.1 GENERAL DESCRIPTION OF DISTRIBUTED GENERATION

This section will cover a general description of DG by presenting it in terms of definition, benefits, central ower station model, growth, and challenges.

8.1.1 Definition

Distributed generations (DGs) or distributed energy resources (DERs) are small-scale energy resources that are located closer to the end users of power in a rather distributed manner. Most common types of DG technologies include modular generators such as kW-scale solar or photovoltaic (PV) panels, small-scale wind turbines, combined heat and power (CHP), and

Distributed Generation Systems
http://dx.doi.org/10.1016/B978-0-12-804208-3.00008-X
369

microcogeneration units in addition to conventional fossil-fuel (e.g., diesel) generators; yet other available technological options include solar PV thermal systems, concentrated solar power (CSP), fuel cells (FCs), and other sources of energy.

8.1.2 Benefits

In many cases, distributed generators can provide lower cost electricity, higher power reliability, and security with fewer environmental consequences compared with traditional power generators. In contrast to the use of a few large-scale generating stations located far from load centers as in most of the traditional electric power systems, a DG system employs a set of numerous but small generators and can provide power on-site without much reliance on transmission and a distribution grid. DG resources can provide power in capacities that can range from a fraction of a kilowatt (kW) to about 100 megawatts (MW). In contrast, utility-scale generation units can produce power in capacities that are often greater than 1000 MW. Large nuclear plants in the United States have capacities in that MW range.

8.1.3 Central Power Station Model

The current model for electricity generation and distribution in the United States is dominated by a model of centralized power plants. The fuel sources for producing power at these plants are typically combustion (coal, oil, and natural gas) or nuclear. The centralized power station model requires transmission of electricity from the generating resources to the final consumers. Most of these generators are located far from the actual users, sometimes tens to hundreds of miles away. This requires transmission of electricity across a long distance.

This system of centralized power plants has many disadvantages. In addition to the issue of long distance transmission, this system with large-scale power plants contributes to greenhouse gas (GHG) emissions, the production of nuclear waste, inefficiencies, and power loss over the lengthy transmission lines, environmental impact near the area where the power lines are constructed, and security-related issues.

Many of these issues can be avoided by a model of DG. By locating the energy source near or at the end user location, the issues related to the construction of transmission lines are rendered obsolete. For example, a set of solar panels can be a DG. As has been demonstrated by solar panel use in many parts of the world, these units can be stand-alone or integrated into the existing energy system grid. Frequently, consumers who have installed solar panels at their premises can contribute more to the grid than they take out of the grid, resulting in a win–win situation for both the power grid and end users.

8.1.4 Growth

All over the world, there is growing interest in increasing the percentage of electricity generation from DGs, particularly from renewable energy sources (RESs), for a variety of reasons. As a consequence, DGs are poised to grow significantly in the near future. Nowadays, the focus is on the implementation of regulations to establish technical requirements for DG interconnection to distribution networks. However, one of the biggest challenges that DGs or distributed energy systems (DESs) are currently facing is the lack of proper market design or some sort of fair compensation scheme that will allow the provision of the types of resources needed to produce economical power and get paid for this at fair market prices. Even though DGs have the capability to produce energy and participate in frequency regulation, current wholesale electricity markets are not designed to properly and fully accommodate such resources and compensate them accordingly.

8.1.5 Challenges

There are many challenges associated with the growing development of DG resources. The key challenges include
(1) Growing composition of renewable energy resources in DGs.
(2) Technical regulations related to interconnection of DGs into the distribution network.
(3) Compensation schemes for grid-connected DGs for their long-term business viability.
(4) DG's ability to participate in wholesale electricity markets.
(5) DG's roles in future distribution system markets.
(6) Market mechanisms to properly account for energy output from DGs and to provide fair market prices.

(1) Growing Composition of Renewable Energy Resources in DGs

Globally, there is a general trend for the steady growth of RESs, particularly wind and solar resources. While the majority of wind power resources are located in windy but remote areas of the system away from the load centers, the other types of RESs, such as rooftop solar panels and smaller wind plants are directly connected at the consumption location of the end users. These resources pose significant challenges for system operation because of their characteristics on variability, controllability, partial unpredictability, and locational dependency. Therefore RESs as DG a resource pose the same challenges while the growing proportion of RESs in DG resources may exacerbate the operational problems. Active power injection by DG units leads to increasing voltage levels at the point of common coupling and in the vicinity of the power grid. Sometimes, a voltage rise at the distribution system can create system-related problems.

(2) Technical Regulations Related to Interconnection of DGs into the Distribution Network

The integration of small-scale DGs with the capacities up to few MWs to the grid poses significant challenges for distribution network operations. DG units are being installed by private owners and connected to the distribution grid at low-voltage or medium-voltage level, depending on the size of installed DG capacity. Various system configurations of DGs are also installed. Currently, the technical regulations to deal with DG interconnection to the distribution grid are either minimal or nonexistent. Letting any DG to be connected to the grid in a haphazard manner is also not a good option. Therefore there must be some sort of technical regulations that would allow DGs to be interconnected to the distribution network in a manner with little or no impact on system operation and system reliability.

(3) Compensation Schemes for Grid-Connected DGs for Their Long-Term Business Viability

Regardless of whether DGs are owned by individual load customers or incumbent electric utilities (distribution companies), DG owners need to earn a steady stream of revenues for power produced by these resources for their long-term economic viability. Because most DGs will be located at a local distribution system, the regulation for DG revenue requirements will be the responsibility of regulatory authorities that have jurisdiction over the distribution system. In the United States, state regulators play such a role. This revenue requirement is done via revenue tariffs that must ensure fair compensation for DGs. The prohibiting barriers to the long-term business viability of DGs should be removed. In some jurisdictions, schemes such as net energy metering (NEM) and feed-in tariff (FiT) are used to provide proper compensation for DGs. Other innovative compensation schemes are possible.

(4) DG's Ability to Participate in Wholesale Electricity Markets

DG units are installed either at the site of consumption or solely installed for the purpose of electricity generation to participate in the electricity market or benefit from existing financial incentives provided by governments, such as FiTs for renewable electricity generation or other forms of subsidies. Electricity generated from DG units tends to cause a major challenge for distribution system operation because the conventional grid structure was designed according to the concept of unidirectional power flow from points of generation toward the customers connected to medium-voltage and low-voltage networks. However, introduction of DG units has challenged the conventional grid structure as these small-scaled generators begin to feed power to meet local demand and cause reverse power flows toward higher levels of the power network in cases when DG generation exceeds consumption at the consumer's premises. Additionally, because these units are most commonly owned by third-party entities, the distribution system operator (DSO) has limited or no control over the capabilities and performance of these units, bringing about safety and protection issues in grid operations. Another issue that arises from the network planning perspective is that, siting DG units is not centrally planned by the DSO and points of interconnection are not always suitably positioned to meet local demand or relieve congestion where grid constraints exist. Initial network planning and sizing have been done according to the levels of peak demand while the impact of DG installations on grid components has not been properly taken into account.

(5) DG's Roles in Future Distribution System Markets

The growing number of DGs to be connected to the distribution system poses challenges not only to the distribution system operation, but also to the future distribution system market. While it is still uncertain in terms of the final structure of the so-called DSO, the distribution system market design should possess certain attributes to properly accommodate the growing penetration of DGs. At least, these DSOs should be responsible for operation and planning of the distribution grid including DGs, as well as providing rules and mechanisms that properly value the benefits provided by DGs.

(6) Market Mechanism to Properly Account for Energy Output from DGs and to Provide Fair Market Prices

As some DGs such as rooftop solar panels will be owned by individual customers and located at the customer's premises, the excess generation by these resources can be sold back to the electric utility. With the potential establishment of DSOs, the market operator must also properly account for the energy output from DGs and provide proper, fair compensation via some sort of market mechanism. With proper market design, DGs will have a level-playing field with any other type of resources in the relevant network.

8.2 TECHNICAL REGULATION OF DISTRIBUTED GENERATION INTEGRATION

Generally, interconnection of any generator to the grid is governed by grid codes. The grid codes are a set of technical guidelines and operational specifications based on which power system should be planned and operated. The grid codes differ from country to country due to the different regulations, requirements, and characteristics of their national power systems.

At the distribution system level, the grid codes are mainly used to specify and design the guidelines that the distribution network operators (DNOs) will apply in the planning and development of the distribution system with the compliance of end users. When the amount of DG to be integrated to the system was small, the required technical guidelines were simple. In this case, the DNO would normally assess the DG integration by conducting typical integration studies such as load flow and basic power quality studies.

Recently, there has been a substantial increase in the amount of DG penetration into many power systems. The increase in the percentage share of DG resources in the power generation systems, especially in the distribution networks, creates a necessity to develop or update the grid codes due to the impact of DG units on system stability and reliability. Output power of some DG units is intermittent and dependent on environmental conditions. Therefore huge DG penetration to the grid can impact system reliability. Additionally, if there are increasing amounts of DG resources, there should be proper regulations to make the connection of DGs cost-effective.

As a result, to maintain stable and reliable power systems, it becomes important to revise and upgrade the DG connection guidelines. There are several standards that exist to enable DG interconnection, communications, and power system in the new smart grid. There were some attempts to harmonize the varying grid codes related to DG integration at the international level, which recommended some standards. Some of these new standards include IEEE 1547, IEC 62109, IEC 62477, and ENTSO-E draft grid code. The IEEE 1547 standards cover the important requirements for DG interconnection to the grid. These standards cover requirements for all equipment used to connect a DER to the distribution system.

There are positive and negative impacts of interconnecting DG into the power system. It is generally accepted that DG may produce a negative impact on the distribution network operation. Among the main issues are voltage fluctuation, thermal capacity congestion, fault-level contributions, frequency variation, regulation, and harmonics.

There are also some benefits associated with the integration of DG to the distribution grid. Key benefits include the reduction of power losses (subject to the level of DG penetration), provision of ancillary services (e.g., reactive power control and energy balancing), the deferral of distribution and transmission system upgrades (especially in constrained areas), improvements in the security of energy supply (via reduction of the dependency on imported fossil fuels), customer bill savings (net metering), and shorter time needed for the construction of DGs (in comparison with that needed for conventional centralized generating plants).

This section will explore the most recent and comprehensive grid codes with respect to the integration of DG within the distribution grid. In this section, the grid requirements of DGs in distribution systems for steady state operation and dynamic operation during grid disturbances will be covered.

8.2.1 Steady State Operation

To maintain power system security and stability, voltage and frequency of DG units during the steady state operation should be within the specified system voltage and frequency. Four states can happen during a steady state operation:

(1) DGs are working in their normal and continued condition
(2) DGs can operate with reduced power
(3) No requirements
(4) DG should be disconnected

The steady state limits for DGs are different in different countries. For example, Fig. 8.1 shows the voltage-frequency ranges defined at a transmission system operator (TSO) in Denmark for steady state operation of DG units in the distribution system. For example, when the system frequency is between 49.0 and 50.5 Hz and the system voltage is between U_{HF} and U_{LF}, DGs can continue their operation. However, if the system frequency is between 50.5 and 51.0 Hz and given the same voltage range, DGs can still operate for 30 min without the need for reducing their generation. However, after 30 min and the system situation continues to be the same, the DGs have to reduce their generation. By the same token, when the system frequency is between 51.0 and 53.0 Hz, the DGs have only 3 min to avoid producing power for the system frequency to revert back to nominal value.

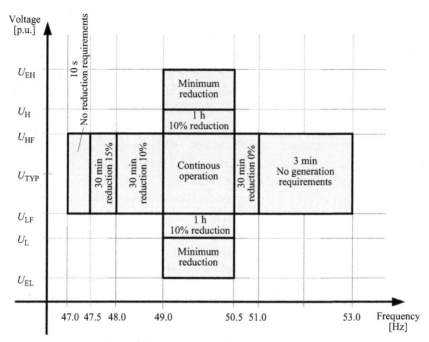

Fig. 8.1 Required voltage-frequency ranges for steady state distributed generation operation for Denmark's transmission system operators. *(Source: T.-N. Preda, K. Uhlen, D.K. Nordgard, An overview of the present grid codes for integration of distributed generation [figure], in: CIRED Workshop, 2012, pp. 1–4.)*

Table 8.1 shows the minimum time periods for DG units to operate at different frequencies deviating from a nominal value without disconnecting from the network. As can be observed in the table, the time periods for operation vary at different frequencies in different countries.

The active power-frequency controllers of DG units are designed to control system frequency within its stable limit by changing active power output. Also the reactive power-voltage controllers of DG units act to keep the system voltage within a stable limit by changing the reactive power set point of DG units. For example, in Spain, a DG unit should exchange reactive power in the range of 0%–120% of its capacity with the distribution system.

8.2.2 Dynamic Operation

Any power system can experience system disturbances at any time for any reason. Disturbances can occur for a variety of reasons such as insulation failure, flashover, physical damage, or human faults. These disturbances can

Table 8.1 Minimum time periods for distributed generation units operation at different frequencies

Synchronous area	Frequency range (Hz)	Time period for operation
Continental Europe	47.5–48.5	To be defined by each TSO while respecting the provisions of Article 4(3), but not less than 30 min
	48.5–49.0	To be defined by each TSO while respecting the provisions of Article 4(3), but not less than the period for 47.5–48.5 Hz
	49.0–51.0	Unlimited
	51.0–51.5	30 min
Nordic	47.5–48.5	30 min
	48.5–49.0	To be defined by each TSO while respecting the provisions of Article 4(3), but not less than 30 min
	49.0–51.0	Unlimited
	51.0–51.5	30 min
Great Britain	47.0–47.5	20 s
	47.5–48.5	90 min
	48.5–49.0	To be defined by each TSO while respecting the provisions of Article 4(3), but not less than 90 min
	49.0–51.0	Unlimited
	51.0–51.5	90 min
	51.5–52.0	15 min
Ireland	47.5–48.5	90 min
	48.5–49.0	To be defined by each TSO while respecting the provisions of Article 4(3), but not less than 90 min
	49.0–51.0	Unlimited
	51.0–51.5	90 min
Baltic	47.5–48.5	To be defined by each TSO while respecting the provisions of Article 4(3), but not less than 30 min
	48.5–49.0	To be defined by each TSO while respecting the provisions of Article 4(3), but not less than the period for 47.5–48.5 Hz
	49.0–51.0	Unlimited
	51.0–51.5	To be defined by each TSO while respecting the provisions of Article 4(3), but not less than 30 min

Source: ENTSO-E (2012), Network Code for Requirements for Grid Connection Applicable to All Generators: Requirements in the Context of Present Practices, Working Document create small or large transients in the system. Usually, DGs with capacities less than 30 kW need to cease operation during faults. Tables 8.2 and 8.3 show one example of DG's response during abnormal condition. Clearing time is the time between starting abnormal condition and DG's ceasing to produce power.

When any voltage is in a range given in Table 8.3, or frequency is in the range of Table 8.2, the DG should cease power generation within the clearing time as indicated. If big DG units were disconnected from the grid for any transient disturbances, it is very probable to encounter a dynamic stability problem in the grid. Therefore to remain stable during and after disturbances, transients standards and requirements are imposed on the DG units.

Fault ride through (FRT) is a capability of large enough DGs to stay connected during disturbances for a specific clearing time. The DG units should inject specific reactive power during the fault. Fig. 8.2 shows the FRT profile of a power generation unit. The figure represents the lower limit of a voltage that is per unit based on rated voltage versus clearing time.

In the figure, U_{ret} is the voltage during the fault; t_{clear} is the instant when the fault has been cleared; and U_{rec1}, U_{rec2}, t_{rec1}, t_{rec2}, and t_{rec3} are parameters

Table 8.2 Distributed generation response during abnormal condition

DG size	Frequency range (Hz)	Clearing time (s)
≤30 kW	>60.5	0.16
≤30 kW	<59.3	0.16
>30 kW	>60.5	0.16
>30 kW	<(59.8–57.0)	Adjustable 0.16–300
>30 kW	<57.0	0.16

Source: IEEE 1547 Standard

Table 8.3 Clearing time for different voltage ranges

Voltage range (p.u)	Clearing time (s)
$V < 0.5$	0.16
$0.5 \leq V < 0.8$	2.00
$1.1 < V < 1.2$	1.00
$V \geq 1.2$	0.16

Source: IEEE 1547 Standard

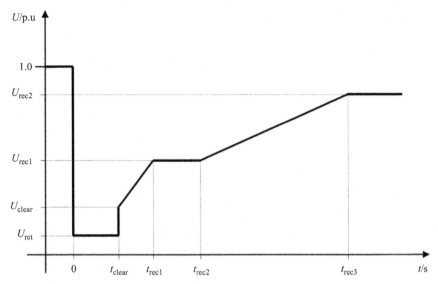

Fig. 8.2 Fault ride through profile of a power generation unit. *(Source: [Johnson, A. Fault Ride through Implementation of RfG and Scope of Study Work, National Grid [Figure]. Retrieved from http://www2.nationalgrid.com/WorkArea/DownloadAsset.aspx?id=30992.)*

Table 8.4 Fault ride through requirements for distributed generation in different national grid codes

Country	Duration of fault (ms)	Voltage fault level (Urated)	Postfault time recovery (s)	Reactive current injection
Hydro-Quebec (Canada)	150	0%	0.18	–
Denmark	50	20%	1	–
Germany	150	0%	3	Up to 100%
Ireland	600	50%	–	–
Spain	500	20%	0.5	Up to 100%
UK	140	15%	1.2	–
ENTSO-E	40	15%	1.5–3	Up to 100%

Source: T.-N. Preda, K. Uhlen, D.K. Nordgard, An overview of the present grid codes for integration of distributed generation [Table], in: CIRED Workshop, pp. 1–4

that depend on the types of generation and protection devices. More generally, these parameters are defined differently in different grid codes. They are certain points of voltage recovery after fault clearance. Table 8.4 shows the FRT capability for seven grid codes.

8.2.3 Country Examples

As country examples, the development of rules and regulations related to DG interconnection to the distribution system in England and Spain are presented further.

8.2.3.1 United Kingdom

The electricity sector in Great Britain is composed of four elements: generation, transmission, distribution, and retail supply. Transmission and distribution are regulated business while generation and retail supply are open for competition. The distribution market is operated by 14 licensed DNOs; 12 in England and Wales and 2 in Scotland as shown in Fig. 8.3. Each DNO is responsible for a regional distribution service area. The 14 DNOs are owned by six different groups (six different colors in the figure) and some DNOs also own generating and supply operations. In addition, there are independent distribution network operators (IDNOs), which own and operate smaller networks within the areas covered by the DNOs. National Grid Electricity Transmission (NGET) owns and operates regulated electricity transmission networks in the United Kingdom. The company owns and operates the high-voltage electricity transmission network in England and Wales (National Grid); and operates two electricity transmission networks in Scotland (Scottish Power Transmission and Scottish Hydro-Electric Transmission).

European Directive 2003/54/EC suggests that the DSOs (DNOs in Great Britain) should consider the integration of DGs in their network planning. This may facilitate the upgrading or replacement of electricity network capacity. However, an independent generator developer and a DNO may have different incentives in selecting specific locations and DG capacities. Generator developers are generally motivated by availability of renewable resources and by the possibility of higher rate of return. However, DNOs are mainly driven by cost minimization, quality of supply standards, and regulatory incentives such as network losses and DG incentives. In Great Britain, a DNO is required to connect generators under the terms of its license, and does not have the option of prioritizing connection at specific sites. Some exceptions may be possible if the DNO would like nonutility developers to propose a DG at a desired location.

Given that DG is connected to the distribution networks, maintaining the traditional passive operation of these networks and the philosophy of centralized control will make it necessary to increase the capacities of both

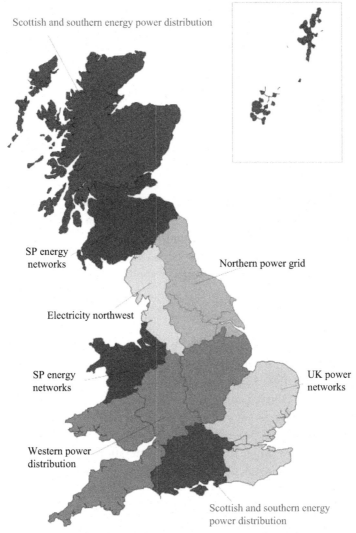

Fig. 8.3 Map of distribution network operators in the United Kingdom. *(Source: http:// sse.com/whatwedo/networks/electricitydistribution/distribution-network-operators/; Public domain.)*

transmission and distribution networks. There are different grid access methods (e.g., deep, shallow) and connection arrangements (including associated charges) for connecting DG facilities.

A DNO in the United Kingdom is used as an example in connecting a DG to the distribution network. Based on the rules set by that DNO, for smaller DGs, such as PV, CHP, wind or hydro generation rated up to 3.68 kW (16A) per phase at a single premise, the owner or user can install and connect it directly to the system by following the guidelines from the DNO. Connecting other generation technologies such as small-scale wind power requires notification and discussion with the DNO as a more complex connection may be required. For larger generators or multiple units, additional network upgrades may be required to handle the new power generation.

For example, UK Power Networks, one of the DNOs in Great Britain, is experimenting with an innovative scheme known as Flexible Plug and Play related to DG integration. Currently, DG developers face many challenges when trying to connect DGs to the electricity network. The company is trying new "interruptible" commercial agreements on suitable and willing customers within the Flexible Plug and Play trial area of Cambridgeshire. This arrangement was developed after having extensive discussions with local and national government, community interest groups, other electricity DNOs, and trade associations.

8.2.3.2 Spain

In Spain, several royal decrees (RDs) have been issued to promote the use of DG. In 1994, *RD 2366* established the rules for the electrical energy production from hydro, cogeneration, and other RESs including units below 100 MVA. These plants can sell the surplus power to the utility that needs to buy it if this power injection does not cause technical problems. In 1999, the Spanish government began a plan to promote renewable energies (PFER) by setting a target for renewable generation penetration. In 2000, *RD 2366/2000* requires all plants over 50 MW to participate in the markets. *RD 314/2006* imposed the incorporation of solar panels for water heating and PV in certain buildings. *RD 1699/2011* regulates the grid connection rules for small DG power plants and the supply of electrical energy produced by the network for their consumption. RESs have priority access to the electrical grid. In addition, they are preferred in the dispatch order in the electricity markets because their marginal cost is zero.

Regarding the technical regulations, three RDs were established. In 1985, the administrative and technical guides were established to allow for the interconnection of DGs with capacities less than 5 MW. In 2000, *RD 1663/2000* established the main technical requirements that regulated the installation and interconnection of PVs to the low-voltage network. This RD was denigrated by *RD 1699/2011*.

RD 436/2004 recommends a methodology to follow for the economical and legal activities of DG production. After the economic recession from 2008 to 2010, *RD 1003/2010* liquidates the bonus provided by *RD 661/2007*. Before 2012 the legislation emphasized the development and promotion of renewable energy. However, after 2012, due to the economic crisis and the increasing tariff deficit, legislation focused on cost reduction. *RD 1/2012* suspended the financial incentives to new installation projects of renewable, cogeneration, and waste-based energy. Also, this decree indefinitely suspended the registration procedures of preallocation for renewable plants.

At the end of 2012, *RD 29/2012* was established to reduce the tariff deficit by correcting or eliminating the special regime for renewable resources that do not meet required obligations for final registration of preallocation. After that, additional regulations were set to ensure the financial stability of the electrical system (*RD 9/2013*). This new regulatory framework establishes a new legal and economic regime for renewable energy, cogeneration, and waste facilities, where the incentive model based on electricity production is abandoned. Afterwards, all facilities are regulated by the same rules and market obligations. Additional remuneration is given based on the installed capacity (€/MW) and operational energy (€/MWh) to compensate for the installation investment.

Finally, *RD 413/2014* updates the regulation related to electricity production from renewable energies, cogeneration, and waste, establishing the remuneration parameters applicable to certain facilities that produce electricity from renewables. Also, this decree sets the investment compensation rules for renewable source facilities. *RD 900/2015* has been a hard blow for generation from renewable sources, penalizing self-consumption.

8.3 REGULATORY AND COMPENSATION SCHEMES FOR DISTRIBUTED GENERATION

The inclusion of DER has a significant impact on the economic decision of the electric system, particularly related to network investments and operation. However, the real impact is still not easy to predict because the implementation of DERs is still new. Several issues can affect the evaluation. Such issues include the DER penetration and concentration level, DER technologies and profiles, network characteristics, and system management. One important factor that has significant influence on the estimation of the economic benefit of DER is the type of regulations under the regulatory environment in which the DER operates. Regulation has a broad meaning that includes aspects of legal, government rules, organizational, and economic issues.

To make proper decisions regarding the inclusion and promotion of DERs, it is important to assess the experiences that are already gained in different countries around the world. In this section, the focus is on the economic regulation of which different experiences are discussed. As the regulation affects the revenue to be received by the owners of the DGs, the current compensation schemes for DG resources are also discussed to the extent applicable.

Given the central role of the behind-the-meter resources, regulators need to ensure that these resources are fairly compensated. By the same token, utilities will continue to provide necessary services to the grid and to the final consumers eventually. Thus regulators need to ensure that utilities are fairly and adequately compensated for providing these essential services. Therefore regulators need to strike a balance to ensure both parties are fairly compensated for the values that are provided to the system. Communications and advanced technologies will continue to evolve and innovate the power sector. Therefore markets, institutions, and regulators have to find a way to align institutional capabilities and regulations with the underlying capabilities of all resources. The main goal of regulators should be to design tariffs that ensure fair compensation for clean DERs ranging from solar PV to CHP.

8.3.1 Some Compensation Examples

There are two well-known compensation schemes that are designed to provide proper revenues for the values provided by the DG resources. The two compensation schemes are known as NEM and FiT.

8.3.1.1 Net Energy Metering

NEM is a special billing arrangement that provides credit to customers with DGs, such as solar PV systems for the full retail value of the electricity that their DGs generate. Under NEM, the customer's electric meter keeps track of how much electricity is consumed by the customer, and how much excess electricity is generated by their DG system and sent back into the electric utility grid. Over a 12-month period, the customer has to pay only for the *net* amount of electricity used from the utility over and above the amount of electricity generated by their DG system (in addition to monthly customer transmission, distribution, and meter service charges they incur). NEM is an important policy for valuing and enabling distributed generation.

At any time of the day, a customer's DG system may produce more or less electricity than they need for their home or business. When the DG system's production exceeds the customer demand, the excess energy automatically goes through the electric meter into the utility grid, running the meter backwards to credit the customer account. At other times of the day, the customer's electric demand may be higher than the amount of electricity that the DG system is producing, and the customer relies on the utility for additional power needs. Switching between the DG system's power and the utility grid power is instantaneous; thus customers never notice any interruption in the flow of power.

NEM can be a good option for optimizing the rate of return on DG investment by consumers. NEM can enable the following:

(1) Allow customers to zero-out their bills.

(2) Credit customer accounts at full retail rates.

(3) Accurately capture energy generated and consumed, providing customers with annual performance data.

Customers that generate a net surplus of energy at the end of a 12-month period can receive a payment for this energy under special utility tariffs. Different utilities develop different tariff rates on net surplus generation as a way to provide incentive for consumers to generate more.

Under a NEM agreement, the customer's utility will continue to read the customer's meter monthly. The customer will receive a monthly statement indicating the net amount of electricity that was consumed or exported to the utility grid during that billing period. Residential or small commercial customers have the option of paying the utility for their net consumption monthly, or settling their account every twelve months.

There are also rebate programs on DG systems, such as solar systems, sized to meet a customer's expected annual electricity needs, with a minimum of 1 kW and a maximum of 1 MW. Peak electricity demand requirements for most residential customers range between two and four kilowatts. The system size will depend on the needs of each individual customer and how much electricity the customer wants to generate using a DG. The DG system can also be built modularly by starting small and gradually expanding over time as long as the system does not exceed 1000 kilowatts.

It is important to keep in mind that the electric rates that the utility charges its customers increase in tiers as the customer uses more and more electricity. A smaller system that satisfies part of the consumer's electric needs may be more cost-effective because the customer will be avoiding the higher tier rates, while the customer is still purchasing the lower cost electricity from the utility for its needs.

Most smaller electric customers have simple bidirectional meters capable of spinning backwards to record energy flowing from their DG system to the utility grid and are currently eligible for NEM. These basic meters are often referred to as "non-time-of-use meters" because they are incapable of recording when electricity was used: only how much was used. Some utilities may want two meters for NEM, one to measure electricity going from the grid to the residential home or business, and another one to measure surplus energy going from the customer's DG system to the grid.

The more sophisticated meters are time-of-use (TOU) meters, which record when electricity is used and allow the utility to charge different rates at different times of the day or week. Currently, TOU meters are optional, but will be required for recipients of incentives in the future if TOU rates are set in the customer's area.

8.3.1.2 Net Metering Credit Examples
The following examples on net metering are referenced from those given by Massachusetts Department of Energy Resources on their website. The examples below will help the students understand better about how net metering works for a typical electricity customer.

(1) Net Metering Facility that Offsets Customer Load for the Month
The following example shows how the net metering credits would apply to the electric bill of a sample residential customer. In the example, it is assumed that the customer uses 600 kilowatt-hours in the month and generates 600 kilowatt-hours in the month, so that all of the customer's use is offset by his or her on-site DG generation and there is no excess generation. Table 8.5 below shows the monthly utility bill for such a customer with and without net metering. With net metering, the customer's bill can be reduced from $101.30 (without net metering) to $8.30 (with net metering).

(2) Net Metering Facility with Net Export for the Month
The following example shows how the net metering credits would apply to the electric bill of a sample residential customer, assuming that the customer uses 600 kilowatt-hours in the month and generates 1200 kilowatt-hours in the month, so that all of the customer's use is offset by the on-site DG generation and there is 600 kilowatt-hours net excess generation for the month.

Table 8.5 Example of the monthly utility bill for a customer with no excess generation

Charge	Rate	Charged without net metering	Charged with net metering
Default service	7.0 ¢/ kW h	$ 42.00	–
Distribution	5.5 ¢/ kW h	$ 33.00	–
Transmission	2.0 ¢/ kW h	$ 12.00	–
Transition	1.0 ¢/ kW h	$ 6.00	–
System benefits	0.3 ¢/ kW h	$ 1.80	$ 1.80
Customer charge	$6.5/ month	$ 6.50	$ 6.50
Total		$ 101.30	$ 8.30

Source: https://sites.google.com/site/massdgic/home/net-metering/net-metering-credit-example; Public domain.

Table 8.6 Example of the monthly utility bill for a customer with excess generation

Charge	Rate	Charged without net metering	Charged with net metering
Default service	7.0 ¢/ kW h	$ 42.00	($ 42.00)
Distribution	5.5 ¢/ kW h	$ 33.00	($ 33.00)
Transmission	2.0 ¢/ kW h	$ 12.00	($ 12.00)
Transition	1.0 ¢	$ 6.00	($ 6.00)
System benefits	0.3 ¢	$ 1.80	$ 1.80
Customer charge	$6.5/ month	$ 6.50	$ 6.50
Total		$ 101.30	($ 84.70)

Source: https://sites.google.com/site/massdgic/home/net-metering/net-metering-credit-example; Public domain

The result is that the customer gets a credit on his or her bill that can be applied to future bills or "assigned" to other customer accounts in the same service territory and load zone. Table 8.6 below shows the monthly utility bill for such a customer with and without net metering. With net metering, the customer will get credit of $84.70.

8.3.1.3 Feed-In Tariff

FiT is an energy supply policy that promotes the increased adoption of renewable energy resources. A FiT scheme offers a guarantee of payments to renewable energy developers for the electricity they produce. FiTs generally pay customers a premium price for generating power from renewable energy resources like solar or wind. They are different than net-metering rates, which are calculated differently and are generally at a lower rate. Payments can be composed of electricity alone or of electricity bundled with renewable energy certificates. These payments are generally awarded as long-term contracts set over a period of 15–20 years.

FiT policies are successful around the world, particularly in Europe. Currently there are six states in the United States, namely California, Hawaii, Maine, Oregon, Vermont, and Washington, that mandate FiTs or similar programs as shown in Fig. 8.4. Electric utilities in a few other states also have voluntary FiTs. There is a growing interest in FiT programs in the United States, especially because there is growing evidence about their effectiveness as a framework for promoting renewable energy development and job creation.

For example, in California the FiT is offered a 10- to 25-year contracts. For contracts starting in 2012, systems are reimbursed for generation at a base

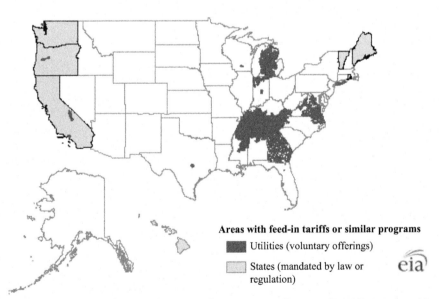

Areas with feed-in tariffs or similar programs

▣ Utilities (voluntary offerings)

▢ States (mandated by law or regulation)

Fig. 8.4 U.S. map with feed-in tariff programs. *(Source: https://www.eia.gov/todayinenergy/detail.cfm?id=11471; Public domain.)*

rate of 7.7 cents per kilowatt-hour produced under 10-year contracts. Under 25-year contracts, they are reimbursed for generation at a rate of 9.2 cents per kilowatt-hour produced. While customers generating under an FiT arrangement can sell the energy they generate to the utility, they are not eligible for other incentives, like net metering or the California Solar Initiative. For most residential consumers with small energy-generating systems, opting for an incentive program is probably going to be a better deal.

FiT policies can be implemented to support all renewable technologies including, wind, PV, solar thermal, geothermal, biogas, biomass, FCs, tidal, and wave power. So long as the payment levels are differentiated appropriately, FiT policies can increase development of a number of different technology types over a wide geographic area. At the same time, they can contribute to local job creation and increase clean energy development in a variety of different technology sectors.

Some of the benefits and impacts of FiT policies include

(1) The rapid renewable energy development seen in jurisdictions with FiT policies has helped reduce the environmental impacts of electricity generation, while providing valuable air quality and other environmental benefits.

(2) Fixed prices created by FiTs for renewable energy sources can also help stabilize electricity rates, which can entice new business and attract new investment.

(3) Due to the guaranteed terms and low barriers to entry offered by FiT policies, they have been highly successful at driving economic development and job creation.

(4) Data from countries like Germany and Spain demonstrate that well-designed FiT policies can positively impact job creation and economic growth. A growing body of evidence from Europe and Ontario, Canada, demonstrates that FiT policies have on average fostered more rapid developments of renewable energy projects than other policy mechanisms.

8.3.1.4 Feed-In Tariff Examples

In the UK system, there are three savings elements related to the FiT scheme:

(1) *Generation tariff* (G) is a fixed rate payable to households for the total amount of electricity generated, calculated per unit. The rate that the customer will receive is determined by when the customer first registered to join the FiT scheme and the type and size of installation.

(2) *Export tariff (E)* is payable on the units of electricity the customer exports back to the national grid because the customer hasn't used them in his or her own property. In the case of most small-scale technologies—or unless the customer already has a two-way smart meter installed—the level of electricity exported is currently "deemed" by assuming the customer exports 50% of the electricity that is generated.

(3) *Savings on the customer's electricity bills (S).* Because the customer will be generating a portion of his or her household electricity by his or herself, the customer's energy bills will be lower.

Fig. 8.5 shows the savings elements explained previously related to the United Kingdom's FiT. Renewable resource technologies that are eligible in the United Kingdom's FiT program include solar PV panels, small-scale wind turbines, hydroelectricity, anaerobic digestion, and micro combined heat and power (micro CHP). The following example is drawn from an example given on the website provided by a company named "Feed-In Tariffs Ltd," which was set up to provide information about the United Kingdom's FiTs.

If a household installed a 2.5 kW solar system, FiTs would provide the following benefits:

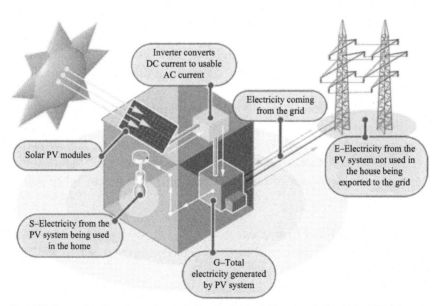

Fig. 8.5 Example of savings elements in the United Kingdom's feed-in tariff scheme. *(Source: http://www.which.co.uk/energy/creating-an-energy-saving-home/guides/feed-in-tariffs-explained/feed-in-tariffs-qanda/; Public domain.)*

In a typical location in South Central England, this system should generate about 2125 kW h each year, earning a generation tariff of about £330 a year, tax-free. In addition, if, say 1500 kW h is used in the home, this would save a further £210 per annum if electricity costs are 14 pence per kW h. Plus, the remaining 625 kW h would be exported, earning about £30 under the export tariff. Therefore the total benefit would be around £570 to £590 per year. This is because most houses without export meters receive the export tariff for 50% of the generation or about £50.

We provide another example to illustrate the FiT scheme in the United Kingdom. A typical home uses 3300 kW h (units) of electricity annually. A south-facing 2 kWp solar PV system at 30 degrees in Sheffield (zone 11) and with no shading would generate around 1770 kW h per year. For systems installed between October 1, 2015. and December 31, 2015, the generation tariff would be 12.47 pence per kW h. Generation tariff income will be

$$0.1247 £/kW h \times 1770 \text{ kW h} = £220.72 \tag{8.1}$$

The estimated annual income from the tariff would be £220.72. Assume that a typical household would use around 50% of the generated electricity (885 kW h) and export the remaining 50% to the grid.

$$1770 \text{ kW h} \times 50\% = 885 \text{ kW h} \tag{8.2}$$

The export tariff is 4.85 pence for every unit that the consumer exports back to the grid.

$$885 \text{ kW h} \times £ 0.0485 = £ 42.92 \tag{8.3}$$

Therefore the customer can receive £263.64 as the sum of generation tariff and export tariff, in addition to the savings on his or her electricity bills.

8.3.2 Country Examples

As country examples, schemes and policy options used in Germany and Spain are described below.

8.3.2.1 Germany

Germany took a big step to include and expand DERs in the resource mix by making a decision to promote the electricity generation based on hydro power, wind power, solar energy, biomass, and other renewable resources as well as to reduce the utilization of nuclear power for electricity generation. This decision led to the approval of the Renewable Energy Sources Act (EEG) in 2000. This act contains three key features:

(1) The investment of renewable energy resources are protected by the FiTs and the requirement of connections.

(2) The charges will not impact the users.

(3) Promoting innovation by reducing the FiTs.

Based on these rules, the system operators must give preferential feed-in inclusion into their network to renewable energy. In addition, renewable resources are guaranteed a 20-years payment for their production. Also, small distributed resources are promoted by giving them access to the market.

Up today, it is necessary to support the inclusion of renewable resources due to the cost. It is important to remark that renewable resources do not produce external costs such as environmental or human health problems. Renewable energy does not receive subsidies because it is not paid by consumers in their bills or by federal taxes. FiTs are designed to pressure energy producers to invest in new and more efficient technologies. This objective is achieved by reducing the FiT for newly installed plants on a regular basis.

The Combined Heat and Power Act, which was recently modified, is another piece of legislation that affects the DER inclusion. CHP resources are subsidized regardless of voltage levels. For the CHP plant, the operator pays an extra charge not only for the CHP plant production but also compensation in case its output is not injected into the electrical grid. The CHP Act also includes the regulation of storage devices.

The cost of photovoltaic (PV) generation is still higher than that of other types of electricity sources. Also, the small local rooftop PV resource cannot compete with other larger-capacity types of generating resources. To encourage the use of local PV resources, based on the Renewable Energy Act, a fixed value is set to purchase, to remunerate the PV generation, and to use it as the self-consumption.

8.3.2.2 Denmark

The competitive market in Denmark is relatively new since its inception in 2003. The Danish market is also part of the Nord Pool, which is the leading market in Europe and comprises nine Northern European Countries. Large-scale CHP generation units represent more than half of the total generation capacity. However, the renewable energy resources increased substantially up to 26% of the total capacity. Given this background, the regulation that sets the objectives and goals in Denmark is the Danish Energy Agreement established in 2012. The primary goals of this regulation are to set three main targets by the year 2020:

(1) Reduce emissions by 34%.

(2) Increase renewable energy penetration by 35%.

(3) Improve energy efficiency by 7.6%.

Given the resource mix, DG is critical as most of the renewable resources are connected to the grid at the distribution level. In 2009, more than half of the DG was wind power and CHP while PV sources made up the rest. Nowadays, the PV resources are substantially taking an increasing share in the resource mix.

The regulation sets nondiscriminatory access to the grid. Therefore there is no guarantee for renewable resources to receive priority for the grid connection. However, in the case of network constraints, renewable energies have priority over conventional resources. Injection of power by the renewable energy can only be reduced for power balance purposes.

The cost method for DG is based on the shallow approach, where the DG units pay only the connection charge and the investment costs are included within the system's use fees. Therefore new generators need to consider only the cost of connection to the distribution grid. Renewable DGs are not required to pay for the system use. However, they are charged for meter installation and administration costs.

8.4 THE ROLE OF DISTRIBUTED GENERATION IN ELECTRICITY MARKETS

In this section we will describe the development of electricity markets, including energy markets and capacity markets, and the roles that DGs will play in these markets. Again, as most DGs will be connected to the distribution grid, market design for the distribution system will be needed and how DGs can participate in these markets will be elaborated.

8.4.1 Electricity Markets

Current electricity markets in the United States, Europe, and Latin America have been in operation for about two decades. Electricity markets were developed around the world as a result of restructuring in the electric power industry. Technological advances, high cost of producing electricity, economic inefficiencies, and stricter environmental regulations are some of the drivers for industry restructuring. One of the major outcomes of this restructuring effort was the separation of transmission and generation businesses from the integrated system into two different and distinct business functions. This process is known as unbundling. Because the transmission business is a natural monopoly, it is still under strict regulation. In other words, all aspects of the transmission business are regulated by relevant regulatory authorities. However, the generation business is exposed to competition in a market setting. This market setting is essentially known as the electricity market.

These electricity markets are wholesale markets in which sellers and buyers of electricity meet, exchange, and trade electricity on a large-volume basis. Sellers are generally owners of generators and buyers are entities that are responsible for meeting or delivering the electricity needs of final consumers. Wholesale buyers can also be third-party brokers who buy electricity from the wholesale market and then sell it to other buyers or final consumers. Electricity markets include the energy market, capacity market, ancillary service market, and other types of emerging markets. The sellers, or resource owners, rely on the final prices coming from these individual markets to cover their fixed and variable costs. The buyers pay at the final prices for the purchase of electricity and other system costs. The pricing mechanism used in these markets is either zonal pricing or nodal pricing.

In general, the electricity market is operated by a market operator that can be an independent entity with the primary purpose to conduct market clearing and determine market prices along with generation scheduling. In some markets, such as those in Europe, the market operator clears the market based on offers/bids by generators/load without considering the network constraints and passes the generation schedule to the TSO. Based on this generation schedule, the TSO ensures that this particular dispatch is possible given the network capacity. If there are network violations, the TSO would change this dispatch by dispatching some generations up and other generations down to ensure that the transmission system is operated in a reliable manner. In other words, the flow on each transmission facility should be within its limits (both precontingency and postcontingency). The dispatch schedule for each participating generator should also be within its minimum and maximum operating limits.

In the United States, the dual functions of market clearing and system operation are done by a single entity, known as an independent system operator (ISO) or regional transmission organization (RTO). These entities also conduct transmission system planning and manage the interconnection of proposed generators for the future. Fig. 8.6 shows a map of the operating ISOs/RTOs in North America. As can be seen in the figure, there are nine such entities in North America (seven in the United States and two in Canada). Even though each entity operates in a geographically distinct

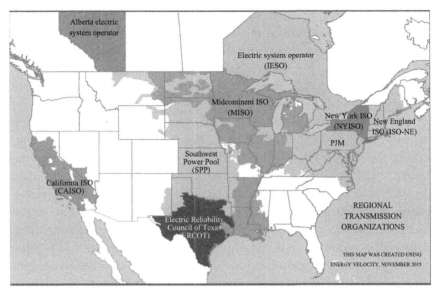

Fig. 8.6 Map of regional transmission organizations in North America. *(Source: http://www.ferc.gov/industries/electric/indus-act/rto.asp; Public domain.)*

region, the responsibilities and functions of these entities are more or less the same. The major functions of these entities are

(1) System operation to maintain reliability and security of the power system.

(2) Market operation to clear the various electricity markets and post the market results.

(3) System planning to plan the transmission system and manage the generator interconnection projects.

In addition to these major functions, these entities are responsible for other functions such as market settlement, transmission service reservation, customer service, compliance with appropriate industry rules and regulations, and so on.

Electricity market is a general term that is generally comprised of the energy market, capacity market, ancillary service market, and other types of markets. These markets, along with the roles that DGs will play, are further explained in the following sections.

Traditionally the majority of electricity in a market system is supplied by large-scale generators. As the amount of DG eligible to participate in the wholesale electricity market is small, the participation by DG in the wholesale market is not necessarily an issue. This is because wholesale markets operate at the transmission or subtransmission system levels and the majority of DGs are located at the distribution system. Electricity markets at the distribution system do not exist. DGs are essentially out of the wholesale electricity markets. If DGs are directly connected to the transmission/subtransmission systems, those DGs can freely participate in the wholesale electricity markets.

However, the current wholesale market status will be challenged when there are a significant number of DGs connected to the distribution system. A large number of DGs at the distribution system not only reduces the amount of electricity traded at the wholesale market system by reducing the wholesale load to serve but also can act as an additional source of electricity supply when DG outputs are greater than load requirementa in the area in which DGs are located. These are some of the key challenges that significant amounts of DGs will impose on the electricity market regime. As electricity markets comprise both the energy market and capacity market, the roles of DG in each type of market are further discussed.

8.4.1.1 Energy Markets

Among different types of electricity markets, the energy market is the most important market. The term *energy* is quite broad, which here literally means "energy produced by electricity." Other source of energy such as oil and gas are beyond the scope of this chapter. In the energy market, the offers by generators and bids by loads are matched based on the economic merit order. Economic merit order means the generating unit with the lowest offer is dispatched first before dispatching the next generating unit with a more expensive offer. This selection of generating units continues until a sufficient level of generation is committed and dispatched to meet the forecasted load. In this dispatch selection process, the generators that have binding contracts for bilateral transactions are also committed and dispatched if operationally feasible.

An energy market generally consists of a day-ahead market and a real-time market. Intraday markets (hourly or subhourly markets) also exist in some markets. In the day-ahead market, the market bidding is open until a certain time period on the day before the actual market date. Market participants such as generators, demand bidders, and financial entities submit their offers/bids before that deadline. Then, market bidding is closed after which the market engineers try to clear the market during next few hours. By a certain deadline on the same day, the market operator clears the market and produces results of market prices and a generation schedule including information on the binding constraints in some markets. Market prices from the day-ahead market are typically made public while the generation schedule for each committed generator is made known only to each individual generator. The results from the day-ahead market are financially and contractually binding. The day-ahead market also takes into account bilateral contracts and self-schedules. Both security-constrained unit commitment (SCUC) and security-constrained economic dispatch (SCED) are done in the day-ahead market. Generator uplift issues are also taken into account. The day-ahead market is typically cleared for each hour the next day. This is equivalent to solving a set of 24 constrained optimization problems. Half-an-hour market clearing is entirely possible for the day-ahead market.

In contrast to the day-ahead market, the real-time market clears generally every three to five minutes—market clearing time interval—at least in the U.S. electricity markets. The real-time market clears on a real-time basis for the next market time interval. Since this market clears in real-time, the SCUC step is not done because a real-time market is only concerned with the dispatch of committed generating units. However, the system operator may decide to turn on any quick-start unit as the system condition requires. This kind of unit start-up is generally beyond the market-clearing framework. As in the day-ahead market, the cleared real-time market prices are made public by posting them on the Internet for market participants to view. Again, the dispatch decision from the real-time market clearing is sent only to the individual generators on a shorter time interval such as every 4–6 s. Because the volatility in the real-time market is higher than that of the day-ahead market in U.S. energy markets, the energy volume traded in the real-time market is typically smaller than that in the day-ahead market. Typically, the real-time market makes up about 5%–10% of the entire energy market volume while the day-ahead market makes up about 40%–50% of the same energy market volume. The rest is made up of a combination of bilateral contracts and self-scheduling. This percentage composition is described here for illustrative purpose as this percentage composition can vary widely among different electricity markets as well as over time.

8.4.1.2 Distributed Generation Participation in the Energy Markets

As the existing wholesale energy markets operate at the transmission or sub-transmission system levels, the DGs that wish to participate in these markets have to be connected at that level. In other words, a DG cannot participate in the wholesale market if the DG is connected as a behind-the-meter generator.

There are two possible ways for a DG to participate in the electricity market. The first possibility is that when the wholesale electricity market extends into the retail market area in which the DG is located or a new retail market, a so-called distribution system market, is created in which the local DG can participate. Another possibility is that the DG, originally connected to the distribution system, has to be reconnected to the transmission system level. In other words, the behind-the-meter DG has to become a DG that has direct connection with the main transmission grid. In this case, the DG can directly participate in the wholesale electricity market.

A typical wind farm consists of a group of wind turbines at the same location used to produce electricity. Wind farms have typically larger capacities and are designed to provide large amount of power. Therefore if wind farms are connected to the main transmission grid, they can participate in the wholesale electricity market. It will be challenging for a wind farm to participate in the electricity market if it is not connected to the transmission grid regardless of the size of the wind farm.

For example, a large number of wind farms were built in western Texas because the wind resource potentials in that area are tremendous. However, unless there is a sufficient capacity of transmission lines built to interconnect these wind farms to the main grid, these wind farms cannot easily participate in the Texas electricity market. It can cost billions of dollars to construct the transmission lines needed to supply the state's central cities with wind power from western Texas.

The situation with the smaller wind resource or PV resource located at the premise of individual consumers is vastly different. A typical household can install a small wind turbine in the backyard or a small PV on the rooftop of the house. The household consumer can use the power output from that DG to meet its own load. If the power output from the DG is less than the consumption requirement, then the consumer will still need additional power from the grid (wholesale market). If the power output from the DG is greater than the consumption requirement, the consumer can sell that excess power back to the grid. That arrangement will be nearly impossible without the explicit market rules set up for DG participation in the electricity market. This applies for any other types of DGs that are located behind-the-meter. The DG owner can still strike a contract with the local electric utility for power purchase and sale but direct participation by that DG in the wholesale power market is out of the question.

To remedy this situation, a possible approach is to aggregate the dispersedly located DG resources by a third-party aggregator. Aggregation of DG resources can make the total size of the aggregated resources larger, which in turn can help make it possible to participate in the wholesale electricity market. This approach is based on the assumption that there are market rules in place for such aggregated DGs to participate in the wholesale electricity markets.

8.4.1.3 Capacity Market

As stated earlier, the energy market makes up the majority of the electricity market. In other words, the majority of revenue for generators comes from the energy markets. The other market complementary to the energy market is known as the capacity market. A capacity market is designed and implemented in some electricity markets to provide additional revenues to the generators, as it is believed that the energy market alone is not sufficient to provide revenues to the generators for their long-term business viability. The problem of the participating generators not making sufficient money in energy markets alone is also known as a missing money problem. Interesting enough, this philosophy is not shared in all electricity markets in the United States. As an example, both ERCOT (Texas market) and Cal-ISO (California market) do not have capacity markets while PJM and ISO-NE (New England) have such capacity markets. Indeed, the capacity market provides additional revenue to the participating generators, which are cleared in the capacity market. This extra revenue is borne by the final electricity consumers in return for maintaining resource adequacy.

8.4.1.4 Distributed Generation Participation in the Capacity Market

Participation in the capacity market for DG resources is similar to their participation in the energy market. Most of the capacity-type resources are required to participate in the capacity auction market, with some exceptions. In some markets, intermittent resources, such as wind and solar, are not required to participate in the capacity market. If they do and get cleared in the market, these types of resources are obligated to offer into the energy market. The nature of highly variable output from these resources makes it very challenging to maximize their output and maximize their revenues because their power output is highly subject to changing weather patterns. If a DG resource that is connected to the transmission system decides to participate in the capacity market, this resource is supposed to follow the existing capacity market rules. In other words, if this resource is cleared in the capacity market for future X-years, that resource must be available at all times in that time frame. This resource may face some penalty for any period when it is not available. Some capacity discounts of these resources are also implemented to account for their significantly uncertain availability.

8.4.1.5 Ancillary Service Markets

The other, smaller market, but important from the system operation perspective, is called the ancillary service market. This market generally comprises a regulation market and a primary reserve market, which includes synchronous and nonsynchronous reserve markets. The regulation market is designed to provide a market-based solution to the fine adjustment of generation up or down to smooth out the instantaneous power variation caused by instantaneously changing load in real-time. In addition to the generation sources, demand-side resources can also be used to provide regulation. Certain types of resources, such as hydro power plants and energy storage devices (batteries), are well suited to providing such smooth regulation. Similar to the regulation market, the synchronous and nonsynchronous reserve markets make up a small portion of overall revenue for generators.

8.4.1.6 Distributed Generation Participation in the Frequency Regulation

Some of DGs such as wind and solar units have unique characteristics; that is, their power output varies randomly based on weather conditions, and the electric system operation cannot control this output profile. The power fluctuation requires additional energy to balance the demand and the ancillary services such as frequency regulation and voltage support. Fig. 8.7 shows an example of wind variability effect on system frequency.

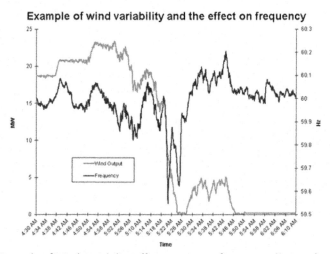

Fig. 8.7 Example of wind variability effect on system frequency. *(Source: https://www.mauielectric.com/clean-energy-hawaii/clean-energy-facts/wind-energy-integration; Public domain.)*

In real-time operation, the system operator must deal with these fluctuations in both frequency and voltage caused by such DG resources. The frequency regulation is performed on a very small time scale such as on a second or minute basis. Once a frequency variation is detected, the electric power needs to redispatch regular generators to compensate for the variations. These changes are detected by the automatic generation control (AGC) function.

The presence of DG resources does not necessarily produce a new problem because compensation for the frequency variation is already in place in the initial stages of the control system history. However, the significant penetration of DG resources into the system adds more variability to the network, whose impact is new. The challenge now is how to manage and control this variation, which is difficult to do at a local or device level. Therefore new grid control functions, technologies, and strategies are needed. Moreover, the fact that the demand shifts are coincident with weather changes further complicates this issue.

There are several ways to diminish the effect of some DGs on power system frequency. The compensation can be performed by the inclusion of fast-acting load control devices or flywheels by controlling the ramping of the power. Another solution is given by the integration of energy storage systems or energy capacitor systems. In addition, the combination of wind or PV resources with FC storage devices can mitigate the effect at a local area. The important characteristics of the storage devices used for frequency regulation is related to the quick response they need to have. Some tests have been done using redox flow batteries for supplementary control in addition to the power quality compensation. Finally, curtailment of the DG output power, rather than absorbing the entire DG output, can be adopted for frequency regulation during emergency conditions.

8.4.1.7 Demand Management in Energy Markets

While the electricity market is cleared based on the supply (generators) and the demand (load), the participation by the generators is the predominant activity. Much focus is put on the supply side (i.e., how generators offer), which generators have market power, and whether they make a sufficient level of revenue from various electricity markets. Just recently, demand response (DR) resources have gained more attention. DR resources are types of resources that can modify or influence the demand consumption pattern. DR resources can be used to reduce the previously anticipated demand in case of system emergency such as supply shortage. DR resources can also participate in both energy and capacity markets as an economic resource, known as price responsive demand (PRD). The objective of PRD is to forgo the prescheduled consumption if the market price (day-ahead or real-time) exceeds a certain threshold level. For example, if the market price exceeds USD100 per MW h for a specific market interval, a particular DR resource feels that it is too expensive so that resource can bid into the market, saying "I will forgo my consumption if the price goes above USD100 per MW h."

Energy markets represent an economic agreement between generation and consumption (demand). DSOs/TSOs are generally in charge of operating the DR in an efficient way with less possible interruptions of electrical supply. To operate and control the demand, it is common to use direct load control (DLC), which is a classical tool based on the assumption that load behavior is fully under control. The controller uses load models to take decisions. Currently, an automated demand response system named "OpenADR" that is an open-source tool was developed for distribution model simulation oriented to DR in which clients can subscribe to programs to reduce their bill and a utility or system operator can send a message to the clients in case of being required to participate in an "emergency program." In energy markets, DRs can be distinguished as

(1) *Incentive-based DR:*

    (a) DLC: utility, DSO or TSO has free access to demand processes.

    (b) Interruptible rates: customers are subsidized in case of a continued interruption of the service.

    (c) Emergency DR programs: voluntary response to emergency scenarios.

    (d) Capacity market programs: customers have a specific contract to allow supply interruption as necessary.

    (e) Demand bidding programs: customers can bid to reduce the price of their bills.

(2) *Time-based rates DR:*

    (a) TOU: it is used as a static price schedule.

    (b) Critical peak pricing: it is similar to TOU, being less predetermined.

    (c) Real-time pricing: wholesale market prices directly submitted to end users.

To optimize the energy system, the energy management uses demand side management (DSM) by operating the energy system at the side of the demand. DSM can be classified according to the timing and the impact on the customers as follows:

(1) *Spinning reserve (seconds, temporarily reduced)*. Distributed spinning reserve acts by imitating the behavior of a traditional ancillary service. Thus, when the frequency in the grid rises or drops, the load can be increased or reduced as required.

(2) *Physical DR (seconds-minutes, temporarily reduced-optimized schedule)*. Includes grid management and emergency signals.

(3) *Market DR (days, optimized schedule)*. Involves real-time pricing, price signals, and incentives.

(4) *Time-of-use (weeks-months, optimized schedule)*. A change in a price schedule of a TOU implies a change in a supply contract/tariff, so it does not occur frequently.

(5) *Energy efficiency (permanent, optimized)*. Energy efficiency is the most valuable method because it implies permanent changes in the equipment or buildings for energy saving. The objective is to reduce peak loads and the analysis can be done by automation equipment, such as an energy information system (EIS) that implies a data acquisition system, user interfaces, calculation, analysis algorithms, and so on, or by applying classical calculations, such as baseline versus peak load comparison, weekly comparison of time series, benchmarks, and process correlations.

Energy controllers are devices used, at the energy meter side of the consumer, to carry monitor the consumption trend in energy markets. When power consumption rises above the desirable threshold, defined according with predetermined settings, the device will switch off the equipment.

Demand shifting, or load shifting, is another issue to take into account in current energy markets. Because the load curve has a typical shape, it would be necessary to flatten it to have a more efficient system with reduced losses. In addition, consumers can also plan ahead in case of prediction of emergency scenarios in the grid. In this sense, plug-in electric vehicles (PEVs) can be a new load on existing distribution grids and their deployment would support demand shifting.

Virtual power plants (VPPs) are commonly a group of renewable energy sources, or loads, acting only as virtual storage by demand shifting or as a power plant to the grid management. This equipment needs to be controlled from a central dispatch through supervisory control and data acquisition (SCADA) with standards such as IEC 61850 for substation automation.

All the concepts mentioned above are crucial and must be taken into account in energy markets to accomplish an efficient operation of an electrical grid, improving power quality, reducing power losses, and obtaining reduced cost for consumers.

Table 8.7 Value of reduced load in the wholesale electricity market by distributed generation

Item	Quantity (MW)	Market price ($/MW h)	Pool revenue ($/h)
Load	50,000	40.0	2,000,000
Reduced load	49,000	37.0	1,813,000
Difference	1000		187,000
Value of load reduction			187,000

Authors' own illustration

8.4.1.8 Example of Distributed Generation as a Demand Control Tool

DG offers potential benefits to the electricity market by acting as a demand response by reducing load. Especially on a local basis, there are opportunities for electric utilities to use DG to reduce peak loads, to provide ancillary services such as reactive power and voltage support, and to improve power quality. Table 8.7 shows an example on the value of reduced load in the wholesale market due to the presence of DGs. Assume the system load is 50,000 MW. Participating DG have 1000 MW, causing the load to reduce to 49,000 MW. The market price before the DG participation is assumed to be 40.0 $/MW h and the price after the DG participation is assumed to be 37.0 $/MW h. Note that reduction in load can cause the reduction in market price. In other words, lower demand requires lower supply and hence lower market price. From this example, the saving or value of load reduction caused by DGs is 187,000 $/hr.

8.5 MARKET DESIGNS FOR DISTRIBUTED GENERATION

There are different motivations behind the inclusion of DGs into the competitive electricity market. For example, competition, electric reliability, cost reduction, energy independence, and emission reduction are some of the key drivers. Based on these motivations, different market mechanisms have been established. RESs also make up the majority of the DGs. In Europe, a compensation scheme for various types of renewable resource generators is implemented. A fixed price plan is set and modified every two years. For some systems, this amount is paid as a subsidy and in other systems the price is paid directly by the customers.

In the United States, some states have implemented special programs in which a minimum energy production from renewable resources is defined and must be provided by utilities. Other systems require large electricity purchasers to be responsible for the energy produced by renewable resources. The market mechanism in these cases is performed by the issuance of reward certificates. These documents, known as renewable energy certificates (RECs), are proportional to the acquisition of electricity based on renewable resources.

Currently, the only possible mechanism is a bilateral agreement between the DG owner and the utility, where the price is set by the wholesale market. The market mechanisms for the inclusion of DG are under review, and are not yet implemented in many countries. However, another possibility of allowing DGs to participate in the market is to develop what is known as a distribution system market in combination with distribution system operation and planning. This novel design is further discussed below.

8.5.1 Distribution System Operation

In Section 8.4, we discussed the issues and challenges associated with DG participation in the wholesale electricity markets particularly in the United States. However, the majority of the DGs will be connected at the distribution system level. Extending the existing wholesale electricity markets into the distribution system is enormously difficult for most market jurisdictions. One key reason for this is that wholesale markets are governed by the rules issued by the federal energy regulator because wholesale market transactions cross state boundaries and constitute interstate commerce.

On the other hand, the jurisdiction for retail market transactions that will take place in a particular distribution system will fall under the state's authority. In this case, only individual states can issue rules and regulations that would govern the operation of the retail market, also known as the distribution system market. From a market design standpoint, it becomes imperative for the industry to set up such a market in the context of a DSO or DSO market. A DSO market will be different from current wholesale markets.

As a growing number of DERs will be connected to the distribution grid, the distribution system may need a new design to ensure new services from DERs are valued and utilized effectively and efficiently. The new kind of DSO construct is needed to take on the responsibility for balancing supply and demand variations at the distribution level, as well as linking the wholesale and retail market agents. This new design should also include a possible market mechanism at the distribution level in which available, feasible, and cost-effective DER solutions become part of any distribution system planning efforts. These market designs should create opportunities for DER resources with reliable track records to compete equally with more traditional solutions. Planners must become more fully conversant with the capabilities, applications, and costs of DER. Only in this way, progress can be made toward using new solutions for emerging problems in distribution system planning and operation. The following key attributes are identified as essential for successful operation of the future DSO market.

8.5.1.1 Flexibility

In general, flexibility is an ability of a system to adapt and modify in response to continuously changing conditions. The concept of flexibility applies to many aspects of the power system. For example, the modern day power system already shows some level of operational flexibility in response to ever changing conditions of demand and supply. Operational flexibility is especially important for a system that would embrace significant amount of DERs that include variable renewable energy (wind and solar). Operational flexibility can be aided and improved by near real-time operational decisions, more sophisticated short-term forecasting for variable generations, and availability of a wide variety of operational tools. In the end, operational flexibility should be helpful for power system operation in response to a wide range of system operating conditions so as to maintain reliability. Flexibility, if done properly, can reduce overall system costs via more efficient power system operation.

Future DSO designs must ensure providing incentives and investment signals to attract and enhance the flexibility of future power systems. For example, rewarding more flexible resources to come online by means of proper market design or incentive mechanisms should enhance system flexibility. From the resource adequacy standpoint, having a set of more diverse resources in the system certainly increases the system resource flexibility. For example, a system with more diverse resources, such as hydro, nuclear, coal-fired, gas-fired, and other fuel types, is better able to withstand a price shock of a particular fuel than a system with less diverse resources. For a power system with 100% renewable resources, the system flexibility would come from smaller DERs, which should then be diversified.

The flexibility that is relevant to DGs is distribution flexibility. As significant amount of DERs and other resources will be connected directly at the distribution system level, the flexibility of the distribution system becomes even more important. Similar to transmission flexibility, flexible distribution requires a distribution system with ample capacity and minimum congestion so that the majority of DERs can provide power anywhere in the distribution system. At high-priced periods, the load consumers can respond and manage their load consumption with their DRs and load management tools. Other resources such as PVs with reactive power capability and electric vehicles (EVs) with V2G capability can also respond and manage their consumptions/generations as necessary. Instead of one-way power flow as currently designed in the conventional distribution system, the future distribution system should have a capacity and capability to allow for two-way or multiple-way power flows for both active and reactive power. Radial design of the distribution system has to be complemented by meshed design or more intelligent design to allow the power to freely flow from one arbitrary node to another. Increasing the level of flexibility in all key system components in the system would certainly increase overall system flexibility.

8.5.1.2 Market Administration

The future DSO should be responsible for administering and operating a distribution system market similar to the retail market. In this market, the possible market products include energy, ancillary services, and capacity from which DER owners can receive revenues. Well-functioning short-term retail markets are necessary to provide appropriate incentives for investment in and retirement of DERs. Better market design can effectively manage the growth of DERs by properly valuing the energy and other system services provided by these resources. They should also provide necessary long-term signals for both regulated and merchant investments in a distribution network. DSO market design can include day-ahead, intraday adjustment, and real-time balancing distribution markets for energy to achieve efficient unit commitment and dispatch and efficient retail prices. Other responsibilities of the DSO market include distribution congestion management via a distribution locational marginal pricing (DLMP) scheme, contract regime (spot, longer term), distribution interconnection and transit rules and prices, and engineering reliability rules. The DSO market will interact with interconnected the wholesale market at the transmission level by acting as another wholesale market participant by netting out the aggregates of resources and loads at the distribution level. Note that imperfections in competitive retail markets have adverse effects on the wholesale market performance.

8.5.1.3 Operational Authority

Even though it is conceptually possible to have separate entities for distribution system operation and distribution system market operation, it is better to designate a single entity to take a dual role of operating both the distribution system and distribution system market. This is similar to the dual roles played by wholesale electricity market operators in the United States. In this case, any future distribution system operation design must include a DSO that has operational authority to ensure short-term reliability of the distribution system for which they are responsible. Therefore the DSO will also operate the distribution system by matching supply and demand instantaneously while considering possible distribution system contingencies and effectively dealing with fluctuating power output from variable generations. There must be clear rules for distribution system operation that properly take into account the generation output from load consumers who also own DGs.

8.5.1.4 Planning Authority

In addition to operational authority, the DSO must also have planning authority for the distribution system as distribution system planning is equally important as distribution system operation. The DSO will be responsible for efficient planning of the distribution network and DERs located in the area. The new entity will conduct distribution planning and expansion by creating incentives for investment in electric infrastructure with clear planning policies for distribution grid expansion. The distribution network expansion policy should allow for both regulated and merchant distribution network investments. More specifically, the DSO will forecast load for the distribution system and assess requirements for resource and distribution network adequacy. If the capacities of the distribution networks are found to be inadequate in the near future, the DSO will propose network upgrades or expansions in an open, fair, and transparent manner so that all stakeholders involved are aware of the projected network changes. It will also conduct interconnection analysis for DERs that are proposed to connect to the distribution grid. Uncertainties associated with DER expansion and load growth have to be explicitly accounted for in distribution system planning.

8.5.1.5 Independence

Any market operator must be independent of any market participant and resource owners (suppliers) for two major reasons: (1) to operate the market in a fair and impartial manner and (2) to utilize the DERs in the most efficient and least-cost manner. For example, all wholesale electricity market operators in the United States are nonprofit entities that are also independent from any buyers, sellers, and brokers participating in the market. They are also independent from the ownership of any other for-profit entities. These market operators are governed by an independent board of directors and comply with the rules set by federal regulators. By the same analogy, the DSO market operator must be independent from any market participant to operate a free and competitive distribution or retail market. Market participants in DSO market include DER owners, load, and third-party market brokers. No market participant should have any influence on the operation and outcome of any DSO market. The governance structure and incentives for the DSO market should be designed to ensure that no one subset of market participants is allowed to control the criteria or operating procedures of the distribution system operation.

8.5.1.6 Open and Fair Access

One of the impediments to competition in the wholesale power market was the lack of open and fair access to the transmission system on which the market operates. To remove that impediment, in 1996, the U.S. federal regulator issued Order 888, which required public utilities to provide open access transmission service on a comparable basis to the transmission service they provide themselves. This order set the stage for a subsequent increase in competition in the wholesale market. Similar to this rule, the DSO design should be governed by provisions or rules that require open, fair, and non-discriminatory access to the respective distribution system by legitimate users of the system. The users can be DER owners, load consumers, third-party aggregators, and power brokers. Only then, can competition in the distribution system flourish, which will lead to the most efficient utilization of DERs. It would be better if the DSO does not own any DERs operating in their distribution service areas. If they do, there must be some sort of functional separation to avoid conflicts of interest.

8.5.1.7 Transparency

Transparency means the extent to which market participants have ready access to any required system or financial information about the market such as price levels, market depth, and other financial figures. Transparency is one of the prerequisites of any free and efficient market. The transparency of the market should apply to data, rules on market administration and market clearing, operating costs, and system operation procedures. For example, the DSO market should include transparency in data regarding forecasted load, available supply, market clearing prices, and so forth. The availability of such information will reduce uncertainty in the market and increase its efficiency. Transparency in the market clearing process requires that if the DSO market uses uniform-price auction as market clearing mechanism, that auction process has to be as transparent as possible.

Transparency in rules governing market operation requires clarity on (1) criteria and process for DER owners to become participants in the market, (2) rights and obligations of participants under the rules, (3) criteria for suspending or terminating a market participant from participating in the market, and so on. Those rules should be applied by the DSO consistently and without inappropriate bias. Transparency in operating cost of the DSO market is also important. This type of transparency reinforces the accountability of the DSO market for its efficient operation.

8.5.1.8 Interface Between the Independent System Operator, Transmission System Operator, and Distribution System Operator

As the DSO will operate at the distribution level, there must be some kind of interaction between the DSO and the ISO/TSO. Particularly, it is important to have a functional intersection between the wholesale transactions at the transmission system and the retail transactions at the distribution system. One possible way is that those retail products (DERs) can be aggregated and allowed to participate in a market exchange through a wholesale–retail interface with the bulk system. The need for increased coordination and collaboration between the DSO and the ISO/TSO has been addressed in many reports.

We also envision that if the DER is connected to the low-voltage distribution system, then the DSO will get priority in controlling that resource. If the DER is connected directly to the higher voltage transmission system, then the ISO/TSO has priority over the DSO. It is very important to have clear roles and responsibilities between the ISO/TSO and the DSO including the priorities of resource control by the respective entities.

8.6 CONCLUDING REMARKS

As the latest development, the NYISO, the wholesale market operator for the state of New York, proposed rules that would allow behind-the-meter generation with excess capability to take part in its wholesale markets. The new rules can bring 200–300 MW of existing capacity into the New York markets and mesh well with the state's future energy vision. The rules will apply to all behind-the-meter-with-net-generation resources that would like to participate in the markets. Without this rule, the big customers with DG facilities can only sell their excess power to the local utilities under power purchase agreements. Under the new NYISO rule, the eligible DG resources will be treated as other existing resources that participate in its markets. To qualify, the DG will need a nameplate capacity of at least 2 MW, native load of at least 1 MW, and an interconnection letting at least 1 MW be sold into the transmission grid.

The NYISO's proposed tariff changes will help promote DG and provide developers with the opportunity to gain additional revenue through the sale of extra energy and capacity while promoting resiliency. The rules come with flexibility as generators can enroll as a single unit, in aggregations, or where the appropriate metering exists, and one facility can be split into several behind-the-meter generation resources. Additional software, rules, and procedures will be developed to further integrate DGs into the wholesale markets and to align them with the energy policy choices of the state. These DG resources must also be identified, modeled, and studied in the system planning studies.

Energy production from distributed generation, apart from being a handicap as far as power regulation means, also presents an opportunity to increase the diversity of electricity generation, particularly from renewable sources. Renewable energy offers less dependence on sources of fossil origin. As regards PV generation, at least 48.1 GW were installed globally in 2015, representing a growth of 25% over 2014. The largest growth of PV installation was in China followed by Japan and the United States. Asia accounted for 60% of this global growth. Global installed PV capacity reached about 227 GW in 2015. More efforts are necessary to improve the performance of these technologies. Furthermore, as far as DG markets are concerned, opportunities such as the net energy balance should be considered so as not to penalize this type of electricity production technology. It would be more appropriate if there existed a market mechanism that would properly value the benefits brought about by DGs to the system.

FURTHER READING

[1] International Electrotechnical Commission (IEC), in: Grid Integration of Large-Capacity Renewable Energy Sources and Use of Large-Capacity Electrical Energy Storage", White Paper, Geneva, Switzerland, 2012.
[2] Carl Linvill, John Shenot, Jim Lazar, Designing Distributed Generation Tariffs Well: Fair Compensation in a Time of Transition, RAP, Nov 2013.

[3] T.-N. Preda, Kjetil Uhlen, Dag Eirik Nordgard, An overview of the present grid codes for integration of distributed generation, in: Integration of Renewables into the Distribution Grid, CIRED 2012 Workshop, IET, 2012.

[4] IEEE 1547 Standard for Interconnecting Distributed Resources with Electric Power Systems.

[5] ENTSO-E, Network Code on Requirements for Grid Connection Applicable to all Generators (RfG), (March 2013). http://networkcodes.entsoe.eu/connection-codes/requirements-for-generators/.

[6] R. Datta, V.T. Ranganathan, Variable-speed wind power generation using doubly fed wound rotor induction machine—a comparison with alternative schemes, IEEE Trans. Energy Convers. 17 (3) (2002) 414–421.

[7] http://mstudioblackboard.tudelft.nl/duwind/Wind%20energy%20online%20reader/Static_pages/Cp_lamda_curve.htm

[8] http://etap.com/renewable-energy/photovoltaic-101.htm

[9] http://powerelectronics.com/inverters/silicon-carbide-diodes-promise-benefits-solar-microinverters.

[10] V.J. Fesharaki, et al., The effect of temperature on photovoltaic cell efficiency, in: Proceedings of the 1st International Conference on Emerging Trends in Energy Conservation—ETEC, Tehran, Iran, 2011.

[11] The Institute for Electric Innovation, Net Energy Metering: Subsidy Issues and Regulatory Solutions, Issue Brief, (Sept 2014).

[12] National Renewable Energy Lab (NREL) Report on "Flexibility in 21st Century Power Systems", NREL/TP-6A20-61721, May 2014.

[13] K. Knezović, P. Codani, M. Marinelli, Y. Perez, Distribution grid services and flexibility provision by electric vehicles: a review of options, in: Power Engineering Conference (UPEC), 2015, 50th International Universities, Sept 2015. pp. 1–6.

[14] J. Lin, K. Knezović, Comparative analysis of possible designs for flexible distribution system operation, in: 13th International Conference on the European Energy Market (EEM 2016), Porto, Portugal, 6–9 June, 2016.

[15] Renewable Energy Consumer Code, Feed-in Tariff Scheme, UK, (2015).

[16] American Public Power Association, Distributed Generation: An Overview of Recent Policy and Market Developments, (November 2013).

[17] U.S. Department of Energy, The Potential Benefits of Distributed Generation and Rate-Related Issues That May Impede Their Expansion, February 2007.

[18] New York Independent System Operator (NYISO) Report, "A Review of Distributed Energy Resources", (prepared by DNV GL), Sept. 2014.

[19] National Renewable Energy Laboratory (NREL) Technical Report, "Compensation for Distributed Solar: A Survey of Options to Preserve Stakeholder Value", September 2015.

[20] H.A. Gil, G. Joos, On the quantification of the network capacity deferral value of distributed generation, IEEE Trans. Power Syst. 21 (2006) 1592–1599.

[21] D.Q. Hung, N. Mithulananthan, A simple approach for distributed generation integration considering benefits for DNOs, in: 2012 IEEE International Conference of Power System Technology (POWERCON), 2012.

[22] P. Palensky, D. Dietrich, Demand side management: demand response, intelligent energy systems, and smart loads, IEEE Trans. Ind. Informat. 7 (3) (2011) 381–388.

[23] M.H. Albadi, E.F. El-Saadany, A summary of demand response in electricity markets, Electr. Pow. Syst. Res. 78 (2008) 1989–1996.

[24] J. Han, M. Piette, Solutions for summer electric power shortages: demand response and its applications in air conditioning and refrigerating systems, Refrig. Air Condition. Electric Power Mach. 29 (1) (2008) 1–4.

[25] Electric Power Research Institute (EPRI), The Integrated Grid: Realizing the Full Value of Central and Distributed Energy Resources, http://www.epri.com/abstracts/Pages/ProductAbstract.aspx?ProductId=000000003002002733, 2014.

[26] http://www.boe.es/ [visited in 4/29/2016].

[27] http://www.solarweb.net/termosolar.php [visited in 4/29/2016].

CHAPTER 9

Distribution Generation Optimization and Energy Management

Barry Hayes
National University of Ireland Galway, Galway, Ireland

9.1 INTRODUCTION

9.1.1 Background and Motivation

Distribution networks were initially designed as passive systems, which provided one-directional links between the transmission network and electricity end users. The introduction of DG is resulting in many distribution networks becoming energy harvesting systems, with much more variability and bidirectional power flows where there are high penetrations of DG [1].

This creates many new challenges for utilities, and requires distribution networks to be operated as active distribution systems, with real-time control and optimization of multiple distributed energy resources [2,3]. This chapter will introduce the most important concepts around distributed energy management systems, operator systems designed for decentralized management of DG and other energy resources connected to the distribution networks. In addition, the microgrid concept is introduced, and issues related to energy management and optimization of DG in the context of microgrids are discussed.

This chapter is organized as follows. Section 9.2 deals with the impacts of DG on network operation and energy management. This deals with issues related to DG monitoring and communication and distribution network energy management systems. It also introduces the concept of network hosting capacity and the problem of managing congestion in the distribution networks caused by DG.

Section 9.3 discusses DG optimization methods and applications. First, the basic principles and methods used in general optimization problems are introduced, including the problem of optimal power flow in electricity

Distributed Generation Systems
http://dx.doi.org/10.1016/B978-0-12-804208-3.00009-1

415

networks. Examples related to the optimization of DG location, sizing, and voltage optimization in systems with DG are presented.

The final part of the chapter, Section 9.4, deals with issues related to the coordination of multiple DGs in electricity distribution and transmission networks. It introduces the concept of microgrids, and how DGs are managed in the context of microgrids. It also discusses DG system integration issues and potential future electricity networks based on multiple microgrids.

9.2 DISTRIBUTED GENERATION IMPACTS ON NETWORK ENERGY MANAGEMENT

9.2.1 Introduction

The integration of DG presents a number of new challenges for electricity network operators, as in almost all cases, power networks were originally designed for the transmission and distribution of energy from centralized generating systems. Previous studies have shown that the connection of significant numbers of DGs results in an increase in the occurrence of network voltage violations and congestion, particularly in the distribution part of the system.

Traditionally, the design philosophy of electricity distribution networks was based around a "fit and forget" approach. This means that the distribution network is designed and built with enough capacity to cope with any possible changes in the electricity demand and network power flows over a long period of time (e.g., one or two decades) without overloading any system components. In this way, network issues such as congestion and excessive voltage variation are dealt with at the network design stage.

However, with the introduction of DG, there are various local resources in the system, which need to share access to the distribution network. In the fit and forget approach, network operators will only approve the network connection of a DG if it has been validated that the DG will not violate any of the network limits, under all normal operating conditions. DGs are typically given a "firm" connection, meaning that the connection is not restricted, and that the DG can export to the network at all times.

Once the penetration of DG in the electricity network becomes significant, the fit and forget approach may not be valid, as it is no longer economical to build enough capacity in the networks to deal with every possible load and DG output scenario [4]. In some networks, applications for DG network connections have been declined, because the networks may not have sufficient capacity to allow DGs access to the network, or costly network upgrades are required to provide this capacity. This has led to the development of the "connect and manage" approach to distribution network operation. In this approach, network operators allow nonfirm connections to the network, meaning that DGs are allowed to connect to the network, with a restricted connection, where the network operator needs to manage the capacity in the network and share this among multiple DGs. The problem of DG energy management is discussed in more detail, with several examples in Section 9.2.3 of this chapter.

At the transmission level, energy management is generally carried out using an energy management system (EMS), a system of computer-aided tools designed to assist the operator in the monitoring, control, and optimization of the network. The monitoring and communications requirements for control and management of DGs are discussed in Section 9.2.2 of this chapter.

As distribution networks are seeing increasing penetration of DG and other distributed resources, distribution management systems (DMSs) are becoming increasingly important. These are discussed in Section 9.2.4. Finally, the concept of dedicated systems for management of distributed energy resources (DERs), distributed energy resources management systems (DERMSs), and the microgrid concept are introduced in Section 9.2.5.

9.2.2 Distributed Generation Monitoring and Communications

The requirements for monitoring, or telemetry, for DG vary according to the country and region, and according to the network operator. In addition, the requirements from the distribution system operator (DSO) and transmission system operator (TSO) may differ. Communications from DG facilities to the DSO and/or TSO are needed for *metering* purposes (i.e., billing and payments), and in some cases for *monitoring* purposes, so that the network operator has real-time visibility of the various resources on the system, to ensure reliable and secure operation of the grid.

Some examples of typical values that are measured and telemetered to the network operator are

- Directional instantaneous MW and MVAr export/import from DG unit, or group of units in a DG facility
- Directional MW h and MVAr h export/import, averaged at intervals of one hour or less
- Open/closed status of DG interconnection circuit breakers or switches

- Open/closed status of DG generation unit circuit breakers or switches
- Instantaneous voltage V at point of interconnection between DG and network
- Instantaneous current I at point of interconnection between DG and network
- Alarms or warnings indicating trips or communication system failures

Measurement and communications are generally carried out using supervisory control and data acquisition (SCADA) systems. For metering purposes, data is generally required over longer time intervals, for example, hourly or half-hourly. Real-time monitoring for network operation and management purposes typically requires shorter time intervals, for example, on the order of 5–15 min or less, depending on the application.

The exact requirements for DG metering and monitoring depend on many factors, and some countries or regions may require more or less detailed measurement data depending on whether a DG participates in the energy markets, and depending on the impact a given DG has in the local network. However, in general, larger DGs are required to provide more detailed telemetry data, because these larger systems will usually have more influence on the behavior of the electricity system.

Larger DG installations will not only impact the distribution network where they are connected, but will also have an effect on the regional or national transmission system. Table 9.1 shows the requirements for DGs to provide telemetery to the TSO in several European countries [5]. In the United States, the requirements for monitoring of DG varies considerably according to the responsible system operator in each region [6]. Similar information for other counties can be obtained from the national or regional TSO.

Table 9.1 shows that there is significant variation in the requirements for DG telemetered data from one country/region to another. In Germany, all DG facilities with a capacity of 100 kW or larger are required to provide

Table 9.1 Distributed generation requirements for providing telemetered data to transmission system operators in several European countries [5]

Country/region	DG telemetry requirements
Germany	100 kW capacity or larger
Spain	10 MW capacity or larger
UK (Northern Scotland)	10 MW capacity or larger
UK (Southern Scotland)	30 MW capacity or larger
UK (England and Wales)	100 MW capacity or larger

telemetry to the TSO. In Spain, this threshold is 10 MW, which means that there is a large amount of DG on the system (including 3.5 GW of distribution-connected photovoltaic (PV) generation[1]) that the TSO has no direct monitoring or control over. In the United Kingdom, this threshold for providing real-time telemetry to the TSO varies according to the region.

At the distribution level, there may be additional requirements for DGs to provide telemetered data to the DSO. The monitoring of DG is closely related to the issue of DG controllability, that is, whether the DSO can control the output of DG to manage the energy flows in the network. Certain types of DG may be dispatchable, for example, a thermal generation plant owned by the DSO, where the DSO has the ability to dispatch, or schedule the output according to the requirements of the network.

Renewable DG, such as wind and solar, are nondispatchable, because the output depends on weather conditions, although in some cases, curtailment or controlled reduction of energy output is possible. The ownership of DG also depends on local regulations. In many countries, deregulated electricity markets require that the functions of electricity generation and electricity distribution are separated, so it is not permitted for the DSO to own generation assets such as DG.

A summary of the various possibilities for DG dispatchability and ownership is given below:

- *Dispatchable, owned by the DSO.* These are DG units that are dispatchable and can be fully controlled by the DSO; for example, a small thermal generating station.
- *Dispatchable, not owned by the DSO.* This is typically the case in deregulated electricity markets, where the DSO is not allowed to own generating assets. The DG owner may inform the DSO of the expected output from the DG, and the schedule of operation is often agreed with the DSO ahead of time.
- *Nondispatchable, owned by the DSO.* DG units based on renewable energy sources, such as wind or solar, which are owned by the DSO. In these cases, the DSO is able to directly manage and curtail the output of the DG if required.
- *Nondispatchable, not owned by the DSO.* Variable DG, not owned by the DSO; for example, a community-owned small wind farm or rooftop-connected solar PV. The DSO has no real-time monitoring or control of these devices.

---

[1] Based on 2015 figures.

The issues of DG dispatchability and controllability are important for the following sections, which discuss energy management in electricity networks with DG.

9.2.3 Distribution Network Congestion Management

The issue of increasing demands for access to the distribution network has occurred in a number of countries worldwide, due to the growth in renewable energy and a resulting increase in requests for DG connections.

Low penetrations of DG can often affect network operation and performance positively, as DG can reduce the peak demand of the system, and relieve loading of network components at times of high demand, when the system is under the most stress. However, very high penetrations of DG can cause congestion, or overloading of components in the distribution system. A simple example of network congestion is shown in Fig. 9.1.

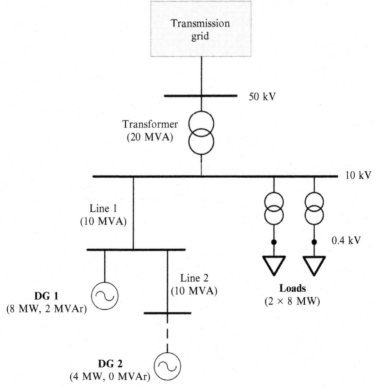

Fig. 9.1 Simple example of network congestion due to distributed generation.

In this distribution network, Transformer 1 is located at the interface between the transmission network (at 50 kV) and the medium voltage distribution system (at 10 kV). Transformer 1 has a capacity of 20 MVA, and Lines 1 and 2 both have a capacity of 10 MVA. This network serves two loads, which are connected at low voltage (0.4 kV) through secondary transformers. The main focus of this example, however, is on DG 1, which has a capacity of 8 MW, and the proposed connection of DG 2, which has a capacity of 4 MW.

The apparent power flow in each line of the network in MVA, S_{line}, is the combination of the real and reactive power flows. It is calculated by

$$S_{line} = \sqrt{P_{line}^2 + Q_{line}^2} \tag{9.1}$$

where P_{line} is the real power flow through the line in MW and Q_{line} is the reactive power flow through the line in MVAr.

The thermal capacity limit of each line is the maximum apparent power flow allowed, according to the line's physical limits, is $S_{MVA\ limit}$. The line flow must be kept within this limit according to

$$S_{line} \leq S_{MVA\ limit} \tag{9.2}$$

In the following examples, the objective is to calculate the line loading and compare this to the limits of each line, in order to determine whether the proposed connection of DG 2 would cause network congestion in the form of line overloading, e.g., exceeding the line limit in Eq. (9.2).

Example 9.1

Calculate power flow in Line 1 and check if the line limit is exceeded when DG 1 is at full output.

In this example, DG 1 is at full output, exporting real power P_1 of 8 MW and reactive power Q_1 of 2 MW. DG 2 is not yet connected. Active and reactive power losses are neglected.

The power flow in Line 1 is given by Eq. (9.1):

$$S_{Line\ 1} = \sqrt{P_{Line\ 1}^2 + Q_{Line\ 1}^2} = \sqrt{8^2 + 2^2} = 8.246\ MVA$$
$$S_{Line\ 1} < 10\ MVA$$

The apparent power flow is less than the Line 1 thermal limit of 10 MVA, and therefore the limit is not exceeded in this case.

Example 9.2

Calculate power flow in Lines 1 and 2. Check if the line limit is exceeded when DG 1 and DG 2 are both operating at full output.

In this example, DG 1 is at full output as before, and DG 2 is also now exporting real power P_2 of 4 MW and no reactive power $Q_2 = 0$ MW.

Starting with Line 2,

$$S_{Line\,2} = \sqrt{P_{Line\,1}^2 + Q_{Line\,1}^2} = \sqrt{4^2 + 0^2} = 4\ MVA$$

The line flow in Line 1 is the sum of the outputs from DG 1 and DG 2:

$$S_{Line\,1} = 8.246 + 4 = 12.246\ MVA$$

In this example, $S_{Line\,1} > S_{MVA\,limit}$, indicating overloading of Line 1. The recommendation from the DSO in this case may be to not allow DG 2 to connect to the network, or to allow both DGs access to the network via a nonfirm connection, where the energy outputs from each DG need to be managed by the DSO during times of network congestion.

The reality of distribution network operation is far more complex than the simple example illustrated above, and one needs to consider the daily and seasonal variations in demand profiles and DG output profiles to fully understand network congestion issues. In addition to line overloading issues, a number of other technical issues such as voltage control, power quality, protection, and fault levels may need to be considered.

Some DSOs have used rules of thumb for determining how much DG can be allowed to connect to the distribution networks. For example, it is estimated that it is possible to connect DGs with a total capacity of up to 15% of the network's peak demand without causing significant problems in most distribution networks [7], and this has been used as a guideline by some DSOs.

The concept of hosting capacity gives a more formal definition. The hosting capacity is defined as the amount of additional DG capacity that can be added to the network without causing unacceptable deterioration in the network performance [8]. This can be calculated for each node in any network through running detailed power flow simulations.

9.2.4 Distribution Network Energy Management Systems

An EMS is a set of computer-aided tools designed to assist the network operator in the monitoring, control, and optimization of electricity networks. EMSs were developed for transmission systems in the early 1970s, and have

Fig. 9.2 Electricity network control center. *(Source: Eirgrid)*

since become a crucial aspect of the transmission network operation, and a central part of the network control room, shown in Fig. 9.2. Some of the main functions of an electricity network EMS are given below:

- Monitoring of the network power flows and voltages
- Forecasting of demand and generation
- Dispatch and control of generation
- Coordination of network protection systems
- Restoration of the network after fault events
- Optimizing network performance and utilization of network assets

Until relatively recently, the application of EMS at the distribution level, or distribution energy management systems (DEMSs), has not been of significant interest. This is largely because distribution networks have traditionally been designed and operated as passive systems, where power flows are one-directional, and are relatively easy to predict and manage (at least under normal conditions). This means that there has not been a need for implementing some of the more advanced EMS functions, such as scheduling and optimization.

However, distribution networks are seeing increasing penetrations of DG and other distributed resources such as flexible demands and devices with energy storage capability. This has led to a requirement for improved observability in distribution systems, and the need for DSOs to take a more active role in monitoring and controlling the operation of the networks. A DEMS is very important in this context.

The following examples also use the simple network schematic shown in Fig. 9.1, Section 9.2.3. The load factor of the network can be defined as the ratio of average loading to the peak loading (in MW h), or maximum loading experienced over a given time interval:

$$LF = \frac{\text{Average loading}}{\text{Peak loading}}$$

The load factor will vary according to the type of load served in the network and according to the season.

The capacity factor of a DG in the network can be defined as the ratio of average DG output to the rated capacity of the DG:

$$CF = \frac{\text{Average output}}{\text{Rated DG capacity}}$$

The capacity factor of a DG depends on the energy source and the generator running schedule. The capacity factors for renewable DGs such as wind and solar generation are limited by the availability of the energy resource, with typical values of 0.15–0.35.

Example 9.3

Calculate expected total monthly load in the network in Fig. 9.1, given the load factor and peak load values.

Each of the two network loads in Fig. 9.1 has a peak demand of 8 MW, and a load factor of 0.55. The reactive power demand and the losses are neglected in this case. It is assumed that there are 30 days in the month.

The total monthly load is then given by

$$\text{Total monthly load} = P_{peak\ demand} \times LF \times N_{days} \times 24\ \text{h}$$
$$\text{Total monthly load} = 16 \times 0.55 \times 30 \times 24 = 6336\ \text{MW h}$$

Example 9.4

Calculate expected total monthly output from DG and the expected percentage reduction in the total monthly load due to the DGs.

The capacity factors of DG 1 and DG 2 are 0.2 and 0.3, respectively. The total outputs of the DG 1 is given by

$$\text{Total output DG1} = P_{rated\ DG1} \times CF_1 \times N_{days} \times 24\ \text{h} = 8 \times 0.2 \times 30 \times 24$$
$$= 1152\ \text{MW h}$$

Similarly, for DG 2:

$$\text{Total output DG2} = P_{rated\,DG2} \times CF_2 \times N_{days} \times 24\text{ h} = 4 \times 0.3 \times 30 \times 24$$
$$= 864\text{ MW h}$$

The total expected monthly output from DG is then

$$\text{Total output} = 1152 + 864 = 2016\text{ MW h}$$

The total reduction in monthly demand due to DG in percentage form is

$$\left(1 - \frac{\text{total load} - \text{total DG output}}{\text{total load}}\right) \times 100$$

$$\left(1 - \frac{6336 - 2016}{6336}\right) \times 100 = 31.82\%$$

Example 9.5

If DG 2 needs to be curtailed (switched off) 10% of the time due to network congestions issues, how much potential generation output is lost?

This can be calculated by simply taking 10% of the expected total output of DG 2:

$$\text{DG 2 curtailment} = P_{rated\,DG2} \times CF_2 \times N_{days} \times 24 \times 0.1 = 86.4\text{ MW h}$$

9.2.5 Coordination of Multiple Distributed Energy Resources

9.2.5.1 Distributed Generation and Distributed Energy Resources in Future Energy Systems

It is anticipated that the continued growth of DG, along with other distributed energy resources, such as electric vehicles (EVs) and energy storage systems (ESSs), will lead to a much more decentralized operation of the electricity system. Traditionally, EMS in the electricity system has been focused on managing several large, centralized generating plant, at the transmission level.

As mentioned previously, DEMSs designed to manage local DG and distributed energy resources in real-time are becoming more prevalent. This means that energy management is increasingly being carried out at the local level, as well as at the system level. In order to coordinate this energy management across multiple areas of the system, hierarchical EMSs have been proposed by a number of authors, for example [9].

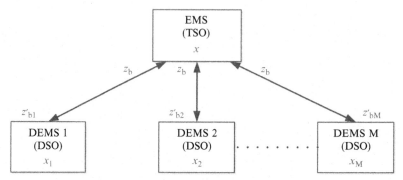

Fig. 9.3 An example of an energy management system hierarchy.

Fig. 9.3 shows an example of an EMS hierarchy, where there is a central EMS (in this case, managed by the TSO), and below this, a number of decentralized DEMSs (see DEMS 1, DEMS 2. etc., in Fig. 9.3), which could be implemented in a DSO control center, or in a distribution system substation. There is a clear hierarchy, or master–slave arrangement in this design.

The advantage of this approach is that it is not necessary for the EMS to know the internal states of all of the DEMSs connected to it (the internal states are represented by x_1, x_1, ..., x_M in Fig. 9.3). Similarly, none of the DEMSs need to know the internal state of the central EMS, x. In this approach, the only measurements that need to be exchanged between the EMS and DEMS are at the borders between their control areas (these border measurements are denoted z_b and $z_{b1}{}'$, $z_{b2}{}'$, ..., $z_b{}'M$ in Fig. 9.3).

9.2.5.2 Virtual Power Plants
Another concept often discussed in the context of DG is the virtual power plant (VPP) shown in Fig. 9.4. The idea behind a VPP is to link together a large number of distributed resources (for example DG units and flexible user loads) in such a way that these can be scheduled or dispatched from a central control room in the same way as a traditional large power plant. This approach can, in principle, replace conventional power plants with aggregations of local resources. However, designing and implementing the necessary communications and control infrastructure to effectively manage large numbers of distributed resources is a complex task.

9.2.5.3 Microgrids
A microgrid is a low voltage (LV) electricity distribution network, with its own energy resources, which is capable of functioning autonomously as a small-scale electricity grid. Microgrids are often implemented in remote areas, where there is no connection to the grid. Recently, there has been much interest in applying

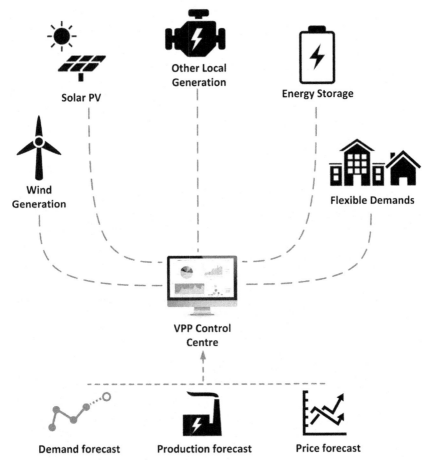

Fig. 9.4 The virtual power plant concept.

microgrids to integrate high penetrations of DG into the electricity system. These issues are discussed later in this chapter in Section 9.4.

9.3 DISTRIBUTED GENERATION OPTIMIZATION METHODS AND APPLICATIONS

9.3.1 Fundamentals of Optimization

An optimization problem can be generally defined as finding the best solution to a mathematical problem from all feasible solutions. The methods used in optimization vary depending on the type of problem and the variables involved. Optimization problems with discrete variables are known as combinatorial optimization problems. If the variables in the problem are continuous, we can use calculus to solve the problem.

A continuous optimization problem can be defined using the following standard form as an *objective function* (the function to be maximized or minimized), subject to a number of constraints [10]:

The standard form of a (continuous) optimization problem is

$$\underset{x}{\text{minimize}} \ f(x) \tag{9.3}$$

$$\text{subject to} \quad g_i(x) \le 0, \quad i = 1, \ldots, m \tag{9.4}$$

$$h_i(x) = 0, \quad i = 1, \ldots, p \tag{9.5}$$

where

$f(x): \mathbb{R}^n \to \mathbb{R}$ is the objective function to be minimized over the variable x,

$g_i(x) \le 0$ are called inequality constraints, and

$h_i(x) = 0$ are called equality constraints.

The process of solving an optimization problem is best demonstrated by example. The following example presents a very simple optimization problem, designed to introduce the main concepts of continuous optimization.

Example 9.6

Maximize the area of a rectangular field, given a limited amount of perimeter fencing.

A farmer has 200 m of fencing material, and wishes to enclose a field using a rectangular fence, as shown in Fig. 9.5. One side of the field is bordered by a straight river, and does not require any fencing.

In order to maximize the area of the field, the optimization function must be defined first, using the formulae for the area and perimeter of a rectangle:

$$\text{Maximize} \quad A = xy \tag{9.6}$$

$$\text{subject to} \quad x + 2y = 200 \tag{9.7}$$

Solving the constraint equation for x gives

$$x + 2y = 200 \Rightarrow x = 200 - 2y$$

Substituting the results into the objective function above allows A to be expressed as a function of one variable:

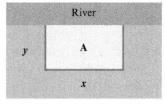

Fig. 9.5 Simple optimization example.

$$A = xy \Rightarrow (200 - 2y)y = 200y - 2y^2$$

The absolute minimum and absolute maximum for y are calculated as follows. The absolute minimum of y is 0 m (consider the fence in Fig. 9.5 with sides x of zero length). The absolute maximum of y is 100 m (a field with two sides of length 100 m, but zero width).

These extreme dimensions of y do not make physical sense, as both would enclose a field with a total area of zero, but they allow us to define the limits of y as follows:

$$y \geq 0$$

$$200 - 2y \geq 0 \Rightarrow 100 \geq y$$

Combining these gives

$$0 \leq y \leq 100$$

Finding the derivative of $A(y)$ yields the critical point(s) of the function

$$A'(y) = (200y - 2y_2)' = 200y' - 2(y^2)' = 200 - 4y$$

Solving gives

$$200 - 4y = 0 \Rightarrow 4y = 200 \Rightarrow y = 50$$

The maximum value of $A(y)$ can be found by evaluating the end points and critical points of the function along the interval $0 \leq y \leq 100$:

$$A(0) = 0, \quad A(100) = 0, \quad A(50) = 200 \times 50 - 2(50^2) = 5000 \text{ m}^2$$

There is only one critical point along this interval, and this occurs at the maximum of the function. The maximum area A is 5000 m^2, and this is found at the *optimal* values of y and x, which are

$$y = 50 \text{ m}$$
$$x = 200 - (2 \times 50) = 100 \text{ m}$$

9.3.2 Optimal Sizing of Distributed Generation

To maximize the return on investment when developing DG sites, the selection of parameters such as the size and location of DG units in an installation needs to be considered carefully [11]. A number of factors and limitations may affect DG sizing, including DG unit costs, the expected market price for energy produced by DG, space restrictions, and availability of the primary energy resource (this is particularly important in the case of weather-dependent renewables such as wind and solar). The network operator, or DSO, may also impose restrictions on the size of the DG installation, to manage technical constraints in the electrical distribution system.

The following example formulates the problem of selecting the best number of DG units to install at a particular site as a simple mathematical optimization.

Example 9.7

Calculate the optimal number of wind turbines to install at a DG site, given the parameters below.

In this example, a developer is planning to build a wind farm in a rural area. The wind farm will be made up of a number of individual wind turbines (WTs), each with a rated capacity of 1 MW. The optimal number of WTs to be installed depends on two main factors: the expected profit that will be made by selling the generated electricity to the market over the lifetime of the wind farm, and on the cost of installing the WTs.

The main parameters are listed below. For simplicity, all mechanical and electrical losses are neglected in this analysis.

- The number of WTs installed at the site will be between a minimum of 1 and a maximum of 18, due to space restrictions.
- The first 1 MW of installed capacity is expected to receive a price of $100 per MW h generated over the entire lifetime of the installation.
- For each additional 1 MW of capacity installed, the effective price received for generated energy is expected to decrease by $5 per MW h.
- The total cost of installing each 1 MW WT is $1,000,000.
- The DG capacity factor is 0.25.
- The average availability of all WTs is 95%.
- The lifetime of the wind farm is 25 years.

The energy output in MW h from each WT unit over the lifetime of the wind farm can be calculated by

$$E_{WT} = P_{rated} \times A \times CF \times h$$

where the rated capacity of each WT, $P_{rated} = 1$ MW, the availability, $A = 0.95$, and the capacity factor $CF = 0.25$.

The number of hours h is calculated by multiplying 8760 (the number of hours in each year, ignoring leap years) by the number of years in the lifetime of the wind farm:

$$h = 8760 \times 25 = 219,000 \text{ hours}$$

This gives

$$E_{WT} = 1(0.95)(0.25)(219000) = 52012.5 \text{ MW h}$$

The selling price of the generated energy, p_{sell}, depends on the number of WT units, N, and can be defined as

$$p_{sell} = 100 - 5N \quad \$/\text{MW h}$$

The *revenue* over the lifetime of the wind farm, $R(N)$, is

$$R(N) = \text{Number of WTs} \times \text{Energy per WT} \times \text{Selling price}$$
$$R(N) = N \times E_{WT} \times p_{sell} = N \times E_{WT} \times (100 - 5N)$$

The *cost*, $C(N)$, depends only on the WT unit cost, and the number of units:

$$C(N) = 1,000,000 \times N$$

The *profit*, $\mathcal{P}(N)$ is generally defined as

$$\mathcal{P}(N) = R(N) - C(N) \tag{9.8}$$

Filling in the terms for $R(N)$ and $C(N)$:

$$\mathcal{P}(N) = NE_{WT}(100 - 5N) - 1,000,000N$$
$$\mathcal{P}(N) = 100E_{WT}N - 5E_{WT}N^2 - 1,000,000N$$

In this problem, the aim is to find the optimal value of N, which maximizes the profit $\mathcal{P}(N)$. From the problem definition, the range of N is

$$0 \leq N \leq 18$$

The critical point(s) of $\mathcal{P}(N)$ are found by calculating the derivative of the profit function:

$$\mathcal{P}'(N) = \left(100E_{WT}N - 5E_{WT}N^2 - 1,000,000N\right)'$$
$$\mathcal{P}'(N) = 100E_{WT} - 10E_{WT}N - 1,000,000$$

There is a single critical point, found at $\mathcal{P}(N) = 0$:

$$-10E_{WT}N + 100E_{WT} - 1,000,000 = 0$$
$$\Rightarrow -10E_{WT}N = +1,000,000 - 100E_{WT}$$
$$\Rightarrow N = \frac{1,000,000 - 100E_{WT}}{-10E_{WT}}$$

Substituting for E_{WT} gives

$$N = \frac{1,000,000 - 100(52012.5)}{-10(52012.5)} = 8.077$$

Therefore the optimal number of turbines, $N_{opt} = 8$.

Example 9.8

Calculate the expected profit over the lifetime of the wind farm if the developer installs (i) 1 WT; (ii) 18 WTs; and (iii) the optimal number of WTs previously calculated.

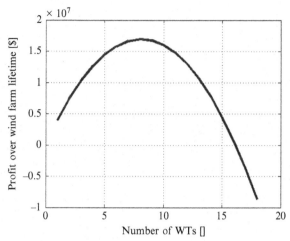

Fig. 9.6 Profit over the lifetime of a wind farm for the range numbers of wind turbines, indicating the optimal value of eight units.

For $N=1$, the profit can be calculated as

$$\mathcal{P}(1) = 100E_{WT}1 - 5E_{WT}1^2 - 1,000,000(1) = \$3,941,187.5$$

For $N=18$, the profit is negative, indicating a loss of $8,637,750:

$$\mathcal{P}(18) = 100E_{WT}18 - 5E_{WT}18^2 - 1,000,000(18) = -\$8,637,750$$

For $N_{opt}=8$:

$$\mathcal{P}(8) = 100E_{WT}8 - 5E_{WT}8^2 - 1,000,000(8) = \$16,966,000$$

The solution to this problem is illustrated graphically in Fig. 9.6 for the range N from 1 to 18.

9.3.3 Practical Considerations in Optimization

The examples above illustrate simple optimization problems, designed to introduce the reader to optimization methods and their applications in DG. For larger and more realistic problems, with large numbers of variables and constraints, it becomes very difficult to calculate solutions manually. Hence computer tools are used for almost all real-world applications of optimization techniques.

There has been a large amount of research devoted to the area of optimization and a variety of techniques for solving optimization problems are available, which depend on the exact nature of the problem. Linear optimization, or linear programming, involves the solution of mathematical models where all of the relationships in the model are linear. Integer programming refers to problems where all of the variables are restricted to be integers. Mixed integer linear programming deals with problems where only some of the variables are integers, while other variables can be nonintegers. Finally, nonlinear programming methods are designed to solve optimization problems some of the constraints or objective functions are nonlinear. For a more detailed introduction to optimization methods and techniques, see Ref. [12].

A range of software tools are available for solving optimization problems. One of the most widely used is the CPLEX solver designed by IBM, which can solve integer programming and very large linear programming problems [13]. Other software tools are available commercially, such as the General Algebraic Modeling System (GAMS) [14]. This software provides the user with access to various solvers for linear, nonlinear, and mixed integer optimization problems. In addition to these examples, many specialist software packages have optimization functions designed for specific applications in a particular field; for example, power systems optimization, building energy management optimization, and others.

9.3.4 Optimization of Distributed Generation in Electricity Networks

This section introduces some of the optimization techniques and applications in electricity systems, and discusses the impacts of DG on energy system optimization.

9.3.4.1 Optimal Power Flow

In the first half of the 20th century, the electrical power system was "optimized" by engineers using a combination of judgment, experience, and rules of thumb developed by the network operators. As computing technology became more widespread and more advanced in the second half of the 20th century, engineers began to utilize computers to improve system operation, and in particular to apply optimization techniques to optimize energy flows in the system.

The optimal power flow (OPF) was first formulated in 1962 by Carpentier [15], and methods for solving the OPF were advanced in the landmark paper by Dommel and Tinney in 1968 [16]. OPF techniques have since become crucial to the operation of modern power and energy systems. The most important application of the OPF is in economic dispatch.

This refers to the problem of scheduling, or "dispatching" all of the energy sources connected to the system in the most economic way, or in other words, minimizing the total cost of generation.

This optimization is usually constrained by the physical limitations of the electricity network, where all of the generators need to be dispatched in such a way that power line flows, voltages, and other technical constraints are not violated. This is a complex problem, especially in large systems.

Fig. 9.7 shows an example of an IEEE test network with 118 nodes, which is often used for research and demonstration of OPF and economic dispatch methods. This is considered a relatively small system; real networks may have up to 1000s or even 10,000s of nodes.

OPF is also applied to other aspects of system operation, for instance, minimization of the energy losses in the networks and minimizing the environmental impact of system operation (i.e., managing the total output from carbon-emitting generators on the system). OPF techniques have also been adopted for planning purposes, including maximizing the amount of DG and other distributed energy resources that can be connected to the distribution network [17,18].

The OPF for minimization of total generation cost (the economic dispatch problem) is formulated as follows:

$$\text{minimize} \quad F_{cost} = \sum_{i=1}^{N_{gen}} f_{cost_i}(P_i) \tag{9.9}$$

where f_{cost_i} are the cost curves (curves describing the relationship between cost of generation and output level) of each system generator. P_i is the generated power at generator i, and N_{gen} is the total set of generators on the system.

The OPF is subject to the constraints of the power network, some of which are given below. The power flow balance constraints in Eqs. (9.10) and (9.11) are

$$P_n = \sum_{n=1}^{N} |V_n||V_m|(G_{nm}\cos\theta_{nm} + B_{nm}\sin\theta_{nm}) \tag{9.10}$$

$$Q_n = \sum_{n=1}^{N} |V_n||V_m|(G_{nm}\sin\theta_{nm} - B_{nm}\cos\theta_{nm}) \tag{9.11}$$

where P_n, Q_n are the real and reactive net power injections at bus n; N is the total number of buses in the system; G_{nm}, B_{nm} are the real and reactive parts

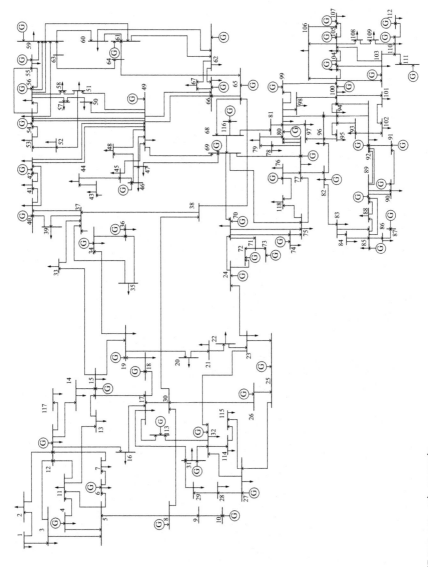

Fig. 9.7 IEEE test network example.

of the elements in the network bus admittance matrix; and Y_{bus} corresponding to the nth row and m^{th} column (the bus admittance matrix, or nodal admittance matrix), Y_{bus}, describes the admittance, or the inverse of the impedance, $Y = 1/Z$, between all nodes in the system that are connected by lines or transformers. The angle θ_{nm} is the difference in voltage angle between buses n and m.

In addition, the OPF needs to satisfy bus voltage constraints in Eq. (9.12) and line thermal constraints in Eq. (9.13).

$$V_{min,n} \leq V_n \leq V_{max,n} \tag{9.12}$$

$$|S_k| \leq |S_{max,k}| \tag{9.13}$$

where $V_{min,n}$ and $V_{max,n}$ are the minimum and maximum allowed voltages at each network bus n, and S_k is MVA power flow through network branch k.[2]

It is also possible to include additional constraints in the OPF to meet the technical requirements of a specific power network, for example constraints on particular power transfers in a part of the system, or transformer tap-changer settings.

Example 9.9

Formulate an objective function for minimizing the losses in the branches of a power network.

The objective function F_{loss} is given by

$$\text{minimize } F_{loss} = \sum_{k=1}^{N_{lines}} f_{loss_k}$$

The function f_{loss_k} for the losses in each branch k in the system is

$$f_{loss_k} = P_{n-m} + P_{m-n}$$

where the branch k connects node n to node m.

9.3.4.2 Optimal Network Location of Distributed Generation

DG affects the operation of the distribution networks, including power flows (described above in Eqs. (9.10)–(9.13)) and voltages. DG also has an effect on system losses. In many cases, the effect of DG on losses is positive, as DG is often located close to the demands where the energy is consumed. This reduces the distance over which energy needs to be transported, therefore reducing losses.

---

[2] A "branch" in the electricity network refers to either a power line or a transformer.

DG can also impact the reliability and quality of the power supply in the electrical system. Previous studies have demonstrated some positive effects of DG on reliability, where DG is able to reduce the loading on the network at critical times, or supply part of the demand in the network during faults and shortages. However, DG can also have a negative impact on reliability, particularly in cases where it has caused problems for the coordination of the network protection systems [19].

Previous research has investigated the optimal placement of DG in the networks; that is, methods for determining the optimal location of DG to deliver system performance benefits [20,21]. The DG placement problem can be formulated as an optimization, with similarities to the OPF problems described above. However, in practice, a network operator may only have a limited influence on the actual location of DG, because this often depends on factors such as site availability, as well as construction and planning permission issues.

9.3.4.3 Distributed Generation and Voltage Optimization

The system voltage constraints, Eq. (9.12), are often the most important limiting factor when connecting DG to the distribution network. In "weak" networks (e.g., rural systems with long, radial electrical lines), problems such as excessive voltage variations and voltage rise (overvoltages) are a major concern. This often occurs when demand is low and the output from DG is high, resulting in reverse power flows and an excess of energy from DG. This causes the voltages at nearby nodes to increase.

Fig. 9.8 shows an example of the variation in output from a PV system, and Fig. 9.9 shows the corresponding voltages in the distribution network,

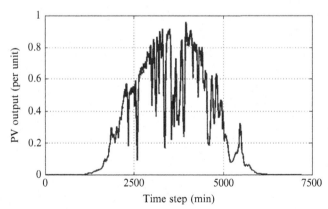

Fig. 9.8 PV output on a cloudy day, showing significant power output variations.

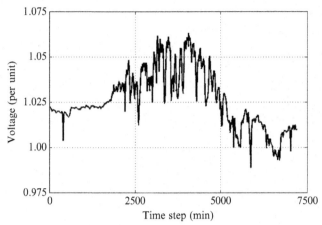

Fig. 9.9 Voltage in the distribution network at the node where the photovoltaic system is installed.

which are affected by the PV output. It can be seen that the voltage output varies significantly during the hours that the PV generation is active and that in particular, there are high voltages during the time periods where PV output is at its maximum.

Traditional voltage control in distribution networks is done mainly using tap-changing transformers, and by switching capacitor banks [22]. These are slow response controls, which were designed to respond to relatively gradual changes in electrical demand over the course of the day. However, these controls may not be fast enough to respond to rapid changes in voltage such as those shown in Figs. 9.8 and 9.9. Many utilities are looking to new voltage control technologies, based on power electronics [23].

Voltage optimization refers to a set of techniques used to coordinate various voltage control resources in the system. The optimization in this case finds the best voltage settings for all of the transformers, capacitors, and reactive power compensation devices on the system. This can improve the voltage profile and the quality of power supply received by customers.

Voltage optimization can also be a very effective energy efficiency measure, because energy savings can be obtained by slightly reducing the supply voltage to certain load types. Studies have shown up to 2–4% energy savings can be obtained by optimizing distribution system voltage [24].

The benefits of voltage optimization include
- Reduction of losses and improvement of efficiency.
- Reduced electrical demand and conservation of energy.
- Improved quality of supply.
- Improved integration of DG and other distributed resources.

9.4 DISTRIBUTED GENERATION AND MICROGRIDS

9.4.1 The Microgrid Concept

A microgrid can be defined as a LV electricity distribution network, with its own energy resources, which is capable of functioning autonomously as a small-scale electricity grid. Microgrids may be isolated from the main grid (this is typically the case in islands, or in extremely remote areas where it is not economical to provide a grid connection), or they may be connected to the main grid. Microgrids that are connected to the main grid have the potential to operate in two modes: interconnected mode, where the microgrid exchanges energy with the main grid through the distribution substation transformer, or in islanded mode, where the microgrid is disconnected from the main grid, and operates autonomously, serving its own local demand using DG and other distributed energy resources.

Fig. 9.10 shows a typical microgrid configuration. There are multiple DGs, which are located close to the loads in the system. There is also an ESS installed. The DGs and the ESS are connected via power electronic interfaces to provide the control to manage the various microgrid resources. All of the distributed resources are controlled by a central microgrid controller, which is designed to schedule DG and maintain reliability and quality of supply when the microgrid is not connected to the main grid.

9.4.2 Advantages of Microgrids

Increasingly, microgrids are being viewed by some experts as a means of effectively integrating DG and other DERs into the energy system. Microgrids have a number of potential advantages over conventional electricity grids:

- *Reduced losses*: In a conventional electricity grid, generation sources are often local at large distances from the demand center and energy is transported over long lines, resulting in electrical losses. The combined energy losses from electricity transmission and electricity distribution can be as high as 4%–5%. In a microgrid arrangement, such as the one shown in Fig. 9.10, the generation is installed very close to the load where it is consumed. This reduces the distance over which electrical energy needs to be transported, thereby reducing losses significantly.
- *Reliability*: Because microgrids have the capability to generate their own energy, and to work independently from the main grid, they can offer a very high level of reliability to customers. A microgrid should be able to operate continuously, even if the main grid suffers a blackout or an energy shortage. For this reason, there has been particular interest in

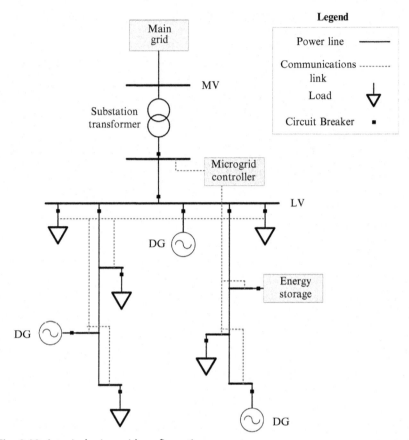

Fig. 9.10 A typical microgrid configuration.

using microgrids for applications where very high levels of reliability are required, such as military installations, hospitals, and jails. In locations where severe weather often leads to outages, microgrids have also been proposed as a means of making the overall energy system more resilient [25].

• *Environmental benefits*: Microgrids make use of multiple DG sources. Usually at least some of these sources are based on renewable energy. The ability of microgrids to operate either in interconnected or islanded mode allows them to use local energy resources when they are available, and rely on importing energy from the main grid when they are unavailable. This can provide the opportunity to integrate a higher penetration of renewable DG than in conventional electricity grids, which in turn leads to reduce carbon emissions and environmental benefits.

- *Energy independence*: As DG and energy storage technologies become more affordable and more reliable, some users may opt to move off-grid, becoming completely independent from the traditional electricity grid, and becoming fully self-reliant in terms of their energy needs. In the future, this could have profound effects on the entire energy industry. This could greatly reduce the market power of large power companies, and provide significant challenges to the business models that have been established for many years.

  If a large number of customers were to become energy independent, or go off-grid, this would reduce the revenue that energy retailers and distributors receive. It would mean that it would be necessary to increase energy prices for the customers that remain connected to the grid, as the fixed costs of managing and maintaining the networks now need to covered by fewer customers. Increasing prices would then increase the incentive for more customers to move off-grid.

  Some industry experts have described this cycle of reduced revenue, and further customer disconnections, as the "utility death spiral," and many utilities are concerned about the impacts that DG and microgrid technology may have on their businesses in the future [26]. Already in European countries with high penetrations of DG, such as Germany and Spain, DSO revenues have been significantly affected. These technological changes may prompt many DSOs to diversify their business models, and become more involved in the development and supply of DG and microgrid technology.

9.4.3 Challenges of Microgrids

There are significant challenges associated with microgrids, some of which are summarized below.

- *Technical issues*: There is, in general, a lack of operating experience with microgrid configurations, and significant technical challenges, in particular around control and protection. Power and frequency control in a small, isolated system is difficult, because changes in local demand and DG output have a much greater relative impact on the system, compared to large conventional systems. This makes maintaining stability a significant challenge. Many microgrids use a frequency-droop control approach, such as the one described in [27].

  Protection is also a major concern, because traditionally, protection systems in LV networks were designed for unidirectional power flows.

Any connected DGs are required to disconnect upon detection of a system fault, to avoid excessive reverse power flows, and to prevent "unintentional" islanding. There are major challenges in implementing a protection system for "intentional" islanding in a microgrid, allowing DGs to operate correctly and maintain synchronism when switching between interconnected and islanded modes.

- *Cost*: Supplying energy from DG can be expensive, as many DG technologies have yet not reached full maturity. Capital costs can be relatively high, without the same economies of scale associated with more established, conventional generation technologies. In addition, microgrids typical require a significant investment in communication and control technologies to implement a centralized microgrid control system and to resolve the control and protection issues discussed above. This can make the cost of implementing a microgrid configuration prohibitive for the vast majority of electricity users. One sector where much progress has been made in the application of microgrids is in the military sector, where the reliability of the power supply is of crucial importance. Making the very large financial investments in new technologies required to achieve this reliability may be feasible for military installation, but in other sectors, these costs may be difficult to justify.

- *Standardization*: As microgrids are a relatively recent development in electrical energy systems, one challenge has been the lack of microgrid standards. Currently there are almost as many microgrid communication and control configurations as there are microgrid projects worldwide. Recently, efforts have been made at introducing new standards for DG interconnection and protection, such as the Institute of Electrical and Electronic Engineers (IEEE) Standard 1547 [28]. At the time of writing, IEEE is in the process of developing specific standards (e.g., P2030.7 and P2030.8) for the specification of microgrid controls [29]).

- *Regulatory barriers*: There are a number of legal and regulatory obstacles to the large-scale implementation of microgrids. The ownership of microgrids is an important issue in this context. For instance, if the microgrid is legally defined as a utility serving multiple customers in a local area, it is unlikely that it would be allowed to operate within another, larger utilities' service territory, connecting and disconnecting to the main grid when required.

DSOs may resist the implementation of microgrids, as they are likely to be in direct competition with conventional utilities, and for the technical reasons highlighted above. Traditional protection and safety regulations were designed to prevent the occurrence of electrical islands at all costs. Intentional islanding of part of an electricity network, as proposed in microgrids, is currently prohibited by regulations in almost all countries and regions worldwide. The microgrid interconnection process will need to be formalized and agreements on many technical and market regulatory issues will need to be made before microgrids can be implemented on a wide scale [30].

9.4.4 Examples of Existing Microgrids

Table 9.2 provides several examples of existing microgrid projects from various parts of the world. The concept of microgrids was outlined by Hatziargyriou in [31], which discusses the structure of a microgrid with a centralized controller. Other examples of microgrid projects worldwide, including some of the examples in Table 9.2 are discussed in [32].

9.4.5 Microgrid Energy Management Systems

Microgrids require an EMS similar to the EMS for distribution systems described earlier in Section 9.2.4, to manage the various energy resources in the microgrid. As can be seen from Table 9.2, these energy resources may include DG (dispatchable or nondispatchable type), energy storage, and flexible/nonflexible loads. The microgrid EMS carries out the task of balancing supply and demand within the microgrid, and meeting technical requirements, for instance, maintaining frequency and voltage within specified limits. A number of different approaches and control strategies for a microgrid EMS have been applied. Most of these have involved using a completely centralized controller, but some have implemented decentralized control, which applied local controls to the various DERs, or a combined centralized/decentralized approach [33].

The best approach to designing an EMS for a particular microgrid depends on the mix of energy resources, the type of load/DG connected, whether the microgrid can operate in both islanded and interconnected modes, and various other technical and economic factors. The following describes a microgrid EMS for the simplified microgrid shown in Fig. 9.11, and provides some examples of energy balancing and optimization of energy storage charging cycles.

Table 9.2 Examples of real-world microgrids

Name	Country	Main characteristics
Kythnos microgrid	Greece	The Kythnos project is a small village microgrid located on a Greek island, and is one of the earliest microgrid projects. It has 12 kW of PV generation capacity, 5 kW of diesel generators, and a total of 85 kW h of battery capacity. It has been used to test a number of communication infrastructures and microgrid control strategies
Bornholm island	Denmark	A larger, isolated grid located on one of Denmark's islands, supplying 28,000 customers. The DG resources are 34 MW of diesel generation, 62 MW of steam turbines powered by various sources, 2 MW of biogas turbines, and 29 MW of wind generation
Sendai microgrid	Japan	A microgrid located on the campus of the Tohoku Fukushi University in Sendai, Japan. It has 700 kW of natural gas generation, 50 kW of PV, and a small ESS. This microgrid gained international notoriety during the 2011 earthquake and Fukushima nuclear disaster, when it supplied the energy needs of the local hospital during a two-day blackout
CERTS microgrid	United States	This microgrid was developed by a collaborative project called Consortium for Electric Reliability Technology Solutions, and is located at the American Electric Power Walnut test facility. It is powered by three 60 kW CHP sources. The technology developed in the CERTS projects has been applied in a number of subsequent microgrid projects in the United States
Santa Rita jail	Unites States	This microgrid supplies a jail located in Dublin, California, which has a peak load of 3 MW. The DG is comprised of 1.5 MW PV generation, a 1 MW molten carbonate fuel cell, and a 4 MW h Lithium-ion battery. It can operate in both interconnected and islanded modes

Fig. 9.12 shows a simplified time series for the demand and DG profiles and Fig. 9.13 shows the resulting net import from the main grid; that is, the difference between the load and DG output. The time series are arranged so that each time series profile is averaged to reduce to four time periods across the 24-h day: from 00:00 to 06:00, from 06:00 to 12:00, from 12:00 to 18:00, and from 18:00 to 24:00.

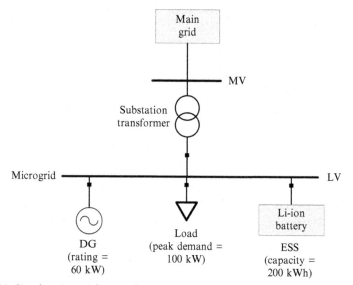

Fig. 9.11 Simple microgrid example.

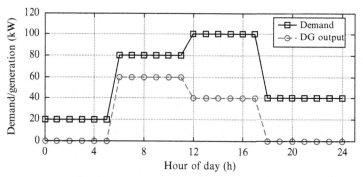

Fig. 9.12 Time series for demand and distributed generation output.

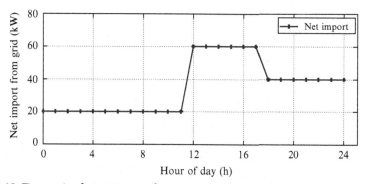

Fig. 9.13 Time series for net import from main grid.

Example 9.10

Calculate the optimal battery charging schedule for the microgrid, with the objective of minimizing the total net import from the main grid, given the demand and DG output time series above and the following restrictions on battery charging.

In this example, a number of assumptions are made around the operation of the ESS in Fig. 9.11. It is assumed that the battery can charge in only one of the four time intervals, or six-hour periods indicated above, storing the energy imported from the grid up to a maximum of the battery capacity of 200 kW h. It can then discharge the stored charge in any one of the four time intervals. The battery is restricted to only one charge/discharge cycle per day, and it is assumed that the initial state of charge of the battery is zero. It is also assumed for simplicity that there are no energy losses incurred during the entire battery charge/discharge cycle, and that there are no electrical losses in the microgrid. The input data for the problem are summarized in Table 9.3.

The number of possible solutions, that is, different options for battery charging and discharging, is limited in this example. The solution that follows applies a complete enumeration approach, which solves the objective function for every possible battery charging option, and selects the optimal solution.

Given the restrictions on battery charging and discharging, there are seven possible solutions to the ESS scheduling problem:

- *Solution 1*: No battery charging
- *Solution 2*: charge T1, discharge T2
- *Solution 3*: charge T1, discharge T3
- *Solution 4*: charge T1, discharge T4
- *Solution 5*: charge T2, discharge T3
- *Solution 6*: charge T2, discharge T4
- *Solution 7*: charge T3, discharge T4

For Solution 1 (no battery charging or discharging), the total power import from the grid in kW h is given by

$$P_{total} = (Demand_{T1} - DG_{T1}) \times 6 + (Demand_{T2} - DG_{T2}) \times 6$$
$$+ (Demand_{T3} - DG_{T3}) \times 6 + (Demand_{T4} - DG_{T4}) \times 6$$

$$P_{total} = 120 + 120 + 360 + 240 = 840 \text{ kW h}$$

Table 9.3 Summary of input data

Time interval	Hours (h)	Demand (kW)	DG output (kW)	Net import (kW h)
T1	00:00–06:00	20	0	120
T2	06:00–12:00	80	60	120
T3	12:00–18:00	100	40	360
T4	18:00–24:00	40	0	240

Table 9.4 Total power import in kW h for each solution 1–7

Solution	1	2	3	4	5	6	7
T1	120	120	120	120	120	120	120
T2	120	0	120	120	120	120	120
T3	360	360	240	360	240	360	360
T4	240	240	240	120	240	120	40
Total import	840	720	720	720	720	720	640

For Solution 2 (charge T1, discharge T2), the battery discharged during T2 to reduce the import from the main grid during this time interval. The total net import during T2 is 120 kW h, which is less than the battery capacity; therefore the battery is charged to 120 MW h during T1 and discharged fully during T2. The total power import from the grid is then

$$P_{total} = 120 + 0 + 360 + 240 = 720 \text{ kW h}$$

The results for all of the possible battery charging/discharging options are calculated in a similar manner and summarized in Table 9.4.

From Table 9.4, the solution with the minimum total power import is Solution 7, charging the battery during T3 and discharging during T4.

Example 9.11

Minimize the total cost of energy in the microgrid over the course of 24 h, where the electricity price is fixed.

In this example, a buy price for importing electricity from the grid is set, and the objective this time is to minimize the total cost of importing energy using the fixed price of 0.2 \$/kW h shown in Fig. 9.14. As before, Solutions 1–7 are evaluated fully, and the results are summarized in Table 9.5.

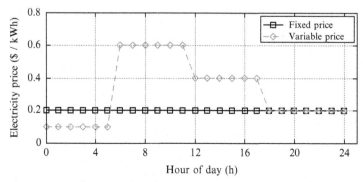

Fig. 9.14 Time series for net import from main grid.

Table 9.5 Total cost over 24 h in $/kW h for each solution 1–7

Solution	1	2	3	4	5	6	7
T1	24	24	24	24	24	24	24
T2	24	0	24	24	24	24	24
T3	72	72	48	72	48	72	72
T4	48	48	48	24	48	24	8
Total cost	168	144	144	144	144	144	128

With the fixed price, the total cost follows the pattern in the previous example in Table 9.4, and again, the optimal solution is Solution 7, with a total cost of $128.

Example 9.12

Minimize the total cost of energy in the microgrid over the course of 24 h, where the electricity price is variable.

In this case the buy price for electricity is set according to the variable price in Fig. 9.14. Again, the objective function is calculated for each solution and the results are shown in Table 9.6.

The results in Table 9.6 show that the variable electricity price impacts the total cost of energy in the microgrid. The optimal charging schedule with the variable electricity price is Solution 2 (i.e., charging in T1 and discharging in T2), with a total daily cost of $204.

In the simplified examples above, the total number of solutions is small, which allows us to easily find the optimal solution by complete enumeration, calculating the objective function for every possible solution to the problem. For larger and more realistic problems, there will be many more variables, and the number of possible solutions may be very large, and computerized integer programming (or mixed integer programming if some of

Table 9.6 Total cost over 24 h in $/kW h for each solution 1–7

Solution	1	2	3	4	5	6	7
T1	12	12	12	12	12	12	12
T2	72	0	72	72	72	72	72
T3	144	144	96	144	96	144	144
T4	48	48	48	24	48	24	8
Total cost	276	204	228	252	228	252	236

the variables are continuous) methods may be more suitable to solve this problem (see the discussion in Section 9.3.3 of this chapter).

For problems such as the microgrid example above, rather than using a complete enumeration approach to find the optimum, it may be possible to limit the number of solutions that have to be evaluated by attempting to identify and terminate nonoptimal solutions early during the solution process, thereby reducing the computational burden. This is often carried out using branch and bound algorithms (these are sometimes called tree search algorithms). A full description of these approaches is beyond the scope of this chapter; for more information the reader is referred to [10,12].

9.4.6 Multiple Microgrids and System Integration

As technologies such as DG, energy storage, and distributed automation become more affordable and widely accessible, many experts have predicted that microgrids will be present in large numbers and will have an important influence on the operation of future electricity systems. A future distribution system where most of the low voltage networks are made up of active microgrids is envisaged in [34]. In this arrangement, multiple microgrids are able to share energy resources with one another, and are coordinated in order to provide services to the system such as frequency and voltage support.

An electricity grid based on multiple microgrids has also been viewed by some as a means to create a more resilient grid, by utilizing the advanced capabilities of microgrids, such as intentional islanding and real-time control of real and reactive power consumption/production. Resiliency in this context means the ability of the electrical system to withstand low-probability, high-impact events, ensuring the least possible interruption to the supply of electricity, and enabling a quick recovery and restoration to the normal operation state [35]. However, there are still many technical, economic, and regulatory barriers that would need to be overcome to develop an electrical power system where microgrids play a prominent role.

REFERENCES

[1] A. Keane, L.F. Ochoa, C.L.T. Borges, G.W. Ault, A.D. Alarcon-Rodriguez, R.A.F. Currie, F. Pilo, C. Dent, G.P. Harrison, State-of-the-art techniques and challenges ahead for distributed generation planning and optimization, IEEE Trans. Power Syst. 28 (2013) 1493–1502.

[2] A. Meliopoulos, E. Polymeneas, Z. Tan, R. Huang, D. Zhao, Advanced distribution management system, IEEE Trans. Smart Grid 4 (4) (2013) 2109–2117.

[3] B.P. Hayes, M. Prodanovic, State forecasting and operational planning for distribution network energy management systems, IEEE Trans. Smart Grid 7 (2) (2016) 1002–1011.

[4] P. Hallberg, Active distribution system management: a key tool for the smooth integration of distributed generation, Eurelectric, Tech. Rep., 2013.

[5] K. Anaya, M. Pollitt, Integrating distributed generation: Regulation and trends in three leading countries, University of Cambridge EPRG Working Paper 1423, Tech. Rep., 2014.

[6] DNV, A review of distributed energy resources, New York Independent System Operator, Tech. Rep., 2014.

[7] S. Papathanassiou, N. Hatziargyriou, et al., Capacity of distribution feeders for hosting DER, CIGRE, Tech. Rep., 2014.

[8] N. Etherden, M.H.J. Bollen, Increasing the hosting capacity of distribution networks by curtailment of renewable energy resources, in: PowerTech, 2011 IEEE Trondheim, June 2011, pp. 1–7.

[9] A. Gomez-Exposito, A. de la Villa Jaen, C. Gomez-Quiles, P. Rousseaux, T.V. Cutsem, A taxonomy of multi-area state estimation methods, Electr. Pow. Syst. Res. 81 (2011) 1060–1069.

[10] S.P. Boyd, Convex Optimization, Cambridge University Press, Vandenberghe, 2004.

[11] B.P. Hayes, A. Wilson, R. Webster, S.Z. Djokic, Comparison of two energy storage options for optimum balancing of wind farm power outputs, IET Generation, Transmission Distribution 10 (3) (2016) 832–839.

[12] D. Bertsimas, in: J.N. Tsitsiklis (Ed.), Introduction to Linear Optimization, third ed., Athena Scientific, Nashua, NH, 1997.

[13] IBM, ILOG CPLEX optimization studio [Online], 2016. Available: http://www.ibm.com/software/.

[14] GAMS, General algebraic modeling system [Online], 2016. Available: https://www.gams.com/.

[15] J. Carpentier, Contribution to the economic dispatch problem, Bull. Soc. France Elect. 8 (1962) 431–447.

[16] H.W. Dommel, W.F. Tinney, Optimal power flow solutions, IEEE Trans. Power Apparatus Syst. PAS-87 (10) (1968) 1866–1876.

[17] G.P. Harrison, A.R. Wallace, Optimal power flow evaluation of distribution network capacity for the connection of distributed generation, IEE Proc. Generation, Trans. Distrib. 152 (1) (2005) 115–122.

[18] B. Hayes, I. Hernando-Gil, A. Collin, G. Harrison, S. Djokic, Optimal power flow for maximizing network benefits from demand-side management, IEEE Trans. Power Syst. 29 (4) (2014) 1739–1747.

[19] A. Girgis, S. Brahma, Effect of distributed generation on protective device coordination in distribution system, in: Power Engineering, 2001 Large Engineering Systems Conference on LESCOPE'01, 2001, pp. 115–119.

[20] S. Ghosh, S. Ghoshal, S. Ghosh, Optimal sizing and placement of distributed generation in a network system, Int. J. Electr. Power Energy Syst. 32 (8) (2010) 849–856.

[21] Z. Wang, B. Chen, J. Wang, J. Kim, M.M. Begovic, Robust optimization based optimal dg placement in microgrids, IEEE Trans. Smart Grid 5 (5) (2014) 2173–2182.

[22] J. Grainger, W.D. Stevenson Jr., Power System Analysis, McGraw-Hill Education, New York, NY, 1994.

[23] J.P. Lopes, N. Hatziargyriou, J. Mutale, P. Djapic, N. Jenkins, Integrating distributed generation into electric power systems: a review of drivers, challenges and opportunities, Electr. Pow. Syst. Res. 77 (9) (2007) 1189–1203.

[24] K. Schneider, F. Tuffner, J. Fuller, R. Singh, Evaluation of conservation voltage reduction (cvr) on a national level, Pacific Northwest National Laboratory, Tech. Rep., 2010.

[25] D.T. Ton, M.A. Smith, The U.S. department of energy's microgrid initiative, Electricity J. 25 (8) (2012) 84–94.

[26] P. Kind, Disruptive challenges: Financial implications and strategic responses to a changing retail electric business, Edison Electric Institute, Tech. Rep., 2013.

[27] E. Barklund, N. Pogaku, M. Prodanovic, C. Hernandez-Aramburo, T.C. Green, Energy management in autonomous microgrid using stability-constrained droop control of inverters, IEEE Trans. Power Electron. 23 (5) (2008) 2346–2352.

[28] IEEE, IEEE Std 1547.2-2008 Application Guide for IEEE Std 1547, Standard for Interconnecting Distributed Resources with Electric Power Systems, IEEE Std., 2008.

[29] IEEE, IEEE P2030.8: Standard for the Testing of Microgrid Controllers, IEEE Std., 2018 (expected).

[30] T.E. Del Carpio Huayllas, D.S. Ramos, R.L. Vasquez-Arnez, Microgrid systems: Current status and challenges, in: Transmission and Distribution Conference and Exposition: Latin America (T D-LA), 2010 IEEE/PES, Nov 2010, pp. 7–12.

[31] N. Hatziargyriou, H. Asano, R. Iravani, C. Marnay, Microgrids", IEEE Power Energy Mag. 5 (4) (2007) 78–94.

[32] N. Hatziargyriou, Pilot Sites: Success Stories and Learnt Lessons, Wiley-IEEE Press, Chichester, UK, 2014, pp. 206–274.

[33] W. Su, J. Wang, Energy management systems in microgrid operations, Electricity J. 25 (8) (2012) 45–60.

[34] N. Hatziargyriou, Operation of Multi-Microgrids, Wiley-IEEE Press, Chichester, UK, 2014, pp. 165–205. Available: http://ieeexplore.ieee.org/xpl/articleDetails.jsp?arnumber=6690579.

[35] A. Khodaei, Resiliency-oriented microgrid optimal scheduling, IEEE Trans. Smart Grid 5 (4) (2014) 1584–1591.

CHAPTER 10

Impact of Distributed Generation Integration on the Reliability of Power Distribution Systems

Mohammad AlMuhaini
King Fahd University of Petroleum and Minerals, Dhahran, Saudi Arabia

10.1 INTRODUCTION

The reliability of distribution systems is an important issue in power engineering for both utilities and customers. The distribution system accounts for almost 40% of the overall power system and 80% of customer reliability problems [1]. Moreover, the power grid becomes more complicated as the demand increases and technology advances. Contemporary loads are often digital in nature, and these loads are frequently sensitive to interruptions and, indeed, many other power quality problems. The customers themselves are perhaps becoming more sensitive to interruptions due to the possibility of industrial manufacturing interruption, commercial loss of sales, and residential nuisance. Competition in power marketing may be impacted as well: industrial customers may seek to locate in places where power system reliability is high. For these reasons, distribution system design and operation is critical for the power industry. One common characteristic of all these industrial and commercial customers is that the cost of downtime is enormous.

Electric power utilities assure a specific average level of power availability to what is called 4 nines. which means power is expected to be available 99.99% of the time. Industrial and commercial customers with sophisticated and sensitive equipment are increasing, such as banks, semiconductor manufacturers, and even some government agencies. These customers are now demanding for more nines in availability and additional reductions in other quality problems such as voltage sags, spikes, and harmonics. As the availability and the reliability of the power system becomes more sensitive to customers and utilities, more research and techniques are needed to evaluate the reliability of the power system.

Distributed Generation Systems
http://dx.doi.org/10.1016/B978-0-12-804208-3.00010-8
453

Moreover, distribution systems are now in a significant transition phase where the system is shifting from a passive distribution system with unidirectional power flow to an active distribution network with bidirectional flow and small-scale generators. Future power systems are motivated by the necessity to diminish the impact of global climate change and reduce the concentration of greenhouse gases in the atmosphere. This can present an extraordinary challenge to the business of electric generation and delivery. Furthermore, another incentive comes from fuel price uncertainty such as oil, natural gas, and other fuel types. These price fluctuations can introduce threats to the economic stability of some countries that depend on imported oil.

Distribution system reliability can be defined as the probability of the distribution system to provide continuous power without failure for a specific period of time. The evaluation of distribution system reliability is the evaluation and calculation of the availability and expected frequency and duration of customer outages. Distribution system reliability is measured by certain count indices, such as the system average interruption duration index (SAIDI) and the system average interruption frequency index (SAIFI).

The reliability assessment of future distribution networks is an important subject due to the increasing demand for more reliable service with less interruption frequency and duration. The connection of a future distribution network may be neither series nor parallel, and analyzing such a network is a complicated process and a time-consuming task. Future distribution systems are often referred to as smart grids where more intelligent technologies are integrated into the system to monitor, control, and operate the system. Therefore, the reliability of future grids is expected to become a more challenging issue in the near future, where the configuration of the system is more complicated and the penetration of the small-scale units is higher.

The diversity and distribution of different types and sizes of distributed generations (DGs) increase the uncertainty and sensitivity of the system, which also makes controlling the power flow much more difficult. Taking into account the deregulation process ongoing in many countries and rapid development in technologies such as DG, there may be a need to reconsider or to extend and enhance the traditional approach to evaluate the reliability of the distribution system. In the way that the power grid is currently operated, power flows from the higher voltage grid to the lower voltage grid. Increased share of DG units may lead the power to flow from the low voltage into the medium voltage grid. Also, the increased demand for more reliability introduces more networked secondary systems and this will add more complexity in evaluating the reliability of distribution systems including DG.

10.2 RELIABILITY OF POWER DISTRIBUTION SYSTEMS

10.2.1 The Configuration of Power Distribution Systems

In general, the power grid can be defined as the complete set of machines, wires, and components that connect between the power plants and load points. Based on the diversity and advance of the sources, controllers, machines, protection devices, and connections, the power grid is considered to be one of the most complicated and scientific grids ever made. Modern power systems are usually subdivided into generation, transmission, sub-transmission, and distribution systems. The energy generated in power plants is transmitted through the transmission network to the substation and from there, distribution systems distribute the power received from the supply points to customer facilities. The distribution network is the local network that consists of connections and transformers to transfer the power and convert it to the final utilization voltage.

The configurations of the distribution system can follow different arrangements based on the cost versus reliability requirements. The distribution system can be in simple radial, primary selective, secondary selective, or secondary network configuration. Each design will provide increasing reliability as well as increasing installation and operational cost. Distribution systems are typically of radial configuration as shown in Fig. 10.1.

The radial configuration is the simplest and least reliable design of load distribution where the power is flowing in one direction from one substation to the loads. The radial system consists of one substation with one or more main feeders and many laterals connecting between the transformers and load points. It is less reliable compared to the secondary networked configuration but it is also less expensive and less complex due to fewer connections and protection devices. Thr radial configuration is usually located in the suburban and rural areas where the density of customers is low and their reliability requirement is not very high. The radial feeders in these areas are overhead lines or underground cables. All system feeders and laterals are designed to operate in their full rated capacity. The redundancy of this arrangement is very low because of the absence of an alternative power supply. If there is any failure in the main feeder, the circuit breaker in the

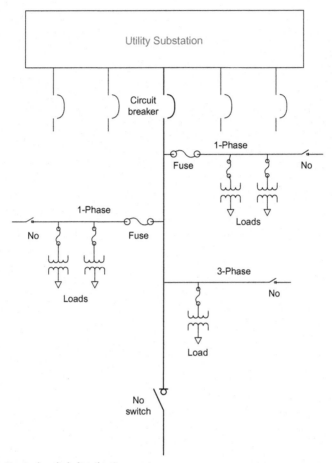

Fig. 10.1 Typical radial distribution system.

transformer side or the reclosers in the feeder will clear the fault and the loads downstream from the protection device will be interrupted.

The secondary network configuration is where all the loads are connected via two or more alternative routes to the main supply. It is designed to provide highly reliable service to the customers. Unlike the radial configuration, there are multiple transformers serving each network. The reliability of this network configuration is very high and every load point in the network is supplied by two or more alternative power supplies. If a fault occurs on one of the transformers or primary feeders, there will not be any interruption to any load point and the network should not experience any interruption.

This type of network configuration usually can be found in urban downtown areas where the density of customers is very high and interruption cost is expensive. Usually the connections of this secondary networks are underground cables. The high voltage side of the transformer is connected to different prime sources and each load point is connected to two or more main feeders to provide a high level of redundancy for each load. When the primary voltage level is interconnected it is usually called a primary network. When the interconnected grid is on the low-voltage side of the transformers, it is called a secondary network. Most networks are in the low-voltage sides (secondary networks). Networks operate at 480Y/277 V or 208Y/120 V in the United States and the network load usually ranges from 5 to 50 MVA.

The protection of the network configuration is more complicated and more expensive than the radial system. The network protectors are used in a network configuration to isolate the faulted section without interrupting the loads. The network protectors are automatic circuit breakers with directional relays connected to the secondary transformers and they are tripped when the power flows from the secondary to the primary side. The installation cost of a secondary network is 175%–200% of the cost of a radial configuration [2]. This increased cost is because of the additional secondary connections, overrated size of lines and transformers, and protection devices.

There are two types of network configurations, spot networks and area (grid) networks. Both networks are served from several primary feeders connected in parallel from different substations. The spot network, usually fed by 3–5 primary feeders and feeds one major load such as a high-rise building or a large commercial customer. The secondary (grid) network is usually fed by 5–10 primary feeders and can serve an area as large as several blocks in a city. A comparison between the spot and area networks is shown in Table 10.1 and typical secondary spot and area networks are shown in Fig. 10.2 [3].

Table 10.1 Comparison between spot and grid networks

	Spot network	Area network
Application	High-rise building	Several blocks
Number of primary feeders	3–5	5–10
Feeders length (mile)	4–6	
Load (MVA)	5–50	
Voltage	208 Y/120	
	480 Y/277	
Typical transformers size (KVA)	300, 500, 750, 1000, 1500, 2000, 2500	

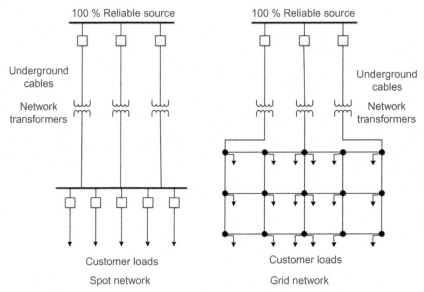

Fig. 10.2 Typical secondary spot and grid network system.

10.2.2 Reliability of Power Distribution Systems

Reliability becomes a subject of great interest in most of the manufacturing and services applications. In an electric power distribution system, reliability is a key issue in the design and operation, especially in view of sensitive digitally controlled loads. The reliability definition based on the IEEE 90 standard is *"the ability of a system or component to perform its required functions under stated conditions for a specified period of time"* [4].

Analyzing and evaluating distribution system reliability is important to improve the operational and maintenance performance of the system and provide highly reliable electricity with high quality. The distribution system is highly complex and contains a large number of connections and components that make it the greatest contributor to the unavailability of power supply to customers. Some sources of power problems are in found in nature, such as tornados, lightning, wind, earthquakes, and snow. Man–made problems include automobile accidents, vandalism, inadvertent contact with overhead conductors, distribution operator errors, and fires. These factors are extremely difficult to predict or control to avoid a power outage. Some factors can be controlled or optimized such as vehicles or construction accidents, overloads, animals' contacts, and equipment failure or wear out. Most power problems can be reduced by implementing underground connections, but then the cost and maintenance inflexibility will increase.

The overall reliability evaluation of power systems should include generation, transmission, and distribution reliability studies. In Ref. [5], the reliability of distribution systems is evaluated considering the impact of the failures from the generation and transmission subsystems. In practice, all the reliability studies have been conducted in relatively small local subsystems because the complete network from the source to the load is enormous. In reality, it is also difficult to collect data that can be used in reliability evaluations. Utilities are conservative or sometimes proprietary in releasing actual reliability data and failure rates. Several references investigated methods to collect and categorize data that can be used in reliability studies [6,7].

The performance of distribution systems may be quantified by measures of voltage regulation and classical power distribution engineering issues. These issues include evaluation of losses. power factor, overhead versus underground designs [8–10], counts of anomalous events [11–15], and power quality at the point of end use [16,17]. Knezevic et al. [12] specifically addresses the value of "count indices" (i.e., counting undesired events such as outages or low voltage cases) for the purpose of standardized distribution system planning. Balijepalli et al. [14] addresses the probabilistic analysis of these indices. In recent years, the move to the use DG resources in the distribution system and the impact of these resources on distribution system reliability has also been considered; for example, in [13,18–23]. These references are only a small sample of the literature as the full literature is voluminous. Refs. [24,25] are samples of distribution system engineering analysis and design; an area of considerable attention for over 100 years.

The impact of protection, interruption, and restoration capability on reliability are also important areas in the literature. In [26], the effect of voltage drop on the reliability evaluation during the restoration process is studied. Sensitive loads can suffer power interruption during the restoration process as the voltage level can be violated. The effect of protection and interruption devices is also discussed in the literature [27–30]. These devices are based on their operation and impact on network reliability. Models and operating characteristics are then incorporated using different techniques to evaluate reliability.

10.2.3 Basic Reliability Models and Analysis Approaches

The reliability evaluation can be divided into two parts: modeling of the reliability characteristics of the components, and the calculation of the reliability of the system. In modeling the component reliability data, it is a usual practice to assume that the failures are independents.

To evaluate the reliability of a system, a mathematical or graphical model of the system should be designed to reflect the reliability characteristics of the system. The models can be categorized into two groups: analytical models and simulation models. Analytical models represent the system by a set of exact or approximate mathematical models and evaluate the reliability based on this mathematical representation of each state. Simulation on the other hand, simulates the reliability indices by repeating the actual process with random behavior. The reliability indices are estimated by counting the failures and summing the duration of each failure in the simulated time. The solution time of the analytical techniques is relatively shorter than the simulation run time. The simulation time can be very high in complex systems and in applications where several reliability indices are required. In applications that include complex systems, the analytical techniques usually include some sort of simplifications or assumptions. The simulation technique, on the other hand, can simulate and include any system behavior without any approximation. The analytical models always give the same numerical results each run because the model contains a fixed mathematical representation for the system, where the results from the simulation models differ in each run because the system characteristics are randomly changing each run [31]. Some basic analytical and simulation models are discussed next.

10.2.3.1 The Reliability Block Diagram

Reliability block diagram (RBD) is the most common model of reliability evaluation. In a RBD, the system is graphically represented, in which its components are connected in series, parallel, and bridge (non-parallel-series) [32]. The system then can be analyzed and the components can be merged based on their reliability characteristics. The typical RBD consists of an input node, an output node, and multiple blocks in series or parallel that represents the physical and logical connection of the actual system. Successful operation requires at least one healthy path from the input and the output.

The main advantage of the RBD is the simplicity of the reliability evaluation, especially for the simple parallel-series systems. The problem with the RBD is that only two states can be represented per component in this model (up and down) and it is difficult to include other states. Additional mathematical notations and formulas must be derived to include more than two states or other redundancy configurations [33,34].

10.2.3.2 Fault Tree Analysis

Fault tree analysis (FTA) is a graphical model that represents the system and shows the failure process and possible combinations of components that can cause a system failure [15,33,34]. The limitations of this model are similar to the limitations of the RBD where it is difficult to model the complex systems with maintenance operation. FTA also does not support the different multiple faults modes or the sequence of the faults in the system.

10.2.3.3 Minimal Cut and Tie Set Method

For any system, there are two subsystems that can represent the connections of the original system in two parallel-series subsystems where the first set is called the minimal tie set (MTS) and the other is called the minimal cut set (MCS). The MTS is the minimal set of components where the components provide a continuous connection between input and output. The MCT is the minimal set of components where the fault condition for all components in the set results in a system outage. The difficulty of these methods lies in the identification of the minimal tie and cut sets and excluding the effect of dependent cut sets.

10.2.3.4 Markov Models

The Markov model can be represented by system states and their transition probabilities between them [15,33,34]. The basic assumption in Markov models about the study system is the lack of memory where it depends only on the current state not on the history of the states. The other assumption is that the transition probabilities are time independent. The states and transition probabilities can be represented in a graph called a state transition diagram (STD) or in a state transition matrix (STM).

In reliability studies of the power systems, each component of the system is modeled with different numbers of states (commonly two or three states). The two states include the up (working condition) and down (repair condition) states and the additional third state can be the planned or scheduled maintenance state. All transition rates between the states are known and the probability of the states then can be easily evaluated. The Markov chain is one of the best models that can represent the dynamic behavior of the system, but it is also very complicated to construct the transition matrix with the large number of components.

10.2.3.5 Monte Carlo Simulation Tools

A widely used technique for reliability assessment in many fields is Monte Carlo (MC) simulation. In MC simulation, reliability is evaluated repeatedly using random parameters from random distributions to simulate the stochastic or deterministic problems [15,33,34]. Usually MC is used when the other deterministic methods failed to apply. It can be useful in evaluating the mean time to failure (MTTF) for very complicated or too large systems. The typical MC consists of the following general steps:

–   For each variable in the model, draw a random number from a random distribution.
–   Evaluate the required functions.
–   Repeat the process for x trail runs and calculate the average for each function under study.

The advantage of this method is that it can simulate almost any system and any failure mode. The disadvantage is that it requires long runs and the accuracy of the output depends on the number of runs and the number of variables in the system.

10.2.4 The Calculation of Distribution System Reliability

Reliability is an important issue in any designed system or products. Customers and users do not expect any failure or interruption of the service as the failure can be expensive or insecure. The question always to raise is "how reliably is this system expected to run in the future?" This question can be answered using two methods, qualitative or quantitative assessment. The qualitative assessment is based on the experience of the engineer and his judgment. Quantitative assessments are based on the evaluation of the components of the system and compute the reliability of the system as a combination of numbers and indices. Quantitative assessment can be used to predict the future performance of the system. It requires knowing the past performance of the system to predict the future. It is important because it shows how the system performs in the future and is a compromise between different alternatives to operate better and evaluate the effect of any future modifications or upgrades in the system.

A typical reliability study focuses on the probability of a component or a system to fail or to operate as intended. This probably does not provide specific definitive information regarding exactly when or how long an outage will occur. For this reason, it is important to introduce other indices that can give information regarding the frequency and duration of outages.

It is important also to indicate the difference between power quality and system reliability. System reliability is more concerned more about the continuity of the service (sustained and momentary interruptions), while power quality contains other power problems such as voltage fluctuations, harmonic distortions, and variations in the wave shape or magnitude.

The IEEE defines a momentary interruption in IEEE Std. P1366 [35] as any interruption of duration of less than five minutes and the sustained interruption as any interruption of duration equal or more than five minutes. In practice, there are two commonly used indices to evaluate the frequency and duration of the interruptions, SAIFI and SAIDI, which are two indices that give information regarding the average duration and frequency of outages that customers experience in the period of study (typically one year). These two indices are related to the configuration of a system and the probability of each component in the system to fail. The indices are used in reliability evaluation to study the effect of components on reliability and to compare different configurations based on their reliability performance. One important route to the examination of reliability relates to the probabilistic modeling of networks and systems in general: as examples, Billinton and others (e.g., [36–39]) have employed the basic properties of the probability of failure of components in series and parallel (including vector-matrix operational analysis) to quantify the probability of failure of a system or network.

Major events such as severe weather conditions are usually excluded from calculating reliability indices because weather conditions can have a major effect on these indices based on the location and configuration of the system. Excluding major events allow the utilities to respond to the real changes of the system reliability.

Utilities used different approaches to define and exclude the major events from the reliability indices. One approach to classify any event as a major event is when the event causes 10% of the utility customers to be out of service for 24 h [40]. Another approach to classify the major events is when 15% of the customers have an outage during the severe weather condition [40]. The IEEE working group proposed an approach to define major events in reliability evaluation [40]. A major event day is any event that causes the outage duration time to be more than major event threshold minutes (T_{MED}) [40]. T_{MED} can be calculated as

$$T_{MED} = e^{(\alpha + 2.5\beta)} \tag{10.1}$$

where

$e =$ exponential constant

$\alpha =$ log average of the daily SAIDI for last five years

$\beta =$ log standard deviation of the daily SAIDI for last five years.

Even though availability and reliability are used interchangeably in several papers in the literature, they are not the same in concept and values.

Reliability basically represents the probability of a component or a system to perform its designed function without any failure in the normal working environment. Reliability does not reflect or contain any time to repair the failed component. It mainly reflects how long the system is expected to work at a specific time before it fails.

Availability, on the other hand, is the probability that the component or the system is working as expected during its operational cycle. It shows the share of time the system is working. Availability depends on both the expected time to fail and time to repair the component or the system. The typical and simple equation to calculate availability is

$$\text{Availability} = \frac{\text{UP TIME}}{\text{UP TIME} + \text{DOWN TIME}}$$

For continuous operating systems such as power systems, it is more informative to study the availability of the components and system to address the quality of service provided to customers. The term *reliability* will be used in this chapter as a general word that represents all aspects of the study (availability, unavailability, failure frequency, duration, etc.) rather than a quantity or a value. Generally speaking, system reliability can be defined as the probability of at least one minimal set of components to work probability between the input and output. This set of components is called tie or path set in graph theory.

In calculating the system's reliability especially complex systems, the concept of permutations and combinations can be used in the calculation process to identify and evaluate different states in the system. The permutations are the number of ways that items can be arranged and can be expressed by this notation $_nP_r$ where n is the total number of elements in the system and r is the number of elements used in each arrangement.

The following equation can be used to calculate the number of ways to arrange a group of components taking into account the arrangement of the components:

$$_nP_r = \frac{n!}{(n-r)!} \tag{10.2}$$

On the other hand, the combinations are the number of ways that r components can be arranged from n components regardless of the arrangement between them. It can be represented as $_nC_r$,

$$_nC_r = \frac{n!}{r!(n-r)!} \tag{10.3}$$

In reliability studies of power systems, a combinations concept is used more than the permutations because the order of failure of different components is not important. What is really important is the combination of the components and not which component failed first.

The life of power system equipment is divided into three intervals: infant mortality, useful life, and wear out periods. In a reliability study of a power system, it is usually the useful life period where the reliability evaluation conducted. Some papers include the wear out period in modeling the components using different probability distributions [41]. It is also common in the literature that the power system components down times and up times are assumed to follow the exponential distribution function. Many components in power systems fail in purely random fashion and the failure rate is assumed to be the same at any time during the components' useful life. Constant failure rate leads to exponential distribution modeling where the failure rate is constant with time.

Most components in power systems are repairable or replaceable. If the component is repaired, it is assumed that it will perform its function as a new component with the same failure rate. The time it takes for each component to fail is called the mean time to failure (MTTF) or simply T_f. Similarly, the time to restore service or to repair the faulted component is called the mean time to repair (MTTR) or simply T_r. Note that both T_f and T_r are the *average* values over a long time and over many cycles of operate/fail-repair/operate/fail-repair/…, and it is assumed that the component has only two states either *up* or *down*. The time it takes for a component to fail and to be repaired is called the mean time between failures (MTBF) or simply the mean cycle time T_{fr} where [42]

$$MTBF = MTTF + MTTR$$
$$T_{fr} = T_f + T_r.$$

(10.4)

As depicted in Fig. 10.3, the MTTF, T_f, and mean time to repair, T_r, and "one average cycle" of time to fail and repair are depicted.

The reciprocal of the mean cycle time is defined as the mean *failure frequency* and denoted as f,

$$f = \frac{1}{MTBF} = \frac{1}{MTTF + MTTR}$$

(10.5)

Note that MTBF, MTTF, and MTTR have the units of time, generally hours, and f has the units of "per hour." The probability of the component to

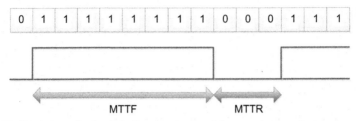

Fig. 10.3 Depiction of a "cycle" of mean time to failure and mean time to repair [42].

operate is called the *availability*, denoted as A or p, and the probability to fail is called the *unavailability*, denoted as U or q. A and U will be used in this chapter as the notation for availability and unavailability. The availability and unavailability are related to MTTF and MTTR as follows:

$$A = \frac{MTTF}{MTBF} = \frac{MTTF}{MTTF + MTTR} \qquad (10.6)$$

$$U = \frac{MTTR}{MTBF} = \frac{MTTR}{MTTF + MTTR}. \qquad (10.7)$$

The frequency and duration of interruptions for a component over one year are defined as the average interruption frequency (AIF) and average interruption duration (AID) [18]. The AIF for a component is defined as the number of failures over one year and can be expressed as

$$AIF_i = 8760f = \frac{8760}{MTTF + MTTR}. \qquad (10.8)$$

where i is the bus or feeder number. Similarly, the AID is the duration in hours for all interruptions in one year and expressed as

$$AID_i = 8760U_i = (MTTR)(AIF_i). \qquad (10.9)$$

For two components connected in series, the system will perform its designed function if both components are working (i.e., they are up). If there is a failure in any one of these two components, the receiving end will experience an interruption or outage (i.e., the load is down). The availability of this system can be expressed as

$$A_{sys} = A_1 A_2 = \frac{T_{f1}}{T_{f1} + T_{r1}} \frac{T_{f2}}{T_{f2} + T_{r2}}. \qquad (10.10)$$

Similarly, for a system of two parallel components, the load will experience an outage if both components fail at the same time. The two parallel components probability is

$$A_{sys} = 1 - U_1 U_2 = 1 - \frac{T_{r1}}{T_{f1} + T_{r1}} \frac{T_{r2}}{T_{f2} + T_{r2}} \quad (10.11)$$

For two simple components in either series or parallel, T_f and T_r are related to the MTTF and the MTTR the entire system, namely T_f^{eq}, T_r^{eq}, respectively. For two components in series, the frequency of failure for the equivalent system equal to

$$f_s = f_1 + f_2$$

In power systems consider the adjustment of the above equation to account for a practical assumption that the second component cannot fail when the first component has already failed. The equivalent frequency will be then equal to

$$f_s = f_1 A_2 + f_2 A_1 \quad (10.12)$$

After substituting all variables from Eqs. (10.6) and (10.7),

$$f_s = \frac{T_{f1} + T_{f2}}{(T_{f1} + T_{r1})(T_{f2} + T_{r2})} \quad (10.13)$$

To find the equivalent failure cycle period (T_{frs}),

$$T_{frs} = \frac{1}{f_s} = \frac{1}{f_1 A_2 + f_2 A_1} \quad (10.14)$$

Then, using Eqs. (10.8) and (10.9), the equivalent time to fail and time to repair can be found as

$$T_{fs} = A_s T_{frs} = \frac{T_{f1} T_{f2}}{T_{f1} + T_{f2}} \quad (10.15)$$

$$T_{rs} = U_s T_{frs} = \frac{(T_{f1} + T_{r1})(T_{f2} + T_{r2}) - T_{f1} T_{f2}}{T_{f1} + T_{f2}} \quad (10.16)$$

A similar procedure can be used to find the equivalent variables in two parallel components. For two parallel components, both components should fail at the same time to cause an outage or service interruption to the customer. The frequency of failures is then equal to

$$f_p = f_1 U_2 + f_2 U_1 \quad (10.17)$$

Table 10.2 Equivalent times of failure and repair of series and parallel components

	T_f^{eq}	T_r^{eq}
Approximate formulas $T_r \ll T_f$		
Series	$\dfrac{T_{r1}T_{f2} + T_{r2}T_{f1}}{T_{f1} + T_{f2}}$	$\dfrac{T_{f1}T_{f2}}{T_{f1} + T_{f2}}$
Parallel	$\dfrac{T_{r1}T_{r2}}{T_{r1} + T_{r2}}$	$\dfrac{T_{f1}T_{f2}}{T_{r1} + T_{r2}}$
Exact formulas		
Series	$\dfrac{\left(T_{f1} + T_{r1}\right)\left(T_{f2} + T_{r2}\right) - T_{f1}T_{f2}}{T_{f1} + T_{f2}}$	$\dfrac{T_{f1}T_{f2}}{T_{f1} + T_{f2}}$
Parallel	$\dfrac{T_{r1}T_{r2}}{T_{r1} + T_{r2}}$	$\dfrac{T_{r1}T_{f2} + T_{r2}T_{f1} + T_{f1}T_{f2}}{T_{r1} + T_{r2}}$

The relationship is shown in Table 10.2 [36,37]. The results in Table 10.2 assume that the power supply is 100% reliable, and outages of components are probabilistically independent. Further, the results show *approximate* formulas for the case that $T_f \gg T_r$. Note that in typical power distribution engineering, T_f is in the order of tens of thousands of hours and T_r is in the order of a few hours. The exact formulas are also shown in Table 10.2 [42].

In many countries, quantified indices are defined to evaluate service reliability. Most of the indices depend on the interruption frequency or interruption duration. Billinton and Allan [43] show how repair time and failure rate may be used in the radial case to find reliability at distribution system buses.

SAIFI is the average interruptions frequency per customer and can be calculated by finding the interruption frequency of all buses divided by the number of customers connected in the system,

$$\text{SAIFI} = \frac{\sum_{i=1}^{B} \text{AIF}_i N_i}{N_T} \tag{10.18}$$

where N_i is the total number of customers connected in each bus, N_T is the total number of customers in the system, and B is the total number of buses. A similar idea can be applied for the average duration of all outages in one year, namely SAIDI, which is simply the summation of the interruption duration of all buses divided by the number of customers connected in the system,

$$\text{SAIDI} = \frac{\sum_{i=1}^{B} \text{AID}_i N_i}{N_T} \tag{10.19}$$

Table 10.3 Equivalent AIF and AID as a function of the AID and AIF of each component

	Series	Parallel
AIF^{eq}	$AIF_1 + AIF_2$	$\dfrac{AID_1 AIF_2 + AID_2 AIF_1}{8760}$
AID^{eq}	$AID_1 + AID_2$	$\dfrac{AID_1 AID_2}{8760}$

It is possible to combine the results of Table 10.2, Eq. (10.8), and Eq. (10.9) to obtain the AID and AIF for a receiving end bus fed by either two series components or two parallel components. This result gives the equivalent AID^{eq} and equivalent AIF^{eq} (as "seen" at the receiving bus) as shown in Table 10.2 [42]. As in Table 10.2, the equivalent AID and AIF of two simple components in series or parallel assume that $T_f \gg T_r$ and the supply bus is 100% reliable. The results in Table 10.3 are simply obtained using the results of Table 10.1 followed by the definition of the equivalent AID and AIF at a power delivery bus being $T_r^{eq}AIF$ and $8760 / \left(T_f^{eq} + T_r^{eq} \right)$, respectively.

Example 10.1

As an example of the use of Table 10.3 and to verify the accuracy of this equation, consider a simple radial distribution system supplied by a 100% reliable source and consisting of two feeders connected in series. Each feeder has $T_f = 5000$ h and $T_r = 5$ h (see Fig. 10.4). Each feeder in Fig. 10.4 is assumed to have the capability to detect and isolate the faults.

The reliability indices of the system in Fig. 10.4 are now calculated. Assuming that there is only one customer at the end of the line then AID and AIF can be found using Table 10.2 and Eqs. (10.8) and (10.9),

$$AIF_1 = AIF_2 = \frac{8760}{T_r + T_f} = 1.75025 f/y$$

$$AIF_s = AIF_1 + AIF_2 = 3.5005 f/y$$

The same approach is used to calculate AID,

$$AID_1 = AID_2 = AIF_1 \times T_{r1} = 1.75025 \times 5 = 8.751 \text{ h}$$

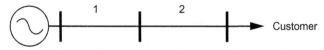

Fig. 10.4 Simple illustrative radial distribution.

$$\text{AID}_s = \text{AID}_1 + \text{AID}_2 = 17.5025 \text{ h}$$

The availability for the above system is equal to

$$A_1 = A_2 = \frac{T_{f1}}{T_{f1} + T_{r1}} = \frac{10,000}{10,000 + 10} = 0.999001$$

$$A_s = A_1 A_2 = 0.998003$$

Example 10.2

Now consider that these two components are connected in parallel as shown in Fig. 10.5 with the same time to fail and repair.

The AIF and AID calculations are

$$\text{AIF}_p = \frac{\text{AID}_1 \text{AIF}_2 + \text{AID}_2 \text{AIF}_1}{8760} = 0.003497 f/y$$

$$\text{AID}_p = \frac{\text{AID}_1 \text{AID}_2}{8760} = 0.0087425 \text{ h}$$

$$A_p = 1 - U_1 U_2 = 0.999999002$$

Table 10.3 implies that the AID and AIF (as well as other derivative indices) may be calculated from circuit topology and values of individual component MTTF or MTTR (T_f, T_r). Equivalently, the AID and AIF at a delivery bus in a network can be calculated from the AID and AIF values of individual components.

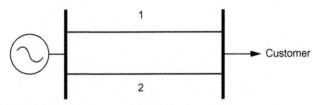

Fig. 10.5 Simple illustrative parallel distribution.

10.2.5 Complex Systems Transformation

In general, network reliability is the ability of the network to perform its designed function, which is delivering power from the input to the output. Evaluating the reliability of power networks (non-parallel-series networks) is a difficult task because of the large size and the complexity of the network. Even by using highly computational computers, an accurate, exact computation of network reliability consumes a lot of effort and time. This justifies the search for some approximate technique to evaluate the reliability of the network using less computational time.

The objective for most techniques used to evaluate the reliability of a complex system is to transform the logical representation of the system into a series-parallel system. Some of these methods are

- Decomposition (factor) method
- Delta-wye conversion
- The MCS method

The decomposition method for system reliability evaluation has been used for a long time, especially for parallel-series systems. The decomposition method (or conditional probability method) can be used basically to find the probability of a system to work, p_s, by assuming that each component can be represented by two states [39]. In the first cited state, the component is assumed to be ideally working (in the case of a line, this ideal state is a short circuit) with a probability p, and then is assumed to be ideally failed (in the case of a line, this is open circuit) with a probability q as shown in Fig. 10.6. Then apply the decomposition method and consider the short circuited and open circuited subsystems until the resulting subsystems are simple parallel and series systems. Then the reduction method can be used to find the availability of each subsystem. The system availability, failure frequency, and duration then can be calculated using Eqs. (10.20), (10.21), and (10.22).

$$A_s = A_{s1}p_5 + A_{s2}q_5 \qquad (10.20)$$

$$AIF_s = AIF_{s1}p_5 + AIF_{s2}q_5 \qquad (10.21)$$

$$AID_s = AID_{s1}p_5 + AID_{s2}q_5 \qquad (10.22)$$

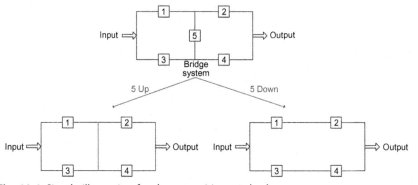

Fig. 10.6 Simple illustration for decomposition method.

where A_{s1} is the system availability when component 5 is short circuit (up) and A_{s2} is the system availability when component 5 is open circuit (down). The decomposition method is very useful in the case of simple and small systems but it will be difficult to apply on complex and large systems where the number of subsystems will increase. It is also difficult to program a general code for this method in digital computers.

Delta–wye conversion also can be used for complex systems where the components are not in simple parallel or series. Using the delta–wye conversion, a reliability study is more complicated than discussed above, and the error in the results can be significant due to the complex conversion equations [32]. Fig. 10.7 shows a simple illustration for delta–wye conversion used in the case of systems that are not simply parallel or series. The probability equations for the delta and wye configurations are shown in Table 10.4 and the MTTF and repair equations to convert from delta to wye configuration are shown in Table 10.5.

Consider at this point a third concept denominated as the cut set method. For any system, there are two subsystems that can represent the connections of the original system in two parallel-series subsystems where the first set is called the MTS and the other is called the MCS (see Fig. 10.8). The MTS is

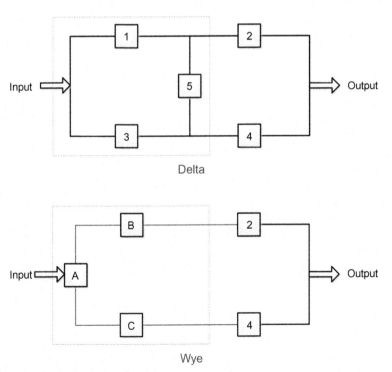

Delta

Wye

Fig. 10.7 Simple illustration for delta-wye conversion.

Table 10.4 Probabilities relations for wye and delta

Delta → Wye	Wye → Delta	
$p_A = \dfrac{\sqrt{X_1 X_2 X_3}}{X_1}$	$p_1 = \dfrac{p_A p_B - p_5 p_3}{1 - p_5 p_3}$	where $X_1 = p_5 + p_3 p_1 - p_1 p_3 p_5$
$p_B = \dfrac{\sqrt{X_1 X_2 X_3}}{X_2}$	$p_5 = \dfrac{p_C p_B - p_1 p_3}{1 - p_1 p_3}$	$X_2 = p_3 + p_5 p_1 - p_1 p_3 p_5$
$p_C = \dfrac{\sqrt{X_1 X_2 X_3}}{X_3}$	$p_3 = \dfrac{p_A p_C - p_5 p_1}{1 - p_5 p_1}$	$X_3 = p_1 + p_3 p_5 - p_1 p_3 p_5$

Table 10.5 Time to failure and repair equations to convert from delta to wye

	A	B	C
T_f	$T_{fA} = \dfrac{T_{f1} T_{f3}}{T_{r1} + T_{r3}}$	$T_{fB} = \dfrac{T_{f1} T_{f5}}{T_{r1} + T_{r5}}$	$T_{fC} = \dfrac{T_{f5} T_{f3}}{T_{r5} + T_{r3}}$
T_r	$T_{rA} = \dfrac{T_{r1} T_{r3}}{T_{r1} + T_{r3}}$	$T_{rB} = \dfrac{T_{r1} T_{r5}}{T_{r1} + T_{r5}}$	$T_{rC} = \dfrac{T_{r5} T_{r3}}{T_{r5} + T_{r3}}$

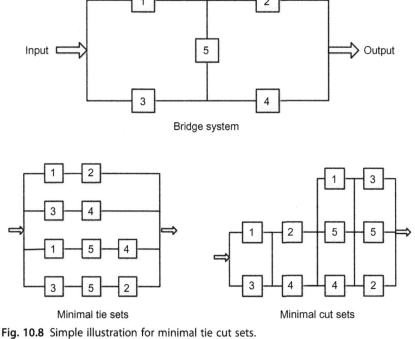

Bridge system

Minimal tie sets   Minimal cut sets

Fig. 10.8 Simple illustration for minimal tie cut sets.

the minimal set of components where the components provide a continuous connection between input and output. The MTS is the minimal set of components where the fault condition for all components in the set results in a system outage. Fig. 10.8 shows an explanation for a complex system and its two minimal sets.

The difficulty of these methods lies in the identification of the minimal tie and cut sets. Allan et al. [44] and Jasmon and Foong [45] indicate that it can be a very difficult task to identify the two sets in large, complex networked systems.

To calculate the reliability of the distribution network using the MCS method, the reliability of each set is computed by calculating the failure probability of all components in parallel. Then, the reliability of the system can be found by evaluating the equivalent MCSs connected in series.

Although the assumption is that it is a series–parallel system, the direct equations to calculate the reliability of the series-parallel components are not accurate when using the MCS method. The reason for this inaccuracy is because in MCSs, components may be repeated and appear in more than one set. For example, consider a system in Fig. 10.8, where Component 1 appears twice in the first and the third cut set. This multiple appearance makes the calculation of reliability imprecise. The only way to avoid this inaccuracy is by solving the MCS subsystem using the basic union and intersection probability calculations to block the effect of the duplicated components. The solution in this case will be complicated and time-consuming, but expected to be as accurate as decomposition or delta-wye methods.

As this method provides nondisjoint events, other methods are used such as inclusion–exclusion and sum of disjoint products to solve the problem with dependent components. Using these methods to calculate the reliability of the network from the cut sets or path sets is very complicated for a large system. To simplify the calculation, approximate equations [Eqs. (10.23) and (10.24)] can be used to solve the reliability, assuming that all the components in the cut sets are independent.

$$U_s = \sum_{i=1}^{n} U_i \tag{10.23}$$

$$\text{AIF}_s = \sum_{i=1}^{n} \text{AIF}_i \tag{10.24}$$

Example 10.3

Consider the bridge test system in Fig. 10.9 to calculate the AIF, AID, and the availability. The data for all components are listed in Table 10.6. For simplicity, it is assumed that there is only one customer at the output of the system.

For the bridge test system, A and U are calculated for each component using Eqs. (10.6) and (10.7). Then, using Eqs. (10.8) and (10.9), the AIF and AID for each component are calculated and listed in Table 10.7.

Table 10.8 shows the AIF, AID, and availability of the system using the three methods described above: the decomposition method, the delta-wye conversion, and cut sets method. Inspection of Table 10.8 shows that similar results with negligible error are obtained using the three methods. The decomposition method is the most flexible technique compared to the other two methods, especially in small networked systems. The MCS method is more powerful in terms of the logical representation of the system and analyzing the reliability of the system. Each cut set represents a different mode of failure.

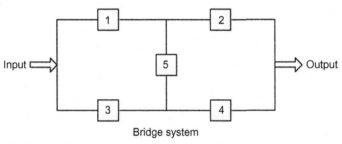

Bridge system

Fig. 10.9 Bridge test system.

Table 10.6 Bridge system data

Component	T_f (h)	T_r (h)
1	10,000	2
2	12,000	3
3	16,000	4
4	10,000	3
5	20,000	2

Table 10.7 Bridge system A, U, AIF and AID results

Component	A	U	AIF (f/y)	AID (h/y)
1	0.9998	0.0001999	0.8758	1.7516
2	0.99975	0.0002499	0.7298	2.1894
3	0.99975	0.0002499	0.5473	2.1894
4	0.9997	0.0002999	0.8757	2.6272
5	0.9999	0.0001	0.4379	0.8759

Table 10.8 Bridge system probability, AIF and AID results

Method	A	AIF (f/y)	AID (h/y)
Decomposition	0.999999875	0.0007661	0.0010945
Delta–wye	0.999999875	0.0007663	0.0010947
Minimal cut set	0.999999875	0.0007662	0.0010944

Example 10.4

For another example to demonstrate the effect of different arrangements of parallel components, consider two radial feeders in parallel with different possible tie connections between them as shown in Fig. 10.10.

For all the lines in this system, the mean times to failure and repair are listed in Table 10.9. The dotted lines in Fig. 10.10, denominated as 1, 2, and 3, are potential locations for *one* added tie line.

AID, AIF, SAIDI, and SAIFI are listed in Tables 10.10, 10.11, and 10.12 for each tie line. These results are found using Table 10.2 for both series and parallel connections and assuming that the number of customers connected at each bus is equal to one.

The results in Tables 10.10, 10.11, and 10.12 show that the minimal AID and AIF for each bus is obtained when the added single tie line is directly connected via the shortest parallel connection and the smallest number of components. From Table 10.3, the equivalent AID and AIF for series connections are the addition of the indices of the series components.

$$AIF^{eq} = AIF_1 + AIF_2$$
$$AID^{eq} = AID_1 + AID_2$$

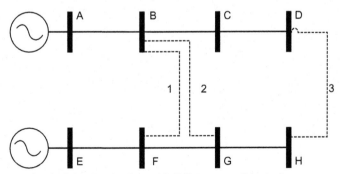

Fig. 10.10 Two parallel radial feeders with different possible tie lines.

Table 10.9 Time to fail and time to repair for all sections

Section	T_f (h)	T_r (h)
AB	10,000	2
BC	12,000	3
CD	8000	2
EF	8000	2.5
FG	9000	3
GH	10,000	3
1,2,3	10,000	2

Table 10.10 AID,AIF, SAIFI and SAIDI for all buses with tie line 1

BUS	AID (h)	AIF
B	0.000898	0.000843
C	2.1904	0.7307
D	4.3787	1.8249
F	0.0011	0.000985
G	2.9201	0.9740
H	5.5456	1.8491
SAIFI	0.672	
SAIDI	1.879	

Table 10.11 AID,AIF, SAIFI and SAIDI for all buses with tie line 2

BUS	AID (h)	AIF
B	0.0011	0.000979
C	2.1906	0.7308
D	4.3789	1.8251
F	0.0020	0.0017
G	0.0023	0.0020
H	2.6295	0.8777
SAIFI	0.429	
SAIDI	1.150	

Table 10.12 AID,AIF, SAIFI and SAIDI for all buses with tie line 3

BUS	AID (h)	AIF
B	0.0029	0.0026
C	0.0055	0.0045
D	0.0070	0.0058
F	0.0042	0.0034
G	0.0068	0.0054
H	0.0075	0.0060
SAIFI	0.003	
SAIDI	0.004	

As a result of the above, it is surmised that series connections increase both indices while parallel connections reduce the indices. The SAIDI and SAIFI values are decreased whenever the tie line is connected farther from the source and connects more sections together. Tie line 3 connects all sections together and provides two different paths from the power sources for all sections and buses. In case of any fault in any section, all the buses can be restored after isolating the faulted line. Fig. 10.11 shows how SAIFI and SAIDI change if the repair time decreases.

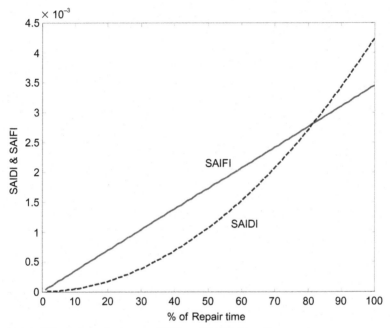

Fig. 10.11 Effect of repair time on SAIFI and SAIDI (Line 3).

10.2.6 Grid Network Reliability

The grid network as explained in Section 10.2.1 differs from the spot network in the secondary side and in the connection points from the primary feeders to the secondary networks. In a spot network, the feeders in the secondary network are connected in parallel and the reliability calculation is done using only the parallel-series concepts. In a grid network, the secondary network is connected in a complex (non-series-parallel) connection and the primary feeders are connected in different nodes in the secondary network. Therefore, the reliability calculation of the loads is not a direct reduction of series and parallel components.

Example 10.5

To calculate the reliability of the secondary grid network, the decomposition and the minimal cut sets methods are used. The system under study is shown in Fig. 10.12 [46]. The two primary feeders are assumed to be identical and the secondary feeders are also assumed to be identical. The time to fail and to repair for the primary feeders and secondary feeders are shown in Table 10.13. To simplify the calculation, one load is specified in this network to calculate the AIF, AID, and availability.

First, AIF, AID, and availability are calculated using the decomposition method. The system is decomposed into two different systems and one feeder is assumed first to be ideally working with a probability (p) then assumed to be ideally failing with a probability to fail (q). This process is repeated until the subsystems are all in series and parallel connections. The connections of these subsystems are shown in detail in Fig. 10.13.

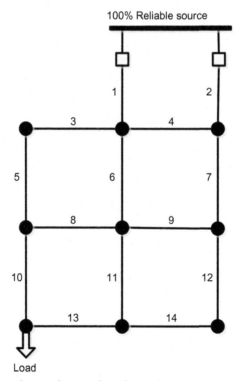

Fig. 10.12 Secondary grid network under study.

Table 10.13 Reliability data for the system under study

	Time to fail (h)	Time to repair (h)
Primary feeders	20,000	10
Secondary Feeders	10,000	10

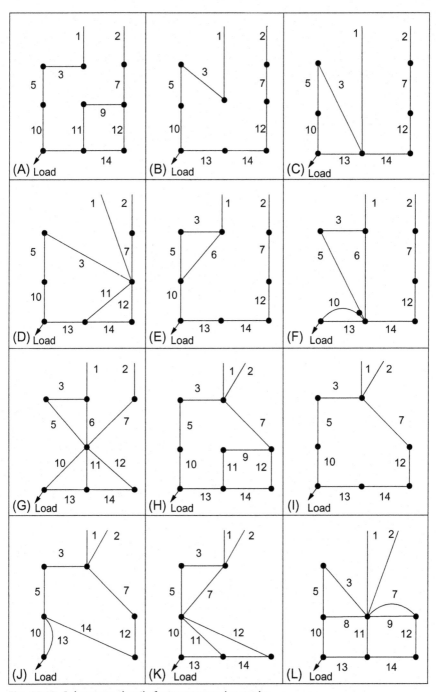

Fig. 10.13 Subsystem details for system under study.

Table 10.14 Reliability calculations for subsystems in Fig. 10.13

Subsystem	A	AIF	AID
A	0.99999127	0.01534	0.0766
B	0.99998433	0.02754	0.1377
C	0.99999526	0.008305	0.04154
D	0.99999625	0.006572	0.03286
E	0.99999327	0.01183	0.0591
F	0.99999725	0.00483	0.02411
G	0.99999825	0.003073	0.01535
H	0.99999376	0.01096	0.05475
I	0.9999878	0.021438	0.1072
J	0.99999277	0.012677	0.0634
K	0.99999675	0.0056886	0.02843
L	0.99999875	0.002197	0.01096

For each of the subsystems in Fig. 10.13, the availability, AIF, and AID are calculated using the successive reduction of the series-parallel components. The result of all the subsystems reliability calculations is shown in Table 10.14.

Using the subsystems reliability results from Table 10.14, the complete system availability, AIF, and AID are calculated using the following equation,

$$X = (A)q^2 + (B)q^4p^2 + (C)q^3p^2 + (D)q^2p^2 + (E)q^3p + (F)q^2p^2 + (G)qp^2 \\ + (H)q^2p + (I)q^3p^2 + (J)q^2p^3 + (K)qp^3 + (L)p^2$$

(10.25)

where X is either availability, AIF, or AID and A, B, C … and so on are the X subsystem values. The complete system availability, AIF, and AID are shown in Table 10.15.

The MCS method is also used to evaluate the reliability of the system under study shown in Fig. 10.12. The MCS is first specified and it is assumed that the maximum number of components that can form a cut set is 3 components. This is a reasonable assumption because in power systems the occurrence of 4 failures at the same time is considered to be rare and the effect of 4 components in parallel considered to be very small and can be negligible. The cut sets for the system are

1-2 10-13 1-4-7 5-8-13 10-11-13 5-6-7 10-11-12 3-6-7 3-8-13

Table 10.15 Reliability calculations using the decomposition method

A	AIF (f/y)	AID (h/y)
0.999998743	0.002202	0.01099

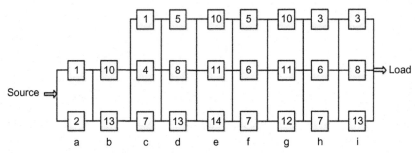

Fig. 10.14 Block diagram for the cut sets.

The blocks that represent the cut sets in parallel and series are shown in Fig. 10.14. Each cut set is represented by parallel connection of its components and the small letters represent each of these sets.

To overcome the complexity in using this method, especially in large complex systems, an approximation to calculate AIF and AID can be applied. This approximation is based on neglecting the dependency between the MCSs. The dependency is a result of the duplicated components in the sets. The MCSs under this assumption will be independent from each other and the reliability can be calculated as simple series-parallel components. To calculate the system unavailability U, AIF, and AID, use

$$AIF = \sum_{s=1}^{n} AIF_s \qquad (10.26)$$

$$AID = \sum_{s=1}^{n} AID_s \qquad (10.27)$$

$$U = 1 - \prod_{s=1}^{n} (1 - U_s) \qquad (10.28)$$

An advantage of the MCS method is that it can be programmed to perform general network reliability analysis. The difficulty of this method lies in the identification of the minimal sets. The MCS can be found by logical inspection of the system and specify the combinations that cause the disconnection between input and output. This logical method can work easily with small systems but with large and complex systems the process is complicated and time consuming.

The calculation of the unavailability, AIF, and AID is shown in Table 10.16 for each cut set.

From Table 10.16, the effect of the two-level cut sets (a and b) in the total reliability indices of the load in Fig. 10.12 is much larger than the effect of three-level cut sets. Almost 99% of the overall reliability indices

Table 10.16 Reliability calculations for cut sets in Fig. 10.14

Cut set	U	AIF (f/y)	AID (h/y)
a[a]	0.00000024975	0.0004376	0.002188
b[a]	0.000000998	0.0017485	0.0087412
c	0.0000000004987	0.00000131	0.00000437
d	0.000000000997	0.00000262	0.000008732
e	0.000000000997	0.00000262	0.000008732
f	0.000000000997	0.00000262	0.000008732
g	0.000000000997	0.00000262	0.000008732
h	0.000000000997	0.00000262	0.000008732
i	0.000000000997	0.00000262	0.000008732

[a]Cut sets a and b are "two-level" cut sets; the remaining cut sets are three level.

Table 10.17 Reliability calculations using the minimal cut set method

A	U	AIF (f/y)	AID (h/y)
0.999998745	0.00000125423	0.002203	0.010987

of the load come from the two-level cut sets. Furthermore, cut set (b) contributes more than cut set (a) due to the shorter time to fail of its components (secondary feeders 10 and 13). The shorter time to fail of cut set (b) components increases the failure rate of the cut set and hence the interruption rate and duration seen by the load also increase.

To calculate the system unavailability, AIF and AID, and Eqs. (10.26)–(10.28) are used and the results are shown in Table 10.17.

10.2.7 Quantification of Reliability

The power industry has become a competitive environment under deregulation and the continuity of power supply to the customers is significant. The basic objective of deregulation is to create a competitive environment in the power area to obtain better service and lower cost. The demand and price are very sensitive to each other and the utilities are under pressure to reduce the cost. The balance between service quality and cost is a key issue in the power market. Therefore, utilities are seeking more accurate data and predictions about the electrical service and its availability to keep their customers satisfied. It is also important to investigate how component failures and repair rates can affect the overall reliability of the system. Several indices are developed to quantify the service performance that measures outage duration, frequency, and system availability. Reliability is basically quantified by the probability of a component or system to operate or not to operate as expected. Moreover, the duration and frequency of misoperation are significant in evaluating the reliability of a device or system. In this chapter, *the event count indices* will be studied, principally the *system average interruption duration index* (SAIDI) and the *system average interruption frequency index* (SAIFI),

$$SAIDI = \frac{\text{Total duration of all interruptions}}{\text{Total number of customers connected}}$$

$$SAIFI = \frac{\text{Total number of all interruptions}}{\text{Total number of customers connected}}.$$

The SAIDI index gives information about the average *time* the customer is interrupted in minutes (or hours) in one year. The SAIFI index gives information about how *often* these interruptions occur on average for each customer. Both indices have been widely used in North America as measures of the effectiveness of distribution systems [1,11]. Note that in [15], the connection between the count indices is addressed, and this area is revisited in this chapter. Both indices are carried out (i.e., averaged) typically over a one-year interval; SAIDI is usually expressed in hours; and SAIFI is unitless.

Another reliability index commonly used with SAIDI and SAIFI indices is the customer average interruption duration index (CAIDI), which is the average duration of each interruption seen by each interrupted customer. CAIDI captures the average time that the utility responds by measuring the average time to restore service. The difference between SAIDI and CAIDI is that in SAIDI the total duration of all interruptions is averaged by the total number of customers connected to the system. CAIDI in the other hand is only averaged by the customers interrupted in each outage event.

Other indices that can be used to evaluate the reliability performance of the distribution system may include [1,47]:
- Momentary average interruption frequency index (MAIFI)
- Customer average interruption frequency index (CAIFI)
- Customers interrupted per interruption index (CIII)
- Customers experiencing multiple interruptions (CEMI$_x$)
- Average service availability index (ASAI)
- Average service unavailability index (ASUI)
- Average system interruption frequency index (ASIFI)
- Average system interruption duration index (ASIDI)

- Energy not supplied (ENS) (also termed energy unserved)
- Average energy not supplied (AENS)
- Average customer curtailment index (ACCI)
- Average energy not supplied per interruption (AENSI).

These indices may not be as widespread in use for utilities and customers as SAIFI, SAIDI, and CAIDI, but they can be useful measures in some complex systems and specific applications. Some indices listed above may also be used to evaluate the quality of power systems in term of momentary low voltage events.

Similar to SAIFI, the MAIFI index measures the momentary average interruption frequency that the customer experienced and it is mostly measured at substation bus. MAIFI is calculated by counting the number of affected customers of any protection device operation and averaged by the number of customers in the system under study. MAIFI can be defined as

$$MAIFI = \frac{\text{Total Number of Customers Momentary Interruptions}}{\text{Total Number of Customers}}$$

$$MAIFI = \frac{\sum DO_i N_i}{N_T} \tag{10.29}$$

where DO_i is the protection device operations, N_i is the number of customers affected per operation, and N_T is the total number of connected customers.

CAIFI is similar to SAIFI and CAIDI. SAIFI measures the average interruption frequency for all customers in the system where CAIFI measures the average interruptions frequency for only interrupted customers.

$$AIFI = \frac{\text{Total Number of Customer Interruptions}}{\text{Total Number of Customer Interrupted}}$$

$$CAIFI = \frac{\sum AIF_i N_i}{\sum N_i} \tag{10.30}$$

where N_i is the number of affected customers in each bus.

The CIII is the customers interrupted per interruption index. It reflects how the customers are connected to the system. It can be calculated as

$$CIII = \frac{\text{Total Number of Customer Interrupted}}{\text{Total Number of Customer Interruptions}}$$

$$CIII = \frac{1}{CAIFI} = \frac{\sum N_i}{\sum AIF_i N_i} \qquad (10.31)$$

Because most of indices are averaged for all customers, it is difficult to identify or truck the number of interruptions (x) for a specific customer, $CEMI_x$ can be used to measure the reliability of the system to a specific number of interruptions. It can be defined as follows:

$$CEMI_x = \frac{\text{Total Number of Customers experienced} > x \text{ interruptions}}{\text{Total Number of Customers}}$$

$$CEMI_x = \frac{\sum N_{xi}}{N_T} \qquad (10.32)$$

The ASAI is the measure of service availability during a given period. It is calculated by dividing the number of hours when service available to the customers by the total number of demand hours for all customers.

$$ASAI = \frac{\text{Total Number of Hours Availability}}{\text{Total Demand Hours}}$$

$$ASAI = \frac{8760 - SAIDI}{8760} \qquad (10.33)$$

Similarly, ASUI can be calculated from the ASAI,

$$ASUI = \frac{\text{Total Number of Hours Unavailability}}{\text{Total Demand Hours}}$$

$$ASUI = 1 - ASAI = \frac{SAIDI}{8760} \qquad (10.34)$$

The above indices basically deal with the frequency or the duration of the interruptions. Other indices can be used to evaluate the power unserved during any interruption. ASIFI is an index where the ratio of connected kVA interrupted to the total kVA connected is calculated as follows:

$$ASIFI = \frac{\text{Total KVA Interrupted}}{\text{Total KVA Served}}$$

$$ASIFI = \frac{\sum S_i}{S_T} \qquad (10.35)$$

where S_i is the kVA interrupted and S_T is the total kVA connected. Similar to ASIFI, ASIDI is the ratio of the total kVA interrupted during the outage by the total kVA connected.

$$\text{ASIDI} = \frac{\text{Total KVA Duration Interrupted}}{\text{Total KVA Served}}$$

$$\text{ASIDI} = \frac{\sum \text{AID}_i S_i}{S_T} \tag{10.36}$$

To report the total energy not supplied by the system during the outages, ENS can be used and calculated as

$$\text{ENS} = \text{Total Energy Not Supplied}$$

$$\text{ENS} = \sum P_{avg,\,i} \text{AID}_i \tag{10.37}$$

where P_{avg} can be calculated as

$$P_{avg} = \frac{\text{Total Annual Energy Demanded}}{8760}$$

AENS is the average of ENS over the total number of customers and can be expressed as

$$\text{AENS} = \frac{\text{Total Energy Not Supplied}}{\text{Total Number of Customers}}$$

$$\text{AENS} = \frac{\text{ENS}}{N_T} \tag{10.38}$$

Similar to AENS, the ACCI is the total energy not supplied but averaged by the total number of customers interrupted,

$$\text{ACCI} = \frac{\text{Total Energy Not Supplied}}{\text{Total Number of Customers Interrupted}}$$

$$\text{ACCI} = \frac{\text{ENS}}{\sum N_i} \tag{10.39}$$

Some utilities use cost indices to evaluate the cost of the interruptions to the utility and customer. The method to calculate these indices differ for each utility based on different survey approaches and approximation methods used to estimate the cost. Two of these indices are

– Customer outage cost (COC)
– Cost of repair of failed equipment (COF).

COCs vary based on the type of customer: industrial, commercial, or residential. Industrial customers' costs usually include loss of products, damaged equipment, maintenance, restarting process, and labor. The same costs may be associated with commercial customers. Residential outage costs are generally far less than the other two sectors. Based on a survey study [1], the estimated outage cost for a residential customer for 1 hour was around $3, for a commercial customer the cost was $1200, and for a large industrial customer the cost was $82,000. It is unknown whether the cited $3.00 figure of value per hour of outage is reasonable (or representative of the average customer). COC can be expressed as

$$COC = \text{Customer Cost Function} \times \text{Average Power} \\ \times \text{Outage duration time}$$

$$COC = \sum \left(CCF_i \times P_{avg,i} \times AID_i \right) \tag{10.40}$$

where CCF_i is the customer cost function for each bus.

Utility outage cost basically consists of the loss of profits for the energy not served and the cost to repair and restore the service. The cost of repair includes labor costs and replacements parts cost. COF can be calculated as

$$COF = \text{Labor Costs} + \text{Replacment Parts Costs}$$

$$COF = \sum (LC_i + RPC_i) \tag{10.41}$$

Most utilities have to report their reliability indices to regulatory bodies. In a 2008 survey, 35 states, including Washington, DC, require routine reporting of SAIDI and SAIFI from the utilities to the public utility commission [48]. Utilities also differ in how to count the number of customers. Some utilities count their customers based on the number of meters connected and some utilities base data on the number of postal addresses. Most utilities also exclude severe weather outages and planned outages from reliability indices because in most storm outages the utility cannot control the incident or severity of the storms. Customers also may be notified before any planned outages (e.g., maintenance) so that the impact of the outage will be minimized.

10.3 RELIABILITY IMPACT OF DISTRIBUTED GENERATION RESOURCES

10.3.1 Overview of Distributed Generation

DG is defined as a small-scale generation unit that is installed in the distribution system and typically connected at substations, on distribution feeders, or at the customer load level [49–51]. DG differs fundamentally from the traditional model of central generation as it can be located near end users within an industrial area, inside a building, or in a community. Different types of DGs have been developed due to the increasing interest in DG in recent years [52].

DG units vary in size, fuel type, and efficiency, and they can be associated with two technologies, conventional energy technology and renewable energy technology. Technologies that utilize conventional energy resources include reciprocating engines, combustion turbines, microturbines, and fuel cells. Conversely, renewable energy resources are based on different forms of natural resources such as heat and light from the sun, the force of the wind, and the combustion value of organic matter [52–55]. A few examples of renewable DGs include photovoltaic (PV) cells, wind turbines (WTs), and biomass. The most promising renewable energies in the United States are wind and solar. The advantages and disadvantages of wind energy and PVs are listed in Table 10.18.

The main difference between renewable and conventional resources is that the output of the renewable resources depends on variable inputs such as wind or solar energy. The power produced from renewable resources may fluctuate more, making it difficult to forecast. In the case in which a DG is connected to the local load to supply the load during interruptions to operate in islanded operational mode, the demand and supply may not match, especially if the DG is renewable. In this case, the DG is either disconnected due to the activation of the frequency or voltage protection devices or the system will shed some loads and only supply critical loads.

Table 10.18 Advantages and disadvantages for several renewable energies

	Advantages	Disadvantages
Wind energy	• Short time to design and install • Low emissions • Different modular size	• Wind is highly variable • Limited resource sites • Audible and visual noise • Low availability during high demand periods
Photovoltaic	• Flexible in terms of size and site • Simple operation • No moving parts and noise • Low maintenance • Short time for design and installation • No emissions	• High capital cost • Large area required • Low efficiency • Low availability during high demand periods • Low capacity factor

There are many potential applications for DG technologies. They can be classified as backup DGs or base load DGs. The DG can be used as a backup generator to replace the normal source when it fails to supply the load, thereby allowing the customer's facility to continue to operate satisfactorily during the power outages. Most backup generators are diesel engines because of their low cost, fuel availability, and quick start time.

A backup DG is connected to the local load, and a manual or automatic switch is installed on the feeder side of the DG local load. During faults on the main feeder or its laterals, the circuit breaker on the substation side will trip to clear the fault, causing the whole feeder to be interrupted. Then, the DG switch can be closed and the DG unit can start supplying its local load.

A base load DG is used by some customers to provide a portion or all of their electricity needs in parallel with the electric power system. It can also be used as an independent stand-alone source of power. The technologies used for these applications include renewable DGs such as wind, PVs, fuel cells (FCs), and combined heat and power (CHP).

The presence of DG in the distribution system may improve system reliability as a result of supplying loads in islanded operation. The islanded operation implies that the loads are disconnected from the substation and supplied from the DG until the utility restores the power from the main supply. The DG may not be able to supply the demand completely during the islanded mode. This is due to the availability and the capacity of the DG especially when it depends on a renewable resource. Note that power balance must occur between the DG, the main supply, and any energy storage capability.

10.3.2 Different Reliability Indices for Distributed Generation Integration

The traditional reliability indices cover sustained interruptions. The time necessary to start-up the DG and switch to islanded operation should be taken into account for the reliability evaluation of the distribution system including the DG. If this time is short enough, the customers will experience a momentary interruption; if not, customers will experience a sustained interruption. The reliability evaluation of the distribution system with the DG and renewable energy resource should be modified to account for both the limitation of the output of the DG and the time it takes to start and switch to islanded operation [18].

The indices in Section 10.2.7 include a number of indices that capture distribution reliability. However, they do not fully capture the energy interrupted, and none fully captures the impact of DG. These subjects are incorporated into several innovative indices discussed below

To find the AENSI the ENS is divided by the total number of interruptions as follows:

$$AENSI = \frac{Total\ Energy\ Not\ Supplied}{Total\ Customer\ Interruptions}$$

$$AENSI = \frac{\sum P_{avg,\ i} AID_i}{\sum AIF_i N_i} = \frac{AENS}{SAIFI} \qquad (10.42)$$

To reflect the insertion of DG into distribution systems, consider the following measures:
- Distributed generation supply during interruptions (DGSI)
- Distributed generation supply duration during interruptions (DGSDI).

The ratio of DG power output share during interruptions can be calculated as,

$$DGSI = \frac{DG\ Power\ Output}{Energy\ Not\ Served}$$

$$DGSI = \frac{\sum P_{DG,\ i}}{ENS} \qquad (10.43)$$

In some cases, the DG cannot supply the full demand to the load, especially when the DG source is renewable such as wind or solar sources. The DG supply duration depends basically on the availability of the fuel or renewable source. The supply duration for DG in this case can be calculated as follows:

$$DGSDI = \frac{DG\ Supply\ Duration}{System\ Average\ Interruption\ Duration}$$

$$DGSDI = \frac{\sum p_{DG} AID_i}{SAIDI}. \qquad (10.44)$$

Example 10.6

The following example will address the impact of DGs on the reliability of an example system, and how this may be evaluated using the SAIDI and SAIFI indices. For this illustration, a test bed system consists of an infinite supply source and a radial distribution feeder is used as shown in Fig. 10.15. For simplicity, one customer per bus is assumed. The T_f and T_r for each section are listed in Table 10.19.

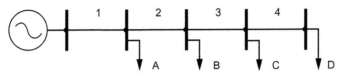

Fig. 10.15 Radial distribution system.

Table 10.19 Time to fail and time to repair for all sections

Section	T_f (h)	T_r (h)
1	10000	2.5
2	9000	2
3	9600	2.4
4	11000	2.3

The AID and AIF for load B and the system are calculated for several cases to study the effect of the DG. The illustrated cases are:

Case 1:
No DG in the system

Case 2:
DG with start/switch time equal zero (momentary)

Case 3:
DG with start/switch time not equal to zero (sustained)

Case 4:
Renewable DG with start/switch time equal zero.

Case 5:
Renewable DG with start/switch time not equal to zero.
These cases are discussed below.

Case 1: No DG in the system

This case captures the basic reliability study of the system before introducing the DG. The AID and AIF are calculated at each bus using the series components equivalent concept.

Case 2: DG with start/switch time equal zero (momentary)

This case illustrates the situation where the DG connected to the load bus B is a standby unit working only during outages. The system with the DG is shown in Fig. 10.16. The DG is assumed to have nearly zero starting and switching time to be considered as *momentary* and not *sustained* interruption. The DG unit is considered to be a nonrenewable unit with infinite available fuel and limited capacity equal to the demand at B. The DG is modeled as an ideal generation unit connected in series with a feeder with the probability of the DG to work or fail as shown in Fig. 10.17. The reliability the DG is represented as T_f and T_r,

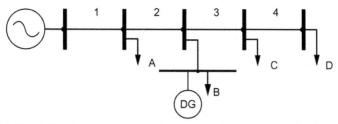

Fig. 10.16 Radial distribution system under study with DG in bus B illustrative Case 2.

Fig. 10.17 DG model used in Case 2.

$T_{\text{f-DG}} = 20,000 \text{ h}$
$T_{\text{r-DG}} = 2.25 \text{ h}$
$T_{\text{ss}} = 0$

In Case 2, the method to calculate AID and AIF for buses A, C, and D is the same as in Case 1: for bus B, where the DG is connected, and the DG forms a parallel source with the substation to supply the load during the interruptions as shown in Fig. 10.18. Zone 1 represent the normal power source to load B and zone 2 represents the standby DG source to the load. In normal operation, zone 1 is connected to load B and supplies the demand. During a fault in any component between the source and load B in the main feeder, zone 1 will be disconnected from the load and zone 2 will be connected to supply the load. Any zone can supply the load and the only case where load B can experience a fault is when zones 1 and 2 are both in fault conditions. Table 10.20 shows the situations for both zones and the status of load B with the repair time included.

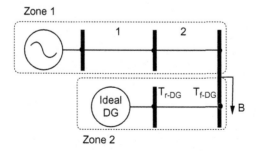

Fig. 10.18 Two parallel zones for Case 2.

Table 10.20 Different operation modes for Case 2

Zone 1	Zone 2	Load B	T_r
UP	UP	UP	0
UP	DOWN	UP	0
DOWN	UP	UP	0
DOWN	DOWN	DOWN	$T_{r-1,2,DG}$

Case 3: DG with start/switch time not equal to zero (sustained)

This case is the same as Case 2 except that the time to switch and start the DG is not zero and assumed to be enough to cause a sustained interruption seen by the load. In principle, the AID and AIF at load B can be calculated using the series–parallel concepts as in Case 2. The presence of the DG and the switching time between the two zones in the system will change the situations where load B can see outages and new modified equations should be introduced to calculate AIF and AID. Table 10.21 shows that load B can experience an outage when a fault occurs in zone 1 alone or when both zones are faulted at the same time (with different time to repair in each case).

When both zones are down, the time to repair will be the same as in Case 2. When only zone 1 is down, the time to repair seen by load B will be equal to the time to start/switch the DG and the T_f seen by load B, when the fault is in any component in zone 1, should be modified as shown in Fig. 10.19. Table 10.22 shows the T_f and T_r used in calculating AID and AIF for both situations in which load B experiences an outage.

Table 10.21 Operation modes for Case 3

Zone 1	Zone 2	Load B	T_r
UP	UP	UP	0
UP	DOWN	UP	0
DOWN	UP	DOWN	T_{ss}
DOWN	DOWN	DOWN	$T_{r-1,2,DG}$

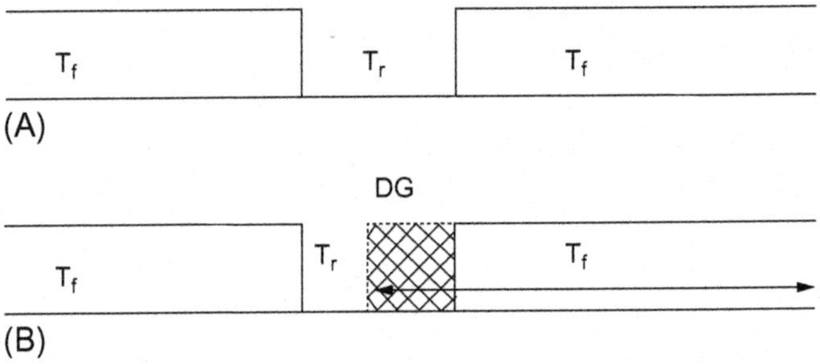

Fig. 10.19 Time to fail and time to repair (A) without DG (B) with DG and start/switch time.

Table 10.22 Time to fail and time to repair for different failure modes for Case 3

	Section	T_f (h)	T_r (h)
Zone 1—DOWN	1	10000	2.5
Zone 2—DOWN	2	9000	2
	DG	20000	2.25
Zone 1—DOWN	1	10002.4167	0.0833
Zone 2—UP	2	9001.9167	0.0833
	DG	20000	2.25

Based on Table 10.22, the new equation used to calculate AIF is

$$AIF_B = AIF_{Z1}.q_{Z2} + AIF_{Z2}.q_{Z1} + AIF_{Z1-ss}p_{Z2} \quad (10.45)$$

where AIF_{Z1}, AIF_{Z2} are the AIF values for zones 1 and 2; q_{Z1}, q_{Z2}, p_{Z2} are the probabilities to fail or operate for zones 1 and 2; and AIF_{Z2-SS} is the AIF for zone 1 when start/switch time is considered as the repair time for each section. AIF is the composite of the two situations where load B can see an outage. The first two terms represent the AIF at load B when both zones are in fault. The third term represents the outage seen by load B caused by start/switch time. AID also can be calculated using

$$AID_B = 8760(q_{Z1}q_{Z2} + q_{Z1-ss}p_{Z2}) \quad (10.46)$$

For this case the probability to operate and fail should be calculated for both zones using

$$p = \frac{T_f}{T_f + T_r} \quad q = \frac{T_r}{T_f + T_r}$$

Case 4: Renewable DG with start/switch time equal zero

This case is similar to Case 2 where is the DG presented at load B with time to start and switch equal to zero. The DG in Case 4 is considered to be a renewable DG (e.g., solar PV) and the output of this DG is based on the availability of the natural source. The output is assumed to be equal to the load in a specific period of the day as shown in Fig. 10.20. In Case 4, the solar unit output is modeled as an ideal DG connected in series with the probability of the source to be available represented as $T_{f-solar}$ and $T_{r-solar}$. The DG is also connected in series with the same T_{f-DG} and T_{r-DG} as in Cases 2 and 3. As shown in Fig. 10.21, the results are summarized below.

Case 5: Renewable DG with start/switch time not equal to zero

Case 5 is the same as Case 4 with time to start/switch the DG at nonzero. This can cause an interruption to load B. The calculation method for AID and AIF is the same as in Case 3 using Eqs. (10.45) and (10.46).

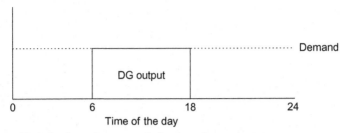

Fig. 10.20 Typical solar unit output during one day.

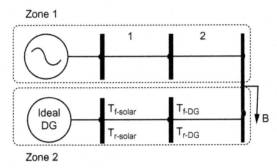

Fig. 10.21 Two parallel zones for Case 4.

The reliability study preformed for the radial distribution system is shown in Fig. 10.15. The AID and AIF are calculated for the five cases discussed in the previous section and the results are shown in Table 10.23. AID and AIF are calculated at load B and the average indices are shown for the whole system.

The effect of the DG in the AID and AIF indices for a specific bus and for the whole system is investigated in this study. It can be seen from the results that in Case 2, the reliability indices at load B and the system will experience considerable improvement compared to Case 1. This is due to the presence of the DG in the system and the assumption that the start

Table 10.23 SAIDI, SAIFI and DGSDI calculation results for Cases 1–5

		Load B			System		
Case		AIF	AID (h)	DGSDI	SAIFI	SAIDI (h)	DGSDI
1	No DG	1.8484	4.1352	0	2.2604	5.2013	0
2	DG $T_{ss}=0$	0.0004	0.0004	0.9998	1.7984	4.1676	0.1987
3	DG $T_{ss}\neq0$	1.8490	0.1642	0.9603	2.2606	4.7326	0.0901
4	Solar $T_{ss}=0$	1.2689	2.0678	0.4999	2.1156	4.6964	0.0970
5	Solar $T_{ss}\neq0$	2.1933	2.1497	0.4801	2.3466	4.6999	0.0963

T_{ss} denotes the switching and start-up time.

and switch time for the DG is equal to zero. Because the DG is assumed to operate in an islanded mode, the impact of the DG is mainly on the reliability of the load bus where the DG is connected.

In Case 3, where the time to start and switch is not equal to zero, the impact of the DG is mainly in the interruption duration and partially in the frequency of the interruptions. The SAIFI will increase slightly as compared to Case 1 due to the slight change of the T_f and T_r for the equivalent system after adding the DG. The SAIDI will be greater than the SAIDI in Case 2 due to the significant increase in the frequency and T_r of the outages.

Cases 4 and 5 show the effect of the limited availability of the DG on the reliability indices. The SAIDI and SAIFI are calculated in Case 4 where the solar source has a limit due to the time of day. The indices are greater than the indices in Case 2 where the DG is assumed to be able to supply the load all day. In Case 5, the SAIFI increased even more than the base case (no DG). In Case 5, the renewable source increases the probability to experience the case of no power production because the capacity factor of the solar resource is low. The increased value of SAIFI is explained as follows: during a repair period in the main feeder, the solar energy afforded at bus B comes "on" after some time at the start of insolation, and then there is an additional outage after insolation.

10.3.3 Adequacy Assessment of the Distributed Generation Islanded System

The main direct contribution of DG to reliability is on the customer side rather than on the utility or system side. The base level of reliability is always provided by the utility, and the DG's role is to boost the level of reliability by supplying the local load during interruptions (assuming that the DG is properly sized to serve at least the critical loads). The duration of interruptions at the load bus are expected to be fewer when a standby DG is connected. Different factors should be considered when evaluating the reliability impact of the DG on the local load, such as fuel availability, power output, unit's failure rate, repair time, and starting time. Many papers have discussed the technologies of DG units and their economical, environmental, and operational benefits [55–59].

The presence of the DG in the distribution system may improve system reliability as a result of supplying loads in islanded operation. Islanded operation implies that the load is disconnected from the substation and supplied from the DG until the utility restores the power from the main supply. The DG may not be able to supply the demand completely during islanded mode. This is due to the availability and capacity of the DG, especially when it depends on renewable resources. The DG also requires protection at the point of common coupling between the utility grid and the DG facility to prevent any unintentional islanding. Generally, the DG cannot be islanded during interruptions with the external loads to the DG facility. This creates quality and safety problems with the utility in the maintenance and restoration processes. In most cases, load shedding is required in the local facility and only the critical loads are restored using the DG.

Analyzing the reliability of future distribution systems, including the DG, is different than analyzing the generation and transmission systems with large-scale central units. The main difference is in the interaction between the generation units, the lines and component network, and the load points. In future distribution systems, higher penetration of the DG will be connected to the local load or at different points of the main feeder. The DG has a smaller capacity-to-load ratio than do the central generation units. This ratio can limit the availability of the DG to supply the demand during interruptions since the probability of the load demand to be greater than the DG power output is high.

As shown in Fig. 10.22, during normal operation the load is connected to the utility supply via the components and the feeders in the distribution network. If the distribution system connection fails to supply power to the load, the load is supplied from the DG in an islanded operation. If the DG is a conventional backup unit, the DG is operated and connected to the load only during emergencies. If the DG unit is a renewable base load unit, it is continuously operated in parallel with the utility supply. But it has to be disconnected and reconnected again during interruptions.

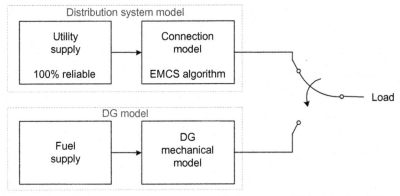

Fig. 10.22 Distribution connection system and DG system reliability models [59].

The DG modeling, DG system adequacy analysis, and integrating the reliability of the DGs in future distribution systems will be covered in this section. Integration of the DG in the reliability evaluation of future distribution systems consists of three main phases: DG unit reliability modeling, DG islanded system adequacy assessment, and DG islanded system reliability integration.

The reliability model of the DG unit is a logical and mathematical representation of the impact of different failure modes of the DG and fuel supply availability on the DG power output. The difference between the DG model of conventional and renewable units, from the reliability point of view, is that conventional units generate the rated power if they are operational. On the other hand, the output power of the renewable units is related to the primary energy source such as wind speed or solar radiation intensity, even if the unit is in a working state. In the DG islanded system, the DG may not be able to supply the local load due to the insufficiency of the generated power to match the demand or internal failure in the unit.

To integrate the DG islanded system's reliability into the distribution system reliability model, all the components (e.g., switches or protection devices) or failure modes that are related to the operation of the DG during interruptions should be incorporated into the DG reliability model. Then, a complete reliability model is proposed to evaluate the reliability of the distribution system including DG, compare different designs and different DG technologies from a reliability point of view, and optimize the size, number, and location of the DG units in the system.

In general, the DG unit reliability model consists of two main models: the fuel supply and the mechanical models. The fuel supply model represents the availability of the fuel supply to the unit during the study period. The mechanical model represents the ability of the unit to operate.

The adequacy of the DG is commonly evaluated by convolving the time varying power output of the DG with the load duration curve to calculate the probability that the DG will supply the load demand. This adequacy analysis can be used to study the impact of the DG on the load and system reliability improvement during interruptions.

The reduction on reliability indices is related to the islanded probability, frequency, and duration as seen by each load. During interruptions, if the total DG available capacity is greater than the load demand, the DG will supply the load. If the load is greater than the generated power, the DG will be disconnected or the DG can supply a curtailed load (e.g., critical components). The annual per unit power output for different DG units and the

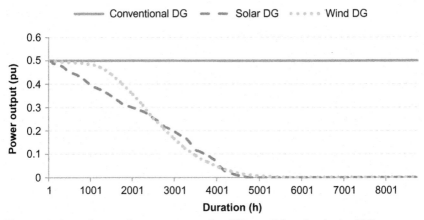

Fig. 10.23 Annual per unit power output for different DG technologies [42].

annual per unit load demand for different customer sectors are shown in Figs. 10.23 and 10.24, respectively.

The most common indices used for adequacy assessment are the loss of load expectation (LOLE), and the loss of load probability (LOLP). The LOLE is the expected number of hours per year during which the generated power is insufficient to supply the demand. If the time used in calculating LOLE is per unit, this index is called LOLP.

The LOLP and LOLE for conventional and renewable DG are given by:

$$LOLP = P(\text{Load} > \text{DG Capacity})$$

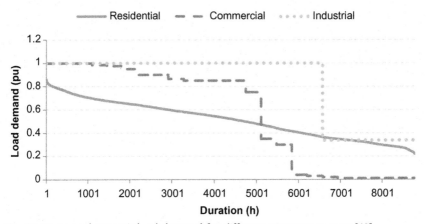

Fig. 10.24 Annual per unit load demand for different customer sectors [42].

$$LOLP = \sum_{i=1}^{n} P_i \sum_{j=1}^{m} P(C_j < L_i) \qquad (10.47)$$

$$LOLE = LOLP \times 8760 \ h/y \qquad (10.48)$$

where n and m are the total number of levels in demand probability table (DPT) and capacity probability table (CPT), respectively, P_i is the probability of the load level i, C_j is the DG capacity of the level j, and $P(C_j < L_i)$ is the probability that the capacity state j is less than load level i.

The adequacy indices are calculated by convolving the CPT of the DG with the DPT. Each power output or load segment is represented by the mean of all the data points in the segment. For each DG capacity level, the percentage of time for which the average dand is higher than the average generated power is used to calculate the LOLP.

The following steps summarize the procedure for evaluating the adequacy of the DG in supplying the load demand:
1. For both the DG and the load under study, the CPT and DPT are generated and the mean is computed for each segment to be used in the adequacy assessment.
2. For each DG power output segment in the CPT, the DPT is used to find the total time (or probability) for which the average load demand exceeds the average generated power for each segment.
3. The probability of each DG power output level is multiplied by the total time (or probability) found in step 2.
4. The cumulative sum of all the products in step 3 yields the LOLE (or LOLP).

Example 10.7

Table 10.24 [42] shows the CPT with the average generated power for each segment for three DG types: conventional, solar, and wind. The DG is assumed 100% reliable and the number of states (n) is 10 (excluding the zero power output level). The rated capacity ratio (RCR) for the conventional, solar, and wind DG is 0.5. As explained in Eq. (10.49), the RCR is the ratio between the nameplate capacity of the DG and the annual peak demand.

$$RCR_{pu} = \frac{DG \ nameplate \ capacity}{Annual \ peak \ demand} \qquad (10.49)$$

Table 10.24 CPT for the conventional, solar, and wind DG (*n* = 10)

State	Segment	Conventional DG (RCR = 0.5)			Solar DG (RCR = 0.5)			Wind DG (RCR = 0.5)		
		Duration (h)	Average Power Output	Probability	Duration (h)	Average Power Output	Probability	Duration (h)	Average Power Output	Probability
0	0	0	0	0	4003	0.000	0.457	3068	0.000	0.350
1	0.0–0.1	0	0	0	1046	0.038	0.119	2226	0.028	0.254
2	0.1–0.2	0	0	0	743	0.151	0.085	662	0.146	0.076
3	0.2–0.3	0	0	0	973	0.255	0.111	510	0.247	0.058
4	0.3–0.4	0	0	0	1027	0.349	0.117	532	0.353	0.061
5	0.4–0.5	8760	0.5	1	968	0.455	0.111	1762	0.475	0.201
6	0.5–0.6	0	0	0	0	0.000	0.000	0	0.000	0.000
7	0.6–0.7	0	0	0	0	0.000	0.000	0	0.000	0.000
8	0.7–0.8	0	0	0	0	0.000	0.000	0	0.000	0.000
9	0.8–0.9	0	0	0	0	0.000	0.000	0	0.000	0.000
10	0.9–1	0	0	0	0	0.000	0.000	0	0.000	0.000

Table 10.25 DPT for the residential, commercial, industrial sectors ($n = 10$)

State	Segment	Residential Customer			Commercial Customer			Industrial Customer		
		Duration (h)	Average Power Output	Probability	Duration (h)	Average Power Output	Probability	Duration (h)	Average Power Output	Probability
0	0	0	0.000	0.000	0	0.000	0.000	0	0.000	0.000
1	0.0–0.1	0	0.000	0.000	2920	0.017	0.333	0	0.000	0.000
2	0.1–0.2	0	0.000	0.000	0	0.000	0.000	0	0.000	0.000
3	0.2–0.3	837	0.276	0.096	365	0.300	0.042	0	0.000	0.000
4	0.3–0.4	1857	0.348	0.212	365	0.350	0.042	2190	0.337	0.250
5	0.4–0.5	1357	0.449	0.155	0	0.000	0.000	0	0.000	0.000
6	0.5–0.6	1793	0.553	0.205	0	0.000	0.000	0	0.000	0.000
7	0.6–0.7	1838	0.650	0.210	0	0.000	0.000	0	0.000	0.000
8	0.7–0.8	916	0.744	0.105	365	0.750	0.042	0	0.000	0.000
9	0.8–0.9	162	0.818	0.018	2555	0.866	0.292	0	0.000	0.000
10	0.9–1	0	0.000	0.000	2190	0.985	0.250	6570	1.000	0.750

Example 10.8

Table 10.25 [42] demonstrates the DPT for the three types of customers (residential, commercial, and industrial). The residential annual peak demand occurs during a short period of time in the summer and winter and can be neglected from the DPT.

Example 10.9

The LOLP for the three DG types (conventional, solar, and wind) and the three customer sectors (residential, commercial, and industrial) are shown in Table 10.26 [59]. The DG unit is assumed to be 100% reliable when it is needed. The smallest LOLP occurs when the residential customer installs a conventional DG to supply the local load. The largest LOLP occurs when a solar DG is used to supply the industrial load during interruptions.

Fig. 10.25 demonstrates the impact of the RCR value on the LOLP of the residential, commercial, and industrial sites if a conventional DG is connected. In Fig. 10.26, the impact of the solar RCR on the LOLP is shown for different customers.

Table 10.26 LOLP for the different DG units and customers in Tables 10.24 and 10.25

$n = 1000$	Conventional (RCR = 0.5)	Solar (RCR = 0.5)	Wind (RCR = 0.5)
Residential	0.5376	0.9128	0.8883
Commercial	0.5833	0.8049	0.7616
Industrial	0.7500	0.9431	0.9345

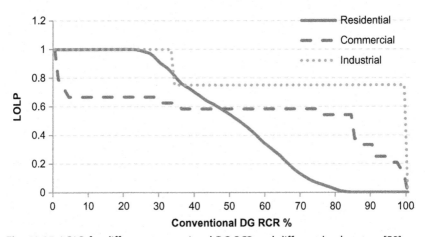

Fig. 10.25 LOLP for different conventional DG RCR and different load sectors [59].

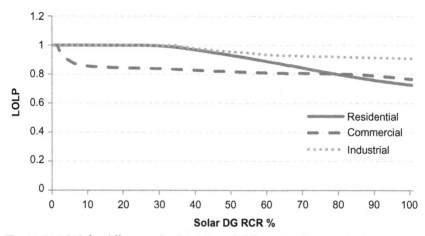

Fig. 10.26 LOLP for different solar DG RCR and different load sectors [59].

REFERENCES

[1] T. Gönen, Electric Power Distribution System Engineering, second ed., CRC Press, Boca Raton FL, 2008.
[2] D.L. Beeman, R.H. Kaufmann, The fundamentals of industrial distribution systems, Trans. Am. Inst. Electr. Eng. 61 (5) (1942) 272–279.
[3] M. Behnke, W. Erdman, S. Horgan, et al., Secondary Network Distribution Systems Background and Issues Related to the Interconnection of Distributed Resources, Report No. NREL/TP-560-38079, National Resnewable Energy Laboratory, July 2005.
[4] IEEE Standard Computer Dictionary: A Compilation of IEEE Standard Computer Glossaries, 1990.
[5] A.M. Leite da Silva, A.M. Cassula, R. Billinton, L.A.F. Manso, Integrated reliability evaluation of generation, transmission and distribution systems, Proc. Inst. Elect. Eng. Gen. Transm. Distrib. 149 (1) (2002) 1–6.
[6] R.L. Robinson, D.F. Hall, C.A. Warren, V.G. Werner, Collecting and categorizing information related to electric power distribution interruption events: customer interruption data collection within the electric power distribution industry, in: IEEE Power Engineering Society General Meeting, October 2006.
[7] V. Werner, D. Hall, R. Robinson, C. Warren, Collecting and categorizing information related to electric power distribution interruption events: data consistency and categorization for benchmarking surveys, IEEE Trans. Power Delivery 21 (1) (2006) 480–483.
[8] S. Mandal, A. Pahwa, Optimal selection of conductors for distribution feeders, IEEE Trans. Power Systems 17 (1) (2002) 192–197.
[9] R.E. Brown, S. Gupta, R.D. Christie, S.S. Venkata, R.D. Fletcher, Automated primary distribution system design: reliability and cost optimization, IEEE Trans. Power Delivery 12 (2) (1997) 1017–1022.
[10] H.L. Willis, Power Distribution Planning Reference Book, Marcel Dekker, New York, NY, 1997.
[11] D.A. Kowalewski, A comparable method for benchmarking the reliability performance of electric utilities, Proc. IEEE Power Eng. Soc. Summer Power Meeting 2 (2002) 646–649.

[12] S. Knezevic, D. Skrlec, M. Skok, The impact of reliability indices in the process of planning radial distribution networks, Proc. Int. Conf. Using Computer as a Tool EURO-CON 2 (Sept. 2003) 244–248.

[13] H. Falaghi, M.-R. Haghifam, Distributed generation impacts on electric distribution systems reliability: sensitivity analysis, Proc. Int. Conf. Using a Computer as a Tool, EUROCON 2 (Nov. 2005) 1465–1468.

[14] N. Balijepalli, S.S. Venkata, R.D. Christie, Modeling and analysis of distribution reliability indices, IEEE Trans. Power Delivery 19 (4) (2004) 1950–1955.

[15] E. Carpaneto, A. Mosso, A. Ponta, E. Roggero, Comparison of reliability and availability evaluation techniques for distribution network systems, in: Proc. Annual Reliability and Maintainability Symposium, 2002, pp. 563–568.

[16] H. Herath, V. Gosbell, S. Perera, Power quality (PQ) survey reporting: discrete disturbance limits, IEEE Trans. Power Delivery 20 (2) (2005) 851–858.

[17] J. Horak, Power quality: measurements of sags and interruptions, in: Proc. IEEE Transmission and Distribution Conference, May 2006, pp. 733–739.

[18] M. Al-Muhaini, G.T. Heydt, A. Huynh, The reliability of power distribution systems as calculated using sys-tem theoretic concepts, in: IEEE Power and Energy Society General Meeting, July 2010.

[19] H. Brown, D.A. Haughton, G.T. Heydt, S. Suryanarayanan, Some elements of design and operation of a smart distribution system, in: IEEE PES Transmission and Distribution Conference and Exposition, April 2010.

[20] A.A. Chowdhury, S.K. Agarwal, D.O. Koval, Reliability modeling of distributed generation in conventional distribution systems planning and analysis, IEEE Trans. Ind. Appl. 39 (5) (2003) 1493–1498.

[21] S. Conti, R. Nicolosi, S.A. Rizzo, Generalized systematic approach to assess distribution system reliability with renewable distributed generators and microgrids, IEEE Trans. Power Delivery 27 (1) (2012) 261–270.

[22] M. Fotuhi-Firuzabad, A. Rajabi-Ghahnavie, An analytical method to consider DG impacts on distribution system reliability, in: IEEE Transmission and Distribution Conference & Exhibition, 2005.

[23] A.C. Nõto, M.G. Dajilva, Impact of distributed generation on reliability evaluation of radial distribution systems under network constraints, in: Proc. of 9th International Conference on Probalistic Methods Applied to Power Systems, June 11–15, 2006.

[24] Westinghouse Electric Corp., The Westinghouse Transmission and Distribution Engineering Handbook, E. Pittsburgh, PA, 1965.

[25] W. Kersting, Distribution System Modeling and Analysis, CRC Press, Boca Raton FL, 2002.

[26] C.L.C. de Castro, A.B. Rodrigues, M.G. da Silva, Reliability evaluation of radial distribution systems considering voltage drop constraints in the restoration process, in: International Conference on Probabilistic Methods Applied to Power Systems, Sept 2004, pp. 106–111.

[27] A.H. Etemadi, M. Fotuhi-Firuzabad, New considerations in modern protection system quantitative reliability assessment, IEEE Trans. Power Delivery 24 (4) (2010) 2213–2222.

[28] K. Jiang, C. Singh, New models and concepts for power system reliability evaluation including protection system failures, IEEE Trans. Power Syst. 26 (4) (1855) 1845–1855.

[29] M.C. Bozchalui, M. Sanaye-Pasand, M. Fotuhi-Firuzabad, Composite system reliability evaluation incorporating protection system failures, in: Proc. 2005 IEEE Canadian Conf. Electrical and Computer Engineering, 2005, pp. 486–489.

[30] P.M. Anderson, G.M. Chintaluri, S.M. Magbuhat, R.F. Ghajar, An improved reliability model for redundant protective systems—Markov models, IEEE Trans. Power Syst. 12 (2) (1997) 573–578.

[31] O. Shavuka, K.O. Awodele, S.P. Chowdhury, S. Chowdhury, Application of predictive reliability analysis techniques in distribution networks, in: 45th International Universities Power Engineering Conference (UPEC), 2010.

[32] M. Ramamoorty, Block diagram approach to power system reliability, IEEE Trans. Power Apparatus Syst PAS-89 (May 1970) 802–811.

[33] R. Billinton, R.N. Allan, Reliability Evaluation of Engineering Systems, Plenum Press, New York, 1992.

[34] R. Ramakumar, Engineering Reliability Fundamentals and Applications, Prentice Hall, New Jersey, 1993.

[35] Guide for Electric Distribution Reliability Indices, IEEE-P1366 standard.

[36] R. Billinton, Power System Reliability Evaluation, Gordon and Breach, New York, 1974.

[37] R. Billinton, R. Ringlee, A. Wood, Power System Reliability Calculations, MIT Press, Cambridge MA, 1973.

[38] D. Midence, S. Rivera, A. Vargas, Reliability assessment in power distribution networks by logical and matrix operations, in: Proc. IEEE/PES Transmission and Distribution Conference and Exposition Latin America, Aug. 2008, pp. 1–6.

[39] J. Endrenyi, Reliability Modeling in Electric Power Systems, John Wiley & Sons, Toronto, 1978.

[40] Lee. Layton, Electric System Reliability Indices. link: http://www.l2eng.com/Reliability_Indices_for_Utilities.pdf.

[41] S. Asgarpoor, M.J. Mathine, Reliability evaluation of distribution systems with non-exponential down times, IEEE Trans. Power Syst. 12 (1997) 579–584.

[42] M. Al-Muhaini, An Innovative Method for Evaluating Power Distribution System Reliability, Ph.D. dissertation, Arizona State University, AZ, Tempe, 2012

[43] R. Billinton, R. Allan, Reliability Evaluation of Power Systems, Plenum Press, New York, 1996.

[44] R. Allan, I. Rondiris, D. Fryer, An efficient computational technique for evaluating the cut tie sets and common cause failures of complex systems, IEEE Trans. REL R-30 (1981) 101–109.

[45] G.B. Jasmon, K.W. Foong, A method for evaluating all the minimal cuts of a graph, IEEE Trans. Reliab. R-36 (5) (1987) 539–545.

[46] M. Al-Muhaini, G.T. Heydt, Minimal cut sets, Petri nets, and prime number encoding in distribution system reliability evaluation, in: IEEE Power and Energy Society T&D Conference, May 2012.

[47] R. Brown, Electric Power Distribution Reliability, CRC Press, Boca Raton FL, 2002.

[48] Joseph H. Eto, Kristina Hamachi LaCommare, Tracking the reliability of the U.S. electric power system: an assessment of publicly available information reported to state public utility commissions, LBNL-1092E report, October 2008.

[49] G. Pepermans, J. Driesen, D. Haeseldonckx, W. D'haeseleer, R. Belmans, Distributed Generation: Definition, Benefits and Issues, Katholieke Universiteit Leuven, August 2003.

[50] W. El-Khattam, M.M. Salama, Distributed generation technologies, definitions and benefits, Electr. Pow. Syst. Res. 71 (2004) 119–128.

[51] M. Werven, M. Scheepers, Dispower—the changing role of energy suppliers and distribution system operators in the deployment of distributed generation in liberalised electricity markets, Ecn Report Ecn-C–05-048, NL, 2005

[52] A. Chambers, Distributed Generation: A Nontechnical Guide, Pennwell, Tulsa, OK, 2001. p. 283.

[53] R. Ramakumar, P. Chiradeja, Distributed generation and renewable energy systems, in: Proc. IEEE Energy Conversion Engineering Conf, 2002, pp. 716–724.

[54] Arthur D. Little, Distributed Generation: System Interfaces, White Paper, Arthur D. Little Inc., 1999.

[55] T.E. McDermott, R.C. Dugan, PQ, reliability and DG, IEEE Ind. Appl. Mag. 9 (5) (2003) 17–23.

[56] P.P. Barker, R.W. de Mello, Determining the impact of distributed generation on power systems: part 1—radial distribution systems, Proc. IEEE PES Summer Meeting 3 (2000) 1645–1656.

[57] R.E. Brown, L.A. Freeman, Analyzing the reliability impact of distributed generation, in: Proceedings of the IEEE Summer Meeting, July 2001, pp. 1013–1018.

[58] E. Vidya Sagar, P.V.N. Prasad, Impact of DG on radial distribution system reliability, in: Fifteenth National Power Systems Conference (NPSC), IIT Bombay, December 2008.

[59] M. Al-Muhaini, G.T. Heydt, Evaluating future power distribution system reliability including distributed generation, IEEE Trans. Power Del. 28 (4) (2013) 2264–2272.

CHAPTER 11

DC Distribution Networks: A Solution for Integration of Distributed Generation Systems

Mehdi Monadi*,†, Kumars Rouzbehi*, Jose Ignacio Candela*,
Pedro Rodriguez*,‡
*Technical University of Catalonia, Barcelona, Spain
†Shahid Chamran University of Ahvaz, Ahvaz, Iran
‡Loyola University Andalusia, Seville, Spain

11.1 INTRODUCTION

In recent years, significant changes in power systems have introduced new technical, economic, and operational challenges. The employment of renewable energy systems (RESs) such as wind energy systems, photovoltaic (PV) systems, biomass power plants, and small hydro turbines is increasing in electrical networks. Integration of these RESs and other types of distributed generation (DG) units is one of the major changes that may impact load flow, stability, and protection of power systems. Most RESs are interfaced with the grid through power electronic converters. Therefore, using direct current (DC) distribution systems, can contribute to reducing the total cost and loss of the system, as the power conversion stages are decreased [1]. Decreasing a power conversion stage also can enhance network reliability.

Moreover, DC networks are not affected by power transfer limitations due to the skin effect that appears in alternating current (AC) lines [2]; therefore they are able to transmit higher power over longer distances. For these reasons as well as the simple integration of most modern electronic loads that are supplied by DC power, the concept of DC distribution systems has attracted considerable attention over the last few years. Until now, using medium voltage direct current (MVDC) systems for common distribution networks remained a research topic and they have been used mostly in special applications like the distribution systems of electrical ships [3]. However, due to their advantages, nowadays there is a major interest in DC grids in research and industrial centers and they are being introduced as an alternative for AC systems in future commercial and industrial grids [4,5].

On the other hand, although MVDC systems can operate as a future alternative for AC distribution systems, there are serious concerns about

Distributed Generation Systems
http://dx.doi.org/10.1016/B978-0-12-804208-3.00011-X

509

their worldwide usage. Protection, control, and network design are the main key issues associated with these grids.

In this chapter, the specifications of DC networks and their advantages for DG integration are described. Furthermore, the main issues related to the worldwide implementation of these networks and possible solutions are discussed. Section 11.2 presents an overview on the specifications of DC distribution systems and the advantages of these grids. Also, the main issues about worldwide implementation of these networks are explained briefly. Section 11.3 introduces the required data for designing DC distribution networks. The issues related to the protection of a DC system are presented in Section 11.4. In this section, the differences between the fault current behavior in AC and DC systems are explained. Then, the challenges associated with the implementation of conventional fault detection methods are explained. Section 11.5 presents the issues related to the control of DC distribution system and introduces suitable control methods for these grids. Finally, Section 11.6 presents the future trends in DC distribution systems.

11.2 OVERVIEW OF DIRECT CURRENT DISTRIBUTION SYSTEMS

AC power networks can be categorized into high-voltage, medium-voltage and low-voltage networks. Nowadays, most of the industrial, commercial, and individual loads are supplied by AC grids; while, the first electrical networks were designed to operate with DC power. However, due to the practical challenges around the implementation of DC networks, in particular the DC voltage levels stepping up/down, utilities and major electrical companies changed their networks from DC to AC systems. In fact, one of the main reasons for the worldwide extension of AC systems is the possibility of the voltage level conversion by the use of AC transformers.

AC transformers, in various levels of power and voltage, are easily designed and implemented; whereas, a change in the level of DC voltage requires implementation of more complex power electronic-based devices that were not available in the past. Indeed, unlike AC networks, the leak of applicable and economic mechanisms to change the level of DC voltage was an important issue that limited the application of DC systems. On the other hand, each cycle of AC current includes two zero crossing points; thus designing and manufacturing AC circuit breakers is easier and cheaper than designing breaking devices for DC currents that do not cross the zero point naturally [5]. In addition, AC motors are cheaper and more robust than their DC counterparts; hence it is more interesting to use these motors in industrial applications. However, as a consequence of recent developments in power electronics devices, new, efficient converters have been developed that facilitate changes in DC voltage level. Using these new developments, almost all the required DC voltage and current levels are available by use of series or parallel structures of new power electronic devices. For example, using high-power convertors, the first high voltage direct current (HVDC) transmission line was energized in 1954 [6].

11.2.1 Motivation for Using Direct Current Distribution Systems

11.2.1.1 Decreasing Power Conversion Stages

The integration of DGs is increasing in power systems. These generators are typically smaller than the conventional power plants and more geographically more distributed [7]. Although most electrical power is still generated by conventional power plants, DGs are the world's faster growing type of power production units [8]. Also, the energy policy of many governments, which is based on liberalism in power markets, encourage power utilities to use, renewable energy-based DGs [9]. Some types of RES-based DGs, like PV systems and fuel cells (FCs), produce DC power; therefore two stages of conversion are necessary to connect them to AC grids. Accordingly, the integration of most types of DGs to the DC systems is simpler and economical than integration of those sources with AC grids. In other words, most of RESs are integrated with AC grids through an interface including a DC stage; therefore using DC distribution systems not only can contribute to the loss/cost reduction, but also may enhance the reliability of the grid as some power conversion stages are eliminated.

Figs. 11.1 and 11.2 show a hypothetical AC distribution network and its DC counterpart, respectively. These figures illustrate that the number of energy conversion stages is reduced when a DC system is used. Therefore, DC systems are an economical and reliable solution for DGs integration. These figures also demonstrate that on the load side, many loads require AC/DC converters, and by using DC distribution systems it is possible to replace individual rectifiers with a centralized rectifier that is located at the DC substation. This station can control the DC voltage of the grid as well.

11.2.1.2 Increase the Transmitted Power by a Cable

The power carrying capability of AC lines depends on thermal limits, required reactive power, and stability issues; while, the capacity of DC lines mainly depends on the thermal limit [10]. In fact, due to the absence of the reactive current component, the current magnitude and cable losses are

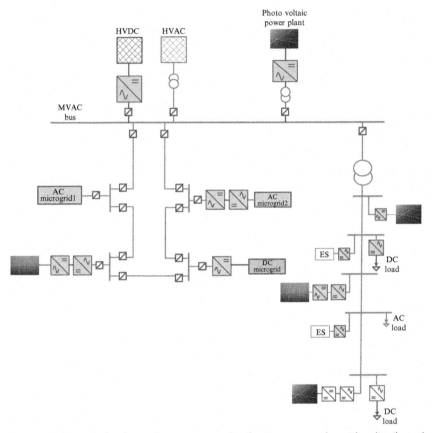

Fig. 11.1 A typical alternating current distribution network with distributed generations.

reduced when DC systems are used. To illustrate this fact, let us consider the current flowing in a line that can supply a given AC and DC voltage. Assume that, in the AC case, the load is a single-phase load that P_{ac}, $\cos(\varphi)$, and I_{ac} are its active power consumption, power factor, and current when it is fed by the AC system; thus the current of the corresponding line is calculated by

$$I_{ac} = \frac{P_{ac}}{V_{ac} \cdot \cos(\varphi)} \tag{11.1}$$

where V_{ac} is the root mean square (RMS) value of AC voltage. The current of this line supplying the load by DC voltage is

$$I_{dc} = \frac{P_{dc}}{V_{dc}} \tag{11.2}$$

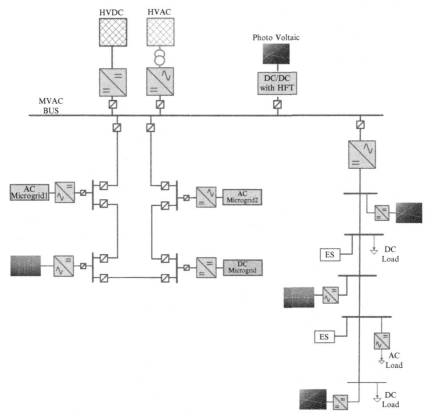

Fig. 11.2 A typical direct current distribution network with distributed generations.

where P_{dc} is equal to the active power when load is supplied by the AC source. Also, V_{dc} is the DC voltage that can be considered as $\sqrt{2}\ V_{ac}$. Indeed, because the insulation of the cable and insulators is designed according to the maximum voltage of AC systems, this equipment can be supplied by DC voltage equal to the maximum value of the AC system. Accordingly, the flowing current of the line when it is connected to the DC source is calculated by [11]

$$I_{dc} = \frac{\cos(\varphi)}{\sqrt{2}} \cdot I_{ac} \tag{11.3}$$

For example, assume that $\cos(\varphi) = 0.87$ and the DC current of the cable is 61% of the current of this line when a load with the same active power is supplied by AC source. Accordingly, from this point of view, using DC systems can provide the following advantages:

(1) A specified load connected to a given cable is supplied by DC systems with a lower value of the current compared with AC systems. Accordingly, the load connected to DC voltage can be fed by a cable with a smaller sized conductor. This, in turn, reduces the cost of the network equipment and the voltage drop. Moreover, resistance of AC cables is affected by the skin effect; thus the resistance of a given cable in a DC system is lower than the AC resistance [2]. Consequently, the use of DC systems also can reduce the total losses of the grid.

(2) Because a given load is supplied by a lower value of current, the total power of loads that can be supplied by a cable increase when DC systems are used. Consequently, by converting an existing AC system to DC, it would be possible to enhance the transmitted power and increase the total power of the connected loads. This is more important when due to the space limitation in a zone or other economic or noneconomic reasons, it is not favorable to construct new feeders while the power consumption of the zone is increasing.

Example 11.1

A single-phase 20 kW load and a power factor of 0.87 is connected to a 220 V AC source using a given cable. If the AC source is replaced by an appropriate DC source, calculate the ratio of the power losses of the cable when it is connected to AC and DC systems. Assume that the AC and DC resistance of the cable are almost the same.

Solution

When load has been connected to an AC source, the current of the cable is

$$I_{ac} = \frac{20000}{220 \times 0.87} = 104.49 \text{ A}$$

The cable losses are

$$P_{loss,\,ac} = 2 \cdot R_L \cdot I_{ac}^2$$

where R_L is the cable resistance. On the other hand, the maximum voltage of the counterpart DC system can be calculated by

$$V_{dc} = \sqrt{2} V_{ac} = \sqrt{2} \times 220 = 311.12 \text{ V}$$

Here, we assume that the selected voltage for the DC line is 310 V. Therefore, the DC current is

$$I_{dc} = \frac{20 \text{ kW}}{310} = 64.52 \text{ A}$$

And the cable losses, in this case, are

$$P_{loss, dc} = 2 \cdot R_L \cdot I_{dc}^2$$

Consequently, the ratio of the cable losses when used in a DC and AC system is

$$\frac{P_{loss, dc}}{P_{loss, ac}} = \left(\frac{I_{dc}}{I_{ac}}\right)^2 = \left(\frac{64.52}{104.49}\right)^2 = 0.38$$

$$\text{or} \quad P_{loss, dc} = 0.38 \times P_{loss, ac}$$

This example shows that using a DC distribution system may result in a significant reduction in the network's losses; this, in turn, increases network efficiency.

11.2.1.3 Multiple Microgrids

AC systems are connected together taking into consideration the synchronizing issues. For example, in the level of distribution systems, two or more microgrids can be connected only if they are synchronized; therefore the connection points should be equipped with the synchro-check devices. However, MVDC links are considered to be a solution of connection between unsynchronized AC microgrids or connection of AC microgrids to DC microgrids. Indeed, one of the main applications of a DC distribution link is to connect microgrids that work with different frequencies. In other words, when DC links are used to connect microgrids, it is not necessary to care about synchronization issues [12]. Fig. 11.3 shows a hypothetical multiple microgrid that is created by the connection of DC and AC microgrids.

It is also worth noting that using DC distribution links between AC microgrids can prevent the effects of faults that happen in a microgrid to the other parts of the system. In fact, DC lines can operate as a fire wall that limits the impacts of a faulty microgrid on other microgrids.

Example 11.2

In Fig. 11.3 assume that a temporary phase-to-ground fault impacts Microgrid1 at $t = 1$ s. Then, CB1 operates and isolates the faulty microgrid after around 80 ms. The temporary fault removes after 200 ms and the current breaker is reclosed at $t = 1.3$ s to reconnect Microgrid1 to the MVDC network. Investigate the effects of this fault on the AC voltage of Microgrid2.

Solution

To investigate this example, first of all, the network of Fig. 11.3 has been simulated in MATLAB/SIMULINK. Fig. 11.4A shows the voltage of

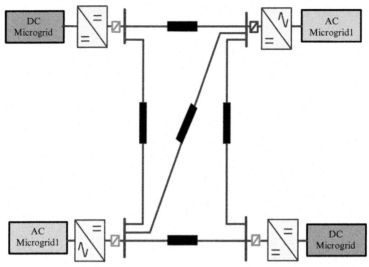

Fig. 11.3 Establishing a multiple microgrid using medium voltage direct current lines.

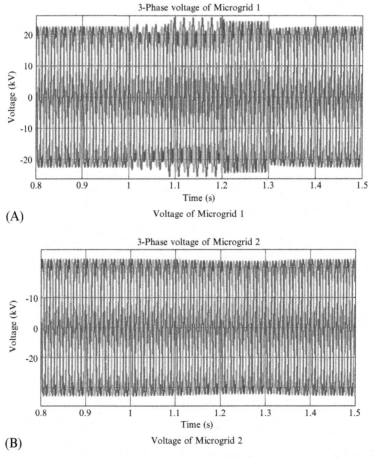

Fig. 11.4 Voltage of Microgrid1 (A) and Microgrid2 (B) before, during, and after the fault.

Microgrid1 before, during, and after the fault. This figure demonstrates effects of the fault and operations of the corresponding current breaker on the voltage of Microgrid1. It is clear that before the fault clearance and network restoration, the AC loads connected to this microgrid were faced with voltage deviations that impacted the power quality of this microgrid.

On the other hand, Fig. 11.4B shows the voltage of Microgrid2 at the same time frame of Fig. 11.4A. This figure illustrates that MVDC lines have been operated as a fire wall between the faulty and healthy microgrids. Hence AC loads connected to Microgrid2 have not sensed any voltage deviation or interruption. Accordingly, by the use of MVDC lines, the effects of faults in each AC microgrid are limited to the borders of that microgrid and the other parts of the network can continue their normal operations.

11.2.2 Voltage Levels of Direct Current Distribution Networks

Distribution networks are classified to medium voltage (MV) and low voltage (LV) grids. Typically, MV feeders receive power from the high-voltage network or DGs and deliver it to the LV grids.

11.2.2.1 Low Voltage Direct Current Networks

Over the last few decades, most of the commercial and individual loads such as office loads, lighting, and so on, are internally supplied by DC voltage. These devices are interfaced with LV AC networks through the appropriate rectifiers; therefore using LV DC networks may result in simpler and more economic connection of these loads and consequently enhance network efficiency and reliability.

Networks that are supplied by voltage below 1 kV are called LV networks. Moreover, to convert an existing AC network to a DC grid, the maximum value of the AC grid can be selected as the nominal voltage of a low voltage direct current (LVDC) system. The LV section of Fig. 11.2, shows a hypothetical LVDC feeder. This figure illustrates that using LVDC grids can provide a simpler and more economic interface for connection of small DGs to the grid by illumination of some of the conversion stages. The comparison of the LV sections of Figs. 11.1 and 11.2 show that most of the distributed converters in low voltage alternating current (LVAC) grid can be merged in a centralized converter when LVDC is used. This change not only can reduce the associated costs of the power losses, but also can provide a more flexible grid with voltage that can be controlled by the controller of the centralized converter. In this case, the voltage conversion and voltage regulation is handled by a centralized converter.

11.2.2.2 Medium Voltage Direct Current Networks

Grids with nominal voltages above 1 kV and up to 35 kV are defined as MVDC grids [13]. Although voltages below or more than these values can be used for distribution systems, the recommended preferred rated voltages are 1.5 kV, 3 kV, 6 kV, 12 kV, 18 kV, 24 kV, or 30 kV [13]. Furthermore, the rated power of each MV feeder is from several kilowatts to several megawatts; thus they are the appropriate grids for integration of DGs that are in the same range of power. Currently, although HVDC and LVDC systems are well-known grids that are used around the world, there is not much experience in the application of MVDC systems and almost all the power distribution grids in the range of MV are supplied by AC voltage. This is mainly due to the lack of effective and economic DC devices in the range of MV. However, following the recent development in the technology of converters, now there is more interest in DC networks in both academic and industrial societies [4]. Note that MVDC systems already have been used in some limited applications; thus the specifications of such systems can be used to facilitate the designing and implementation of future MVDC systems. The following paragraphs describe some of the current MVDC applications.

Medium Voltage Direct Current Systems of Electrical Ships and Aircrafts

As MVDC systems can provide reliable grids, their first application was distribution systems of electrical ships. This is mainly due to the requirement of a higher level of reliability and power quality for the special loads that are used in ships. This level of reliability typically is difficult to be achieved by AC systems. A typical MVDC network for electrical ships is shown in Fig. 11.5. Moreover, the weight and size of the network components of DC grids are less than the components of counterpart AC grids; these are important factors for electrical ships. For the same reasons, MVDC systems are also used for power distribution in aircraft. In both these applications, the main power sources are not connected to a utility grid; therefore those sources can be considered as DGs. Accordingly, MVDC grids already have been used to integrate isolated generators and loads of electrical ships and

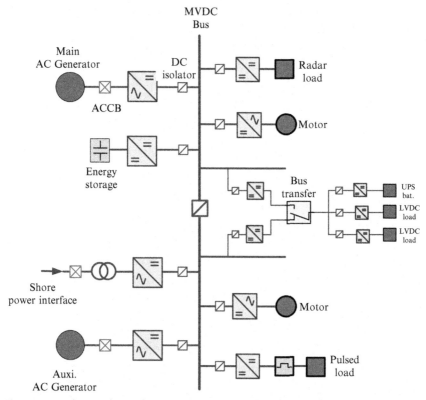

Fig. 11.5 Medium voltage direct current grid used as the distribution systems of an electrical ship [13].

aircraft. In fact, the MVDC grids of electrical ships and aircraft can be considered as the first step in the usage of MVDC grids; thus the investigations on those grids can provide appropriate ideas for designing and implementing future commercial MVDC distribution networks.

Direct Current Collection Grids

As another application, MVDC grids can be used to collect the generated power by the renewable energy units that are distributed in wide-area power plants. For example, in wind turbines, the mechanical torque is produced by wind energy. Because wind speeds are higher at sea and more constant than over land, offshore wind farms are being established. Also, in contrast to conventional land-based wind farms, there is no theoretical limit to the occupied area for the offshore wind farms. Therefore, it is possible to install large wind turbines with huge blades to generate more electrical power. AC collection grids are used to collect and transfer the generated power to onshore stations; however, using a DC collection grid not only can reduce the size of the cable and total losses, but also can increase the maximum possible distance that the generated power can be transmitted [2]. Moreover, the size and weight of the required converters for interfacing to the MVDC system is less than those of required converters for AC grids, which is an important factor in design and implementation of the infrastructure of wind turbines.

On the other hand, large-scale PV power plants are typically low-density power plants; therefore the generated power of such plants is collected from panels that are distributed in a wide area. As PV panels produce DC power, a DC system can be used to collect the power with the minimum number of the power conversion stages. As mentioned above, these collection grids also transfer power with a lower value of losses compared to AC collection grids.

Future Hybrid Alternating Current and Direct Current Grids

Based on the abovementioned advantages of DC networks, MVDC and LVDC grids can be a major part of future distribution networks. AC and DC systems can also contribute to construct hybrid distribution networks in which both AC and DC grids are used simultaneously [14]. Fig. 11.6 shows a conceptual scheme for future hybrid networks, where AC and DC systems work together in HV and MV as transmission and distribution grids. These networks also are suitable for integration of different types of DGs because both AC and DC voltages are available and then each DG can be connected to the proper grid.

It is also worth noting that converters that interface DC feeders to AC feeders can operate as a fire wall between these grids; thus faults in each side do not impact significantly the other side grid of the converter. This can reduce unnecessary power interruptions and hence enhance network reliability and power quality.

11.2.3 Practical Concerns About Future Direct Current Distribution Networks

Despite all the advantages of DC networks, there are limited studies and experiences about protection and control issues of these grids. In this

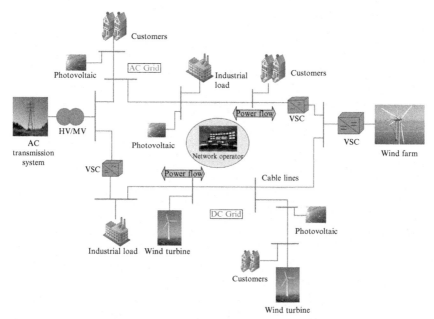

Fig. 11.6 Conceptual architecture of the future hybrid networks [15].

section, some of the main concerns about the implementation of these grids are explained. Note that, before worldwide usage of the DC system it is necessary to find effective solutions for these concerns; thus these issues can be considered as important research topics for interested researchers.

11.2.3.1 Protection Issues

Protection is one of the main concerns about the implementation of DC networks. Protection issues are mainly caused by the particular behavior of the fault current in DC networks. Because the impedance of DC lines are lower than their counterpart AC lines, DC fault current increases very fast; this may disturb the coordination of conventional overcurrent relays [1]. On the other hand, alternating current circuit breakers (ACCBs) clear the fault during a zero crossing point of the fault current waveform; however, there is no natural zero crossing point for DC currents. Therefore, new fault interruption devices are necessary for DC networks. Briefly, issues such as the behavior of converters under faulty conditions, specifications of DC fault current, fault detection, and fault current interruption methods are the main subjects that require more research and development [5,16,17].

11.2.3.2 System Integration

DC distribution networks have to organically grow with time like the development of the AC power system; from a simple initial phase to a desired form, hopefully to arrive at a much more complex DC distribution networks. After the system architecture is defined, the next step is to distinguish clearly what are the objectives and primary functions of network operation, and what are the possible interactions between the network functions. The task of establishing the functionalities inside the system and assuring optimal performance is accomplished by system integration.

11.2.3.3 Power Flow Control

Power flow control capability is one of the main challenges for the successful operation of DC distribution networks. In DC networks, bus voltages do not have a phase angle, and the transmission line impedances do not have an imaginary part. Therefore the variables left for power flow control in those grids are the voltages and currents amplitudes. In point-to-point HVDC transmission lines, one terminal controls the DC-link voltage, while the other terminal controls the current (power) through the DC line. This control philosophy of having only one converter station controlling the DC bus voltages can be extended to DC distribution networks. However, to improve network performance and reliability and due to safety reasons, future large DC distribution networks will require a power flow control strategy capable of sharing the direct voltage control among voltage source converter (VSC) stations of the MVDC grid.

11.2.3.4 Dynamic Behavior

In an AC grid, the main equipment that provides active and reactive power is the synchronous machine. Hence the modeling of the synchronous machine dynamics is the key to successful dynamic behavior assessment of AC power systems.

Similarly, the main component that exchanges power in a DC grid is the power electronic converter. In comparison to synchronous machines, power electronic converters have much faster response due to additional control capabilities and much lower inertia. Therefore, modeling of the power electronic converter dynamics is a key aspect for assessment of the dynamic behavior of multiterminal DC grids. However, due to their switching behavior, the dynamic equations describing the converter operation are discontinuous and complicated to solve. To simplify power electronic converter complexity, the averaged dynamic models can be employed. The advantage of using averaged models is that they simplify the converter analysis while still allowing enough details to understand its dynamics and developed control strategies.

11.2.3.5 Stability

There are no reactive power flow and current/voltage frequency components in DC networks. Additionally, active power flow depends on differences between buses voltages. Therefore, the stability of DC grids, which depends only on the buses voltages, has to be analyzed in a different way than for AC power systems. More importantly, it has to include DC passive components, and power electronic converters and their feedback controllers.

Due to the importance of protection and control issues in DC distribution systems, these issues are considered in the following parts of this chapter with more details.

11.3 DESIGN OF DIRECT CURRENT DISTRIBUTION NETWORKS

11.3.1 Topology of Direct Current Networks

Similar to AC networks that are categorized in single-phase and three-phase networks, DC networks can be established in different topologies. Most common topologies for DC networks that are generally categorized in unipolar and bipolar grids are shown in Fig. 11.7. These two topologies are different in terms of the number of conductors and the voltage level. Unipolar systems include two conductors, as shown in Fig. 11.7A; one of them is connected to the positive/negative pole while the other conductor is connected to the neutral pole. In bipolar DC systems there are two conductors that are connected to the positive and the negative poles. Moreover, this system may include a neutral wire. The absolute voltages of the negative and the positive conductors are normally the same. In the normal operations, current of the neutral wire should be negligible compared to the currents of the other conductors. Fig. 11.7B also illustrates that bipolar networks, indeed, are constructed from two unipolar systems with three conductors.

Both unipolar and bipolar networks can be used to integrate DGs to the DC grids. For example, a bipolar network can be used as a distribution grid

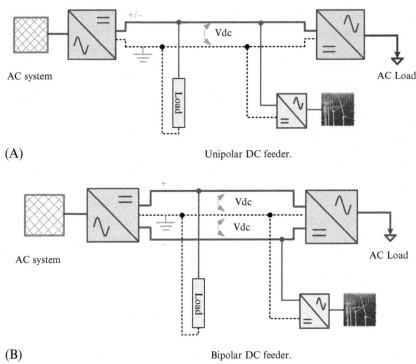

(A)   Unipolar DC feeder.

(B)   Bipolar DC feeder.

Fig. 11.7 Bipolar (A) and unipolar (B) topologies.

to integrate wind turbines (WTs) while PV units can be connected to a unipolar network.

It is worth noting that to convert an existing AC network to a DC grid, the number of cable cores that have been used for an AC system is an important factor in selecting the unipolar or bipolar topology for the DC grid.

11.3.2 Selection of the Direct Current Voltage Level

As mentioned in Section 11.2.2, various voltage levels can be used for MVDC and LVDC distribution networks. However, to convert an existing AC system to a DC system, the DC voltage selection is made according to the specification and limitation of the existing cables. In this section, we investigate the DC voltage selection for a bipolar DC network that is obtained by converting a three-phase AC system to DC, as shown in Fig. 11.7B.

The maximum voltage of each core of the cable that has been used in the AC feeder, is calculated according to the maximum value of the AC network voltage. This voltage is the maximum value that can be selected for a DC network. For a cable that has been designed to be used in a three-phase

AC system with its nominal voltage at V_{ac}, the maximum voltage of each core of the cable is

$$V_{ac,phase} = \frac{\sqrt{2}}{\sqrt{3}} \cdot V_{ac} \tag{11.4}$$

This is the maximum possible voltage of each DC pole; therefore, according to Fig. 11.7B, the nominal pole-to-pole voltage of a bipolar network is calculated by

$$V_{dc} = 2 \cdot \frac{\sqrt{2}}{\sqrt{3}} \cdot V_{ac} \tag{11.5}$$

where V_{dc} determined the nominal voltage of the bipolar network. According to this voltage, we will compare the bipolar DC network with its original three-phase AC network. The transferred power by the bipolar feeder is

$$P_{dc} = V_{dc} \cdot I_{dc} \tag{11.6}$$

On the other hand, the maximum current of a conductor is calculated according to the thermal limits of the conductor. As these limits are related to the power losses, assume that the maximum allowed power losses in a conductor is P_{mloss}. When the cable is connected to the AC voltage, these losses are calculated by

$$P_{mloss} = 3 \cdot R_{ac} \cdot I_{ac}^2 \tag{11.7}$$

When the cable transfers DC power in a bipolar network, the maximum losses are calculated by

$$P_{mloss} = 2 \cdot R_{dc} \cdot I_{dc}^2 \tag{11.8}$$

Accordingly, the maximum allowed currents in AC and DC systems can be presented as

$$\frac{I_{dc}}{I_{ac}} = \frac{\sqrt{3 \cdot R_{ac}}}{\sqrt{2 \cdot R_{dc}}} \approx \sqrt{\frac{3}{2}} \tag{11.9}$$

Therefore, the transferred power in a bipolar system is expressed by

$$P_{dc} = 2 \cdot \frac{\sqrt{2}}{\sqrt{3}} \cdot V_{ac} \cdot \frac{\sqrt{3}}{\sqrt{2}} I_{ac} \tag{11.10}$$

Moreover, the active power of the considered cable, when it was connected to the AC system, is presented by

$$P_{ac,3\varphi} = \sqrt{3} V_{ac} \cdot I_{ac} \cdot \cos(\varphi) \tag{11.11}$$

Consequently, by substituting Eq. (11.10) in Eq. (11.11), the relation between the active power of a three-phase network and the power of its counterpart DC network is

$$P_{dc} = 2 \cdot \frac{P_{ac,3\varphi}}{\sqrt{3} \cdot \cos(\varphi)} \tag{11.12}$$

Example 11.3

Assume that a 2 MW three-phase load with a power factor of 0.87 is fed by a three-core cable. The nominal voltage of the AC system is 20 kV. If the AC feeder converts to a bipolar MVDC feeder, calculate the related AC and DC currents and the ratio of power losses in the cable in AC and DC systems. Assume that the AC and DC resistances of the cable are almost the same.

Solution

The current of the load when it is connected to an AC network is

$$I_{ac} = \frac{2\text{ MW}}{\sqrt{3} \times 20\text{ kV} \times 0.87} = 66.36\text{ A}$$

On the other hand, according to Eq. (11.5), the voltage of the DC system is determined by

$$V_{dc} = 2 \times \frac{\sqrt{2}}{\sqrt{3}} \times 20\text{ kV} = 32.66\text{ kV}$$

Now, we assume that the existing 20 kV AC network is converted to a 32 kV bipolar DC network. Accordingly, the cable current will be

$$I_{dc} = \frac{2\text{ MW}}{32\text{ kV}} = 62.5\text{ A}$$

Moreover, the ratio of the corresponding power losses is

$$\frac{P_{loss,dc}}{P_{loss,3\varphi}} = \frac{2 \times R_L \times I_{dc}^2}{3 \times R_L \times I_{ac}^2} = 0.67 \times \left(\frac{62.5}{66.36}\right)^2 = 0.59$$

which means that the power losses of the DC system are 59% of the AC system's losses.

11.3.3 Interfacing Distributed Generations to Direct Current Distribution Networks

11.3.1 Converter Station

DC switchgears can be connected to several feeders. These switchgears are fed by HVDC/HVAC transmission lines or by MVAC feeders.

The transferred power by DC feeders may reach to several megawatts. Moreover, MVDC/LVDC output feeders, which may include DGs with several hundred/tens kilowatt power, are also connected to DC switchgears. As shown in Fig. 11.2, DGs and AC systems are interfaced to a DC distribution grid using VSCs. These converters also are used to regulate the DC grid voltage. Various topologies can be used for the VSCs; however, the VSCs' topologies are selected according to the level of converted voltage and transmitted power. For example, to connect the MVDC grid to an existing MVAC system a multilevel converter, as shown in Fig. 11.8, can be used.

On the other hand, in MVDC networks the level of voltage would be tens of kV, which is much higher than the nominal voltage of PV/WT units. Therefore, in order to connect these DGs to a MVDC grid, LVDC to MVDC conversions with high output/input voltage ratios are necessary.

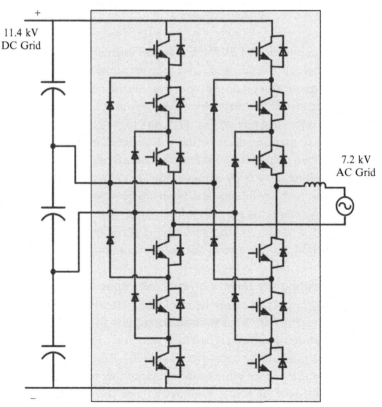

Fig. 11.8 A four-level voltage source converter for medium voltage direct current applications [18].

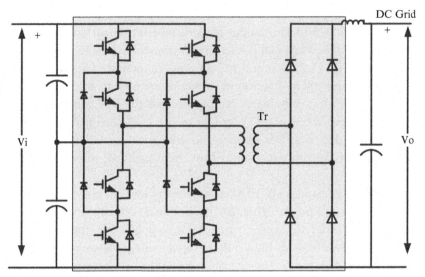

Fig. 11.9 Direct current/direct current topology including a high-frequency transformer for medium voltage direct current applications [19].

Therefore, the connection of PV/WT units to those networks may require a conversion topology including an AC power transformer, as shown in Fig. 11.9. This transformer is necessary to provide the required output/input ratio for integration of DGs that deliver their power at a lower voltage level.

In the converter structure of Fig. 11.9, the high-frequency transformer not only is used to increase/decrease the voltage level, but also provides isolation between two sides of the converter that may facilitate the performance of the fault breaking devices by reducing the fault current level. Due to the existence of the high switching device on the primary and/or secondary sides of this transformer, it is typically selected from high/medium frequency transformers. It is worth noting that, using these types of transformers reduces the weight and the size of the converter station.

11.3.3.2 Calculating the Direct Current Link Capacitor

One of the most important components of VSC stations are DC link capacitors. These capacitors are used to limit the ripples in the DC side of the VSCs and therefore minimize DC voltage variations, which may be caused by significant changes in the converter's load [20]. Accordingly, it is necessary to calculate the proper value for those capacitors to provide an appropriate performance for the VSCs. Various methods and equations have been proposed to calculate the value of DC link capacitors. Each of the already presented methods calculates the capacitor with more attention to one of the network aspects; for example, maximum acceptable voltage deviation and/or converter characteristics. For example, Karlsson [2] has suggested to use the following equation to calculate the DC link capacitor:

$$C_{dc} = \frac{P_n}{V_{dc}^2} \cdot \frac{2\sigma_n^2}{\omega_{lp}} \cdot \frac{1}{(1-\delta_n)\delta_n} \tag{11.13}$$

where P_n is the nominal power of the VSC, the damping factor is $\sigma = 1/\sqrt{2}$, ω_{lp} is the break-over frequency of the VSC's low-pass filter. and $\delta_n = 0.05$ is the maximum desirable voltage ripple.

Example 11.4

According to the network of Fig. 11.2 and Table 11.1, calculate the DC link capacitor for VSC1 that interfaces the DC system to a MVAC grid and VSC2 station of DG1.

Solution

The MVDC network of Fig. 11.2 is connected to the AC network through VSC1. Because this VSC should control the voltage of the MVDC network, the maximum voltage deviation of its output voltage is set at 5%. Thus, according to Eq. (11.13), the value of the DC link capacitor of this VSC is

$$C_{VSC1} = \frac{5 \text{ MW}}{(20 \text{ kV})^2} \times \frac{2 \times \left(\frac{1}{\sqrt{2}}\right)^2}{188} \times \frac{1}{(1-0.05) \times 0.05} = 1.34 \text{ mF}$$

Moreover, DG1 is connected to the LVDC feeder. The voltage deviation of the VSC of this DG is set at 10%. Therefore, the required capacitor to achieve this voltage deviation is

$$C_{VSC2} = \frac{50 \text{ kW}}{(200 \text{ V})^2} \times \frac{2 \times \left(\frac{1}{\sqrt{2}}\right)^2}{188} \times \frac{1}{(1-0.1) \times 0.1} = 73.88 \text{ mF}$$

Table 11.1 Network's parameters for Example 11.4

	VSC1	VSC2
DC voltage	20 kV	200 V
Nominal power	5 MW	50 kW
ω_{lp}	$2\pi30 = 188$ (rad/s)	188
σ	$1/\sqrt{2}$	$1/\sqrt{2}$
Maximum voltage deviation	5%	10%

This example shows that although the nominal power of the main VSC is fairly more than the nominal power of VSC2, the DC link of this converter is smaller than the capacitor of VSC2. Indeed, Eq. (11.13) expresses that the value of the required capacitor for a VSC is mainly related to the ratio of nominal power and the square of nominal voltage.

11.4 PROTECTION ISSUES AND TECHNIQUES IN MEDIUM VOLTAGE DIRECT CURRENT SYSTEMS

A protection scheme can have an acceptable operation if it has been designed based on characterizes and behavior of network parameters and quantities. In other words, protection system must operate based on the network quantities behavior after the fault occurrence. Therefore, to design a general protection method for DC networks, it is necessary to study the impacts and behavior of various types of DC faults. It should be noted that not only are the fault types in DC networks different from AC faults, fault currents characteristics (e.g., maximum current and rising rate) also are not the same for faults in AC and DC networks. Therefore, in this section, we introduce the types of faults that may happen in DC networks. Then, the behavior of the fault currents and the already presented protection methods are investigated.

11.4.1 Fault Types in Direct Current Systems

The main faults that may occur in DC networks are short circuits and open circuit faults. Short circuit faults have some different types, which may happen in a DC network according to the topology of the network. In this section, the common faults that may happen in a bipolar system are introduced. The same explanations also can be presented for unipolar networks.

In bipolar networks, short circuit faults are divided into two main categories, namely, pole-to-pole faults and pole-to-ground faults. In pole-to-pole faults, the positive pole is directly connected to the negative pole or simultaneously connects to the negative pole and the ground/neutral point. These faults are often of low-impedance type that may severely impact the network components. A pole-to-ground fault, on the other hand, occurs when either the positive or negative pole is connected to the ground or the neutral point. This type of fault, which usually occurs due to insulation degradation, is the most common type of short circuit in DC networks;

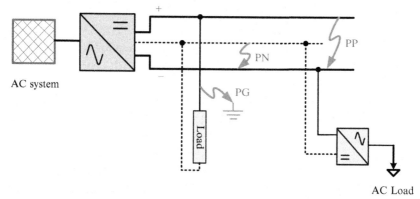

Fig. 11.10 Various types of fault in bipolar direct current networks.

however, it is not as critical as a pole-to-pole fault [21–23]. Therefore, according to Fig. 11.10, different types of faults are summarized as follows:

(1) Positive pole to neutral pole (PP)

(2) Positive/negative pole to ground (PG)

(3) Positive/negative pole to neutral (PN)

(4) Positive/negative pole to neutral pole to ground (PPG)

(5) Disconnection of a pole (open circuit)

The behavior of the fault current is completely different among the above fault types; however, after the occurrence of the first four types, the fault current will increase rapidly. In other words, the first four types are short circuits that may happen due to damaged insulation of cables or direct connection between overhead lines, whereas in the last type no type of short circuit has happened in the system.

The most hazardous fault types are PP and PPG. In these types of faults two conductors with positive and negative voltages, which are carrying power to loads, connect directly. PP faults normally are low-impedance faults. After a PP/PPG fault occurrence, all the capacitors connected to the faulty section discharge to fault path. In this case, a very high current flows through the network components. This type of fault is comparable with three-phase to ground fault in an AC network; however, the rising rate of the fault current is much higher than the fault current in AC systems.

In comparison to PP faults, PG faults have lower effects on the DC grid because, in this case, only one of the poles contributed to the fault. PG faults normally happen after damage to cable insulators or when busbar insulators are broken. Moreover, a PG fault normally happens through a fault impedance; this impedance limits the magnitude and the rising rate of the fault current.

11.4.2 Fault Current Characteristics in Direct Current Networks

Various types of VSCs can be employed to connect DGs to DC grids and to construct DC switchgears. For example, as it was shown in Fig. 11.8, a multilevel VSC can be used to interface an MVDC grid with a MVAC network [18]. Connection of PV systems and/or wind farms to DC networks also requires an energy conversion mechanism, including a medium-frequency interconnection transformer (see Fig. 11.9) [19]. The transformer is particularly necessary for the connection of low-voltage DGs. It is worth noting that, in addition to the network impedance; the operational characteristics of VSCs affect the behavior of DC faults and must be considered in the design of a proper protection system. As this chapter mainly focuses on VSC-based DC distribution networks, the fault characteristics of PP faults in such networks are discussed in this section.

Fig. 11.11 illustrates the components of a typical PP fault that happened in a DC feeder, including a two-level VSC that is used to interface a DG to the DC feeder. In this figure, R and L represent the cable resistance and inductance from the main DC bus to the fault location. The analysis of the circuit in Fig. 11.11 determined that the fault current has three main components, that is, (1) capacitor discharge current, (2) cable inductance discharge current, and (3) the rectified current from the AC side [5]. These components have different behaviors that are explained in the following sections.

11.4.2.1 Capacitor Discharge Component

After the PP fault occurrence in a DC network, the voltage of the main DC bus drops; hence the DC link capacitor is discharged through the fault path. According to Fig. 11.11, the equivalent circuit of this component is a RLC

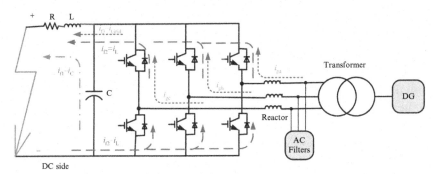

Fig. 11.11 Components of direct current fault current in a voltage source converter-based interface for distributed generations [5].

circuit. Therefore, the DC link capacitor component, i_C, can be obtained by solving the following second-order differential equation:

$$\frac{d^2 i_C}{dt^2} + \frac{R}{L}\frac{di_C}{dt} + \frac{1}{L \cdot C}i_C = 0 \tag{11.14}$$

where R and L denote the DC line resistance and inductance, respectively, and C represents the capacitance of the DC link capacitor. A general solution for this equation is found from the roots of its characteristic equation:

$$s^2 + 2\alpha s + \omega_0^2 = 0 \tag{11.15}$$

where $\alpha = \dfrac{R}{2L}$ (rad/s) and $\omega_0 = \dfrac{1}{\sqrt{LC}}$ (rad/s).

Roots of this quadratic equation are calculated by

$$s_{1,2} = -\alpha \pm \sqrt{\alpha^2 - \omega_0^2} \tag{11.16}$$

Depending on the values of α^2 and ω_0^2, three different responses can be considered for the DC link capacitor current, namely, (i) overdamped response $(\omega_0^2 < \alpha^2)$, (ii) critically damped response $(\omega_0^2 = \alpha^2)$, and (iii) underdamped response $(\omega_0^2 > \alpha^2)$. According to Eq. (11.16), the response of the capacitor discharging current is relative to the values of R, L, and C. For example, if a solid fault impacts a short distribution feeder, the fault current will have an underdamped form, which is obtained as [24]

$$i(t) = \frac{V_0}{L\omega_d}e^{-\alpha t}\sin(\omega_d t) + I_0 e^{-\alpha t}\left[\cos(\omega_d t) - \frac{\alpha}{\omega_d}\sin(\omega_d t)\right] \tag{11.17}$$

where $\omega_d = \sqrt{\omega_0^2 - \alpha^2}$, V_0, and I_0 are the initial values of the capacitor voltage and cable current.

According to Eq. (11.16) and the above-mentioned explanations, the fault impedance can significantly affect the characteristics of the capacitor discharge component of the fault current. In fact, the fault resistance changes both the magnitude and transient response of the fault current. It should also be noted that the capacitor discharge current appears subsequent to a fault in all VSC-based DC grids, whereas the appearance of the two other fault current components depends on the type and control of the VSC supplying the DC bus and the fault impedance.

11.4.2.2 Cable Discharge Component

In all the VSC stations, after the fault occurrence in the DC network, the controller of the converter turns off the main switches of the VSC to prevent the fault current from flowing through the main switches and to protect them against the overcurrent conditions. In this case, the VSC operates as an uncontrolled rectifier; hence once the capacitor is discharged, the energy stored in the cable inductance is discharged through the antiparallel freewheeling diodes. Fig. 11.11 shows that the equivalent circuit of this component is a RL circuit. Therefore, the cable discharge component, i_L, can be obtained by solving a first-order differential equation as follows:

$$L\frac{di_L}{dt} + i_L \cdot R = 0 \qquad (11.18)$$

Solving Eq. (11.18), the cable discharge current is calculated as

$$i_L(t) = I_0 e^{-\frac{R}{L}t} \qquad (11.19)$$

11.4.2.3 Rectified Alternating Current Side Current

As mentioned above, following the blocking of the main converter switches, the VSC operates as an uncontrolled full-bridge rectifier; thus it continues to feed the fault from the AC grid through the freewheeling diode paths. If a three-phase AC/DC converter is used, this rectified current is obtained by [25]:

$$i_{Grid}(t) = i_{ga}(>0) + i_{gb}(>0) + i_{gc}(>0) \qquad (11.20)$$

where i_{gx} (>0) denotes the positive value of phase-x current (i_{gx}), which contribute to the fault current through freewheeling diodes.

Noted that the second and third components of the DC fault currents flow through the freewheeling diodes; thus these diodes may get damaged quickly if proper protections are not put in place [9,26].

The components of a typical PP fault current in a VSC-based DC network are shown in Fig. 11.12. During the first stage of fault current ($t_0 \leq t \leq t_c$), the DC link capacitor discharged while the VSC does not contribute in the fault current. After that, that is, for $t > t_c$, the fault current consists of a cable inductor discharging current and the rectified AC grid current; these two components flow through the freewheeling diodes of the VSC and may damage those diodes [9]. Thus the critical time for the operation of the protection of DC distribution systems is $\Delta t_c = t_c - t_0$ (i.e., during the capacitor discharge period) [28]. Breaking the fault current during this critical time ensures that the smallest possible area of the system is de-energized in response to the fault. This critical time depends on several factors such as fault type, fault location, and fault resistance.

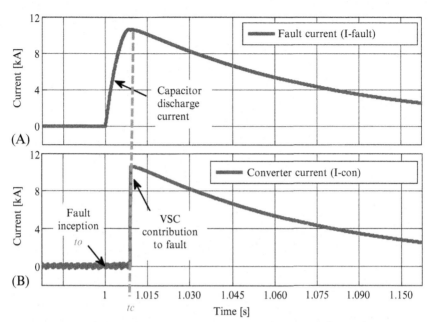

Fig. 11.12 Waveforms of a typical direct current fault current: (A) fault current; (B) converter current [27].

According to the above explanations, the main differences between the protection issues in AC and DC systems can be categorized as the following:

1. Compared to AC fault currents, the rising rate of DC fault currents is very high. Moreover, to protect the semiconductor elements of VSCs, it is necessary to interrupt the fault current in several milliseconds. Accordingly, the protection of DC networks should operate relatively faster than AC systems. Indeed, many conventional protection methods for AC systems are not applicable in DC networks because they are not fast enough; thus the conventional protection schemes in AC grids should be modified to fulfill the fault detection/isolation requirements in DC grids.

2. Conventional AC breakers extinguish the fault current. Considering the sinusoidal waveform of AC currents, such breakers are able to easily interrupt the AC fault currents. However, there is no natural zero crossing point in DC systems; therefore most conventional breakers are not suitable for breaking DC fault currents. Furthermore, the operating time of MVAC breakers is in the range of several tens of milliseconds, which is not fast enough for DC protection.

3. Because the cable inductances of DC networks are negligible, the cable impedance is somewhat less than the counterpart cable in AC networks. Therefore, the peak value of DC fault currents is higher than the peak value of fault currents in counterpart AC networks. Also, the absence of cable inductance cause a higher rate of change for the DC fault currents [29].

Example 11.5

Assume that a PP fault occurs in a DC feeder. The cable resistance and inductance from the fault point to the DC bus, that is, the location of measurement devices, is 0.141 Ω and 0.036 mH, respectively. The DC link capacitor is 2.5 mF.

(A) If the fault resistance is 0.2 Ω, which type of fault current flows to the fault point?

(B) If the fault resistance is 5 Ω, compare the fault current behavior with the previous case.

(C) If the terminals of the VSC in Fig. 11.10 are equipped with 0.1 mH inductors, compare the fault behavior with case A when the fault resistance is 0.2 Ω.

Solution

Case A: To estimate the behavior of the temporary component of the fault current, it is necessary to calculate ω_0^2 and α^2. Accordingly, we can write

$$\alpha^2 = \left(\frac{R}{2L}\right)^2 \approx 3.835 \times 10^6$$

and $\omega_0^2 = \dfrac{1}{LC} \approx 5.556 \times 10^6$.

Therefore the capacitor discharge component of the fault current has an underdamped response. This type of fault current typically has the highest current rising rate.

Case B: The fault resistance impact is on the damping factor of the equivalent circuit of the fault path; thus in this case the value of the damping factor is $\alpha^2 = 1.345 \times 10^9$, which is significantly higher than the damping factor of the previous case. Therefore, in this case, the fault current has an overdamped response. This example shows that fault resistance not only reduces the maximum value of the fault current, but also may change the behavior of this current.

Case C: In this case the values of ω_0^2 and α^2 are

$$\alpha^2 = \left(\frac{0.282}{2 \times 2.72 \times 10^{-4}}\right)^2 \approx 268,720$$

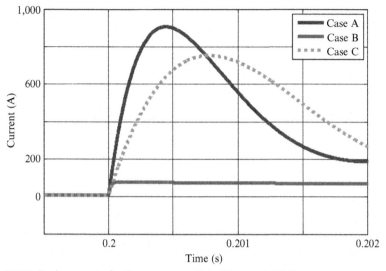

Fig. 11.13 Fault currents for three case studies of Example 11.5.

$$\text{and } \omega_0^2 = \frac{1}{2.72 \times 10^{-4} \times 2.5 \times 10^{-3}} \approx 1.470 \times 10^3$$

Thus, when this inductor is installed, the response of the capacitor discharge current still remains in the underdamped form. However, these inductors have other important effects that are explained here according to Fig. 11.13.

Fig. 11.13 shows the fault current of the three above cases. The comparison of these currents determines that equipping the VSC terminals with the external inductors effectively increases the critical time of the VSC. This is, in particular, important for the VSCs with small critical time because, for those VSCs, not only it is very difficult to coordinate the main and backup protection, but also it is not easy to find fast enough breakers. Moreover, by use of the external inductor, the rising rate and $\frac{di}{dt}$ ratio of the fault current are decreased. Consequently, installing the external inductors can be considered as a part of an effective protection scheme. In some recent papers, researchers have suggested to use these inductors as a part of protection systems of DC networks.

11.4.3 Protection Challenges and Requirements

According to the aforementioned specifications of faults in DC networks, there are some specific challenges that are related to the use of conventional protection devices in these networks. These issues are explained in the following subsections.

11.4.3.1 Fault Detection Methods

Overcurrent relays are the typical protection for conventional distribution systems. The operation of the consecutive time-inverse overcurrent relays are coordinated according to the discrepancy between the fault currents measured by these relays. In AC networks, the fault current in consecutive lines is different due to the difference in the fault paths impedances; however, because of the low value of the line reactance in DC networks, the currents of faults that occur in consecutive short distribution lines are not significantly different. For this reason, as well as the high rising rate of DC fault currents, it is not easy to coordinate the overcurrent relays in DC distribution networks [3,30].

On the other hand, overcurrent relays must be able to effectively detect both pole-to-pole and pole-to-ground faults, whereas there is a significant difference between the current magnitudes of these two types of fault. If the relay settings are selected based on PP fault currents, it may result in the delayed operation of the protection scheme for PG faults. On the other hand, selection of a low pick-up for the relay can cause protection miscoordination for PP faults. Fletcher et al. [1] has studied the impacts of various faults for a typical DC feeder. The results of that study show that overcurrent relays cannot provide a selective protection for DC distribution systems. Table 11.2 provides a summary of the main challenges related to the implementations of protection methods in DC distribution networks with DGs.

11.4.3.2 Impacts of Distributed Generation Integration on Network Protection

In addition to the specifications of DC networks that impact the performance of existing protection methods, DGs integration may also disturb the effectiveness of conventional protection methods. Indeed, the traditional methods, like OC relays, fuses, and reclosers are designed for the conventional distribution networks that are typically passive grids with radial topology. Therefore, these protection devices operate based on the constant settings that are determined by the protection coordination process. However, by integration of DGs, distribution networks not only operate as active grids but also bidirectional currents may flow in their feeders. Consequently, the following issues may happen in a power system after the DGs integration.

Table 11.2 Specifications and challenges of protection methods in direct current networks with distributed generations

Fault detection method	Challenges of implementation in DC distribution systems
Overcurrent	• Fast method • Should be equipped with other methods or communication links to provide selectivity • Applicable in fault interrupt methods that are based on the blocking of converters when selectivity is not desired
Differential	• Only the magnitudes of DCs are compared; thus they can operate faster than differential methods in AC networks • Insensitive to high raising rate of DCs and fault resistance • Are able to provide a selective protection • Requires a reliable communication link
Distance	• Simple algorithm (compared to distance methods in AC networks, simpler fault location estimator can be used) • Faster than AC relays • More sensitivity to fault resistance (as compared to AC systems)
Fault current limiting	• Can be done by external inductors, or current limiting modes/devices of special converters • Requires high-speed semiconductor switches • Requires devices with ability to withstand DCs with high raising rate
Intelligent methods such as ANN, Fuzzy	• It is not easy to use the frequency-based transformations to obtain the required input data from the DC quantities
Directional OC	• Requirement to have accurate and fast methods for detecting the current direction • Not applicable for large DC distribution networks

Disturb the relay coordination. One of the main impacts of the DG connection is missing the coordination between neighbor relays as well as the coordination between fuses and recloses. This disturbance reduces the reliability of the power system due to the disconnecting of healthy loads when the fault is in another feeder or zone. However, it is noted that, in high-impedance faults, the fault current is not much higher than the nominal current, and these protective devices can be remain coordinated [21].

False tripping. False and nuisance tripping can happen due to the increment of the flowing current through an overcurrent (OC) relay when DG is connected to the protected feeder. In this case, in the normal condition of the system, the fault current of a feeder may exceed the threshold of its corresponding relay and consequently cause unnecessary trips.

Blinding of protection. The DG connection can reduce the fault current contribution of the main feeder. For example, in the LV feeder in Fig. 11.2, when all the DGs are connected, after the fault occurrence in the end of the feeder DGs contribute in feeding the fault current. Feeding fault current with more than one passage may reduce the fault current fed from the main feeder. Hence, in some cases, the relay of the main feeder cannot detect the fault or operate after an undesirable time delay.

Changes in the fault behavior due to the change in network topology. In radial feeders, by increasing the distance of the fault position from the main bus, the fault current level decreases. However, the DG integration may increase the fault current level of faults that occur far from the main DC bus.

According to these challenges as well as challenges related to the specification of DC systems, it is almost impossible to provide a complete protection scheme using the conventional current-based relays; therefore it is necessary to develop new methods for DC distribution networks or modify the existing methods. Some recent researchers have suggested the use of communication-assisted methods to overcome the abovementioned issues [27,31]. In fact, using communication links can provide effective and selective protection methods for DC networks. One of the communication-assisted methods is explained in the next subsection.

11.4.3.3 Communication-Assisted Protection for Medium Voltage Direct Current

As mentioned above, conventional OC relays cannot provide selective protection for DC networks. Thus in [31] an effective communication-assisted method has been presented that works based on the OC relays. In this method, the operation of the OC relays is coordinated using an adequate communication link. Fig. 11.14 shows the trip logic and the required communication media for this method. According to this figure, to locate the fault position, the relay installed at each end of a DC feeder receives two signals from another relay installed at the other end of the feeder. These signals confirm that the fault has been detected by the relay and the fault current is flowing from the bus to the line. As shown in Fig. 11.14, each proposed relay has a main protective zone. For example, Line 12 is the main zone of R1; the relay operates instantaneously when a fault impacts this zone.

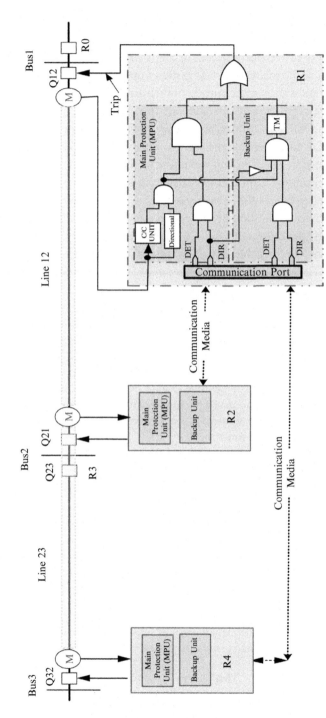

Fig. 11.14 Communication-assisted overcurrent method [31].

In the protection scheme of Fig. 11.14, the main protection unit (MPU) of each relay (e.g., R1 in Fig. 11.14) sends the trip signal when the relay installed at the other end of the feeder (e.g., R2 in Fig. 11.14) confirms that the fault has happened inside the main zone. Using these signals prevents the operation of relays for the faults that occur outside of their protection zone; however, this method blocks the backup operation of the OC-based relays. Indeed, in this case, the OC relay cannot operate as a backup for the adjacent relays. To fix this issue and to enable the relay to operate as a backup protection for the adjacent feeder, a backup unit was embedded into the proposed relay. This unit communicates with another relay that has been located at the border of the adjacent zone (see Fig. 11.14). The trip logic of the backup unit is the same with the main unit; however, their protected zone and their operating time are not the same. Using this communication-assisted method, the OC-based relays can provide selective and reliable protection for DC networks.

Example 11.6

Assume that MV busses of Fig. 11.2 have been equipped with the communication-based OC relays shown in Fig. 11.14. Also, assume that DC busses have equipped with solid-state DCCBs that interrupt the fault current in less than 1 ms. Considering the existence of the adequate inductor on the terminals of the VSCs, the critical time of the main VSC is 7 ms. Accordingly, calculate the time-margin between the operation of the main and backup relays to guarantee the selective protection. The delay in communication link is assumed to be less than 3 ms. Ignore the delay associated with current measurements.

Solution

Both main and backup protection should operate during the critical time of the VSC unit/station (i.e., Δt_c). On the other hand, since fault detection in both main and backup protection requires the adequate communication links, the fault detection time, t_{det}, is heavily dependent on communication delays. Therefore, as it is shown in Fig. 11.15, the main and backup direct current circuit breakers (DCCBs), should operate within a t_{mb} time frame [27].

If we assume that the communication delay is 3 ms and if we ignore the delay in the data process in measurement devices and relays, we can write

$$t_{mb} = 7 \text{ ms} - 3 \text{ ms} = 4 \text{ ms}$$

This value of t_{mb} is the maximum time delay that both main and backup relays should operate. Therefore, the maximum value of TMD is calculated as:

$$\text{TMD}_{max} = \frac{t_{mb}}{2} = 2 \text{ ms}$$

On the other hand, the minimum value of TMD should be larger than the operating time of DCCBs and the delay associated with the current

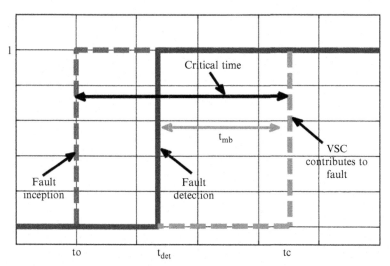

Fig. 11.15 Typical time frames for communication-assisted protection methods in direct current systems [27].

transducer. Therefore, to provide selective protection, the time margin between the operation of the main and backup relays of this example can be set on 1 ms to 2 ms.

This example shows that the time margins in DC networks are significantly smaller than the typical time margins of the protection schemes in AC grids.

11.4.3.4 Fault Breaking Strategies in Direct Current Networks

Fault current breaking is another important part of protection schemes. In AC networks various types of current breakers with various ratings of breaking capacity have already been designed and used. These current breakers typically interrupt the AC fault current during a zero crossing of the current waveform. However, DCCBs are not as common and available as their counterpart ACCBs. Moreover, as DC currents do not cross a zero point naturally, DCCBs require relatively more complex technology to break the fault currents. In addition, typical ACCBs have a mechanical operating mechanism that is capable of clearing the fault in several tens of milliseconds; however, due to the requirement of relatively faster fault interruption devices for VSC-based DC systems, mechanical current breakers are not fast enough for these systems. Thus, solid-state and hybrid DC breakers have been proposed by researchers/engineers. Using these technologies, however, increase the price of the DCCBs, leading them to be more expensive than their counterpart ACCBs.

Accordingly, although for HVDC transmission systems it is acceptable to equip all the transmission lines with the DCCBs, installing these current breakers at terminals of all the distribution (i.e., MVDC/LVDC) feeders is not an economical solution. In fact, for most DC distribution networks it is recommended to use the minimum possible numbers of DCCBs.

Therefore, according to the importance of the DC stations, one of the following strategies can be selected for the fault current breaking:

(1) *Use ACCBs and DC isolator switches.* A DC station protected based on this strategy is shown in Fig. 11.16A. In this strategy, after the fault occurrence in the DC network (e.g., when F1 occurs in Line 1), the fault current is interrupted by operation of the AC side current breaker (i.e., ACCB1). After the operation of this current breaker and interruption of the fault current, the corresponding isolator (i.e., SW2) is opened and isolates the faulty line. Then, ACCB1 is reclosed and the rest of the DC network is re-energized. This strategy provides economical protection; however, typically, it is not a fast enough method to protect the freewheeling diodes of the VSCs. To overcome this issue, it is necessary to modify the topology of the VSC or add external inductors on the VSC's terminals.

(2) *Use DCCBs and DC isolator switches.* This strategy is shown in Fig. 11.16B. Due to the interruption of fault currents by use of the adequate DCCBs, this strategy provides a more effective and reliable protection than the previous case. In this method, after the fault occurrence in point F1, DCCB1 operates and interrupts the fault current after several milliseconds. Then, similar to the previous strategy, the faulty line is isolated by opening the SW1. Finally, DCCB1 closes and restores the rest of the DC grid. In compare to the previous strategy, this strategy can be implemented without requiring any modifications of the VSC's topology or adding external inductors. Moreover, the network restoration time of this strategy is somewhat less than the previous case; thus network power quality is enhanced.

(3) *Use DCCBs.* This strategy, shown in Fig. 11.16C, can provide a selective, reliable, and effective fault current interruption scheme for DC networks. In this method, after the fault occurrence in point F1 of Fig. 11.16C, only, DCCB2 operates and consequently interrupts the fault current and isolates the faulty line. The main feature of this strategy compared to the previous ones is that each fault is cleared by the operation of the local current breaker; thus the other feeders and loads do

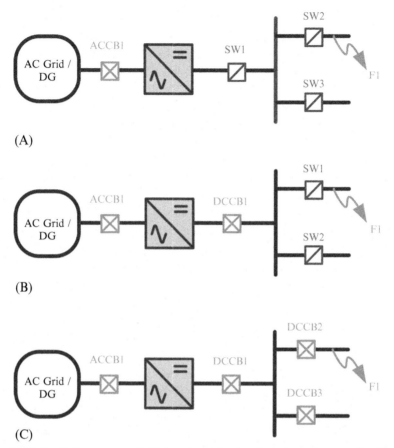

Fig. 11.16 Fault clearance and isolation strategies for direct current networks: (A) use ACCBs and DC isolator switches; (B) use DCCBs and DC isolator switches; (C) use DCCBs.

not face any momentary interruption. This, in turn, can enhance the power quality and the reliability of DC distribution networks. Besides the attractive abilities of this strategy, it is significantly more expensive than the two other strategies.

11.5 CONTROL OF DIRECT CURRENT DISTRIBUTION NETWORKS

Voltage and frequency are two quantities that can be controlled in AC networks. In an AC generator, frequency can be controlled through increasing or decreasing mechanical input power and voltage is controlled by injecting or absorbing reactive power.

In DC networks, however, the frequency does not exist and consequently reactance will not form; hence there is no phase-shift between DC voltage and current and reactive power does not appear in DC lines. Accordingly, voltage is the unique quantity that can be controlled in DC systems. DC voltage can be controlled by the injected power to the DC network; for example, if the injected/generated power is more than the power of the connected loads, the voltage of DC buses will increase. In other words, if we assume that all the generators and loads are interfaced to a DC network through individual converters, the summation of the converters power must always be zero. Therefore, an effective voltage controller in a DC network should continuously monitor the power balance and accordingly, send the appropriate control signals to the related network components (e.g., converters, generators, and loads).

Various methods have already been presented to adjust DC network voltages [32–34]. In these methods, the critical problem is which of the converters of the DC network are able to contribute in voltage control.

Two famous voltage control methods that are found in the literature are master/slave and droop control. In the first method, only the converter of one of the network terminals operates as the voltage controller while the other converters may only control their AC terminals quantities (e.g., active power, reactive power, or AC voltage or frequency). The main advantage of this method is its simplicity. Moreover, high-speed communication links are not needed between the converters. Furthermore, in this method, the network is protected against the undesirable interactions between multiple voltage controllers that might lead to instabilities.

Besides the abovementioned advantages, an obvious disadvantage of this method is that because voltage control is achieved by a controller that is installed on only one of the network terminals, if this terminal is impacted by an internal/external fault, the network will lose its voltage control. In this case, the entire DC network will shut down if the responsibility of the voltage control cannot be transferred to another converter.

On the other hand, the droop control method has been proposed to avoid dependency on a single power converter. In this method, the voltage control is shared among multiple converters that are located in various busses of the network. Compared to the master/slave method, in the droop control method, the network continues to operate normally even if one of the voltage regulating terminals trips. In this case, the remaining terminals are still able to control the DC voltage within its specified boundaries.

11.5.1 Modeling of Voltage Source Converter Stations

Fig. 11.17 shows a single-line diagram of a VSC station. The main task in modeling a converter station is developing an appropriate model for the corresponding VSC. Performing detailed, switching, or average modeling are well-known techniques that are used for VSC modeling. However, the most appropriate model for dynamic and stability studies of DC distribution networks is the average model [35]. On the other hand, modeling of the DC circuit includes deriving a set of algebraic and differential equations to represent the relationship between currents and voltages inside the DC distribution networks.

11.5.1.1 Detailed Modeling of Voltage Source Converter Stations

A detailed model of a VSC station is derived based on the modeling of the switching elements, usually insulated gate bipolar transistors (IGBTs), which generally includes ideal, or nonideal (series and antiparallel) switches, diodes, and a snubber circuit. The nonideal diodes can be modeled as nonlinear resistances using the classical diode function [6].

In this modeling method, a power converter is modeled using the models of the actual power semiconductor devices. Therefore, as this simulation is mainly performed at the circuit level, a very detailed presentation of the VSC can be obtained. For instance, the harmonics generated by the VSC station can be precisely represented. However, in some of the simulator tools (e.g., PSpice) this type of VSC model is described by means of functions that contain exponential terms. This results in slow execution times, large amounts of generated data, and a convergence problem. Moreover, the detailed modeling in MATLAB/SIMULINK involves differential equations that are resolved through the state space method. Consequently, for large DC networks, a large number of differential equations will appear; thus the computations are time-consuming.

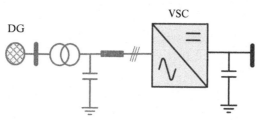

Fig. 11.17 Single-line diagram of a voltage source converter station.

11.5.1.2 Switching Modeling of Voltage Source Converter Stations

In this method, instead of the circuit topologies, the switching functions are used to describe the performance of power converters [35] Fig. 11.18 shows an equivalent model where switches are replaced by three voltage sources on the AC side and a current source on the DC side. The voltage sources are a function of the DC link voltage (v_{dc}) and the switching functions as is shown in Eq. (11.21).

$$\begin{bmatrix} v_{ga} \\ v_{gb} \\ v_{gc} \end{bmatrix} = \frac{1}{3}(v_{dc}) \begin{bmatrix} 2 & -1 & -1 \\ -1 & 2 & -1 \\ -1 & -1 & 2 \end{bmatrix} \begin{bmatrix} s_{ga} \\ s_{gb} \\ s_{gc} \end{bmatrix} \tag{11.21}$$

where s_{ga}, s_{gb}, and s_{gc} are the switching functions that use the 6-IGBT valve pulses as the control input. Sinusoidal pulse width modulation (SPWM) control strategies are used to generate the IGBT valve pulses. The DC link controlled current source is defined as a function of the AC side currents and the switching functions as Eq. (11.22):

$$i_{dc} = \begin{bmatrix} s_{ga} & s_{gb} & s_{gc} \end{bmatrix} \begin{bmatrix} i_a \\ i_b \\ i_c \end{bmatrix} \tag{11.22}$$

It is worth noting that the switching model of the VSC stations can also properly represent the main components of the electromagnetic transients and harmonics generated by the VSC station.

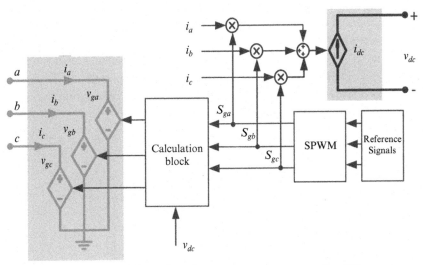

Fig. 11.18 Switching modeling of a voltage source converter station [35].

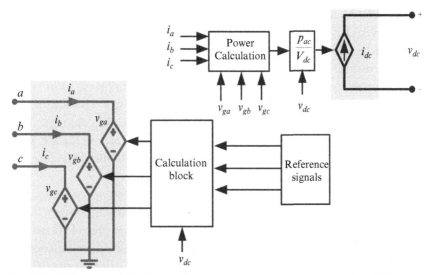

Fig. 11.19 Average model diagram for a voltage source converter station [35].

11.5.1.3 Average Value Modeling of Voltage Source Converter Stations

Although detailed models can describe the accurate behavior of VSC stations, they are very complex and time-consuming methods. Moreover, such levels of detail are not necessary for most network studies. Therefore, it is necessary to develop more efficient models to reduce complexity of detailed models; it is clear that the new models should still provide a similar dynamic response.

Average value models or simply average models are from the simplified models that represent the average response of VSC stations by using controlled sources and switching/averaged functions.

Fig. 11.19 presents an average model of a VSC station. The three reference signals, instead of using the switching functions, are used to represent three average voltage sources on the AC side. Accordingly, the values of the controlled current source are calculated based on power balance by neglecting the internal loss of the VSC.

The DC side quantities of the VSCs are obtained assuming that the power on the AC side must be equal to the power on the DC side plus the converter losses.

11.5.2 Control of Voltage Source Converter Stations

The control of DC distribution networks is directly related to the VSC stations control. Several control approaches for VSC stations have already been proposed in the literature. Among them, vector current control and power angle control have been the most investigated.

The principle of power angle control is fairly straightforward. The active power is controlled by the phase angle shift between the VSC station and the AC system, while the reactive power is controlled by varying the VSC voltage magnitude [33]. The main disadvantage of this control technique is that the control system cannot limit the current flowing into the converter. This issue is a serious problem because after a fault or disturbance in the DC side, if the controller cannot limit the valve current, the VSC should be blocked to prevent damage of the VSC and network components.

On the other hand, vector current control is a current control-based approach. This method can limit the current flowing into the converter during disturbances. Vector current control or simply vector control is a mature and probably the most commonly used technique for VSC station control. In this method, the three-phase voltages and currents of the converter are transformed into the rotating dq synchronous reference frame. These transferred quantities are synchronized to the AC grid voltage by use of a phase-locked loop (PLL). Then, the VSC reference voltage is determined by the inner current controllers of the control system on the dq axis. Finally, the reference signal is transformed back to three-phase using the inverse Park transform. Vector current control is an effective technique to control the instantaneous active power and reactive power independently.

11.5.2.1 Voltage Source Converter Vector Control

The general architecture of the vector control at a VSC station is illustrated in Fig. 11.20. The outer controllers in this figure are responsible for generating the reference currents for the inner current controller. The inner controller, in turn, determines the voltage reference of the converter in the dq frame.

The inner current controller (ICC), shown in Fig. 11.21, usually comprises fast proportional integral (PI) controllers. This PI controller tracks the reference currents that are set by the outer controllers and produces the voltage reference for the converter. Moreover, the outer controllers include active and reactive channels, as is shown in Fig. 11.22.

To derive the structure of the ICC, the point of common coupling (PCC) voltage (i.e., e_s) and the converter side voltage (i.e., v_c) are related by

$$e_s - v_c = R_T i_c + L_T \frac{di_c}{dt} \tag{11.23}$$

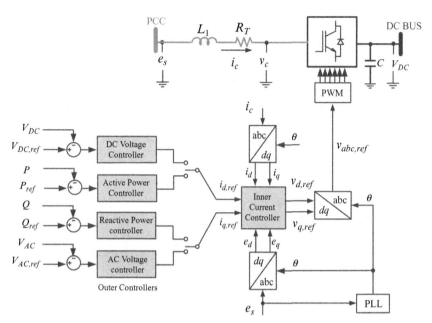

Fig. 11.20 Decoupled vector control of a voltage source converter station [6].

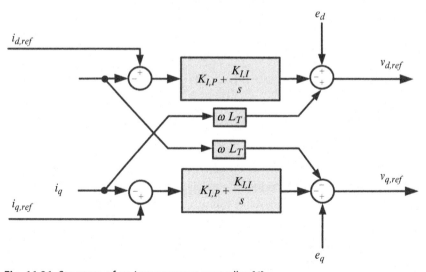

Fig. 11.21 Structure of an inner current controller [6].

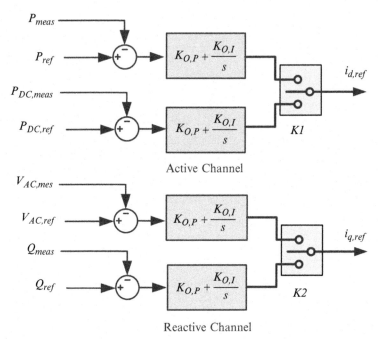

Fig. 11.22 Structure of VSC outer controllers [34].

where i_c is the current flowing from the AC grid to the converter and R_T and L_T represent the equivalent resistance and inductance between the PCC and the converter, respectively.

Then by applying the Park transformation, Eq. (11.23) can be expressed in the dq reference frame as

$$e_d - v_d = R_T i_d + L_T \frac{di_d}{dt} - \omega L_T i_q \tag{11.24}$$

$$e_q - v_q = R_T i_q + L_T \frac{di_q}{dt} + \omega L_T i_d \tag{11.25}$$

where ω is the angular frequency of the AC voltage at the PCC.

The structure of the ICC is obtained using Eqs. (11.24) and (11.25), as shown in Fig. 11.21. The reference voltages ($v_{d,ref}$ and $v_{q,ref}$) are then transformed back into the abc reference frame and used to generate the switching signals for the IGBTs of the converter.

On the other hand, the active channel regulates the active power or DC voltage level by calculating the proper d-axis current. Also, the reactive channel sets the q-axis current to control the reactive power or AC voltage amplitude at the PCC. The possible functions of the active and reactive channels are illustrated in Fig. 11.22.

For active power control, the power equations in the dq reference frame can be written as

$$P = v_d i_d + v_q i_q \tag{11.26}$$

$$Q = v_q i_d - v_d i_q \tag{11.27}$$

Note that the d-axis of the dq frame is aligned with the AC grid voltage phasor, detected by a PLL (i.e., $e_q = 0$) and hence

$$P = v_d i_d \tag{11.28}$$

$$Q = -v_d i_q \tag{11.29}$$

Based on Eqs. (11.26) and (11.27), the current in dq axes can be employed to control active and reactive powers, respectively. The AC voltage controller is intended to regulate the amplitude of the PCC's AC voltage. This can be accomplished by injecting the required amount of reactive power such that the AC voltage at the PCC matches the given reference value. Likewise, the control of AC voltage is carried out by modifying the q-axis current.

To maintain the DC voltage at its reference value, the active power exchanged with the AC grid must be properly regulated. Hence the modification of the d-axis current (i_d) allows us to control the DC voltage within permissible limits.

To control VSC stations using this method, the vector control should be implemented on all the VSC stations. Depending on the control strategy, each VSC may control active power (or DC voltage) and/or reactive power (or AC voltage at PCC). Generally, to control a VSC station, the following operating modes may be implemented for active-reactive channels [32]:

(A) *Constant P-V_{AC} control*: In this case, the converter station maintains the active power and the voltage of the AC grid at the PCC at predefined levels. This control mode is usually applied when the VSC station is connected to a weak AC grid that requires a constant amount of power and support for the AC voltage. The references signals may come from a supervisory control system or can be manually set by operators.

(B) V_{DC}-V_{AC} *control*: This control mode is also usually used when the VSC station is connected to a weak AC grid and it is necessary to maintain voltages of both AC and DC terminals at fixed levels. In this mode, AC voltage is controlled by reactive power compensation and DC voltage is regulated by active power compensation.

(C) *P-Q control*: This control mode is usually applied when the VSC station is connected to a stiff AC grid. Therefore, in this case, the VSC station controls the active and reactive powers while it is not required to regulate AC or DC voltage. It is noted that, if reactive power supply is not needed, the reference signal for Q will be set on zero.

(D) $V_{DC} - Q$ *control*: The $V_{DC} - Q$ control is used when the VSC terminal is connected to a stiff AC grid and the VSC station should operate as a slack bus for the DC distribution network. In this mode, the voltage of the DC grid is regulated by the VSC station. Similar to the previous mode, if reactive power supply is not needed, the reference signal for Q will be set on zero.

11.5.2.2 Direct Current Voltage Control and Power Sharing in Direct Current Distribution Networks

DC voltage control plays a crucial role in the stable and reliable operation of DC distribution networks. If the DC bus voltage rises and exceeds the predetermined threshold, the protection systems may operate. Moreover a large voltage drop may lead to the nonlinear phenomena and results in various difficulties for connected loads and VSCs. Therefore, the voltage control system should be able to provide at least some minimum requirements for DC networks. From this point of view, the most important control requirements of DC distribution networks can be listed as follows [32]:

- The bus voltages should be kept within their steady state limits.
- The system should be able to withstand for a specific time against the disturbances in voltage or power.
- The primary control should not rely on communication among terminals.
- The system should be flexible to connection/disconnection of a VSC station (plug and play characteristic).

An appropriate DC voltage control strategy can guarantee the power balance within the DC distribution network. As discussed in the previous sections, for a DC distribution system comprised of several VSCs, at least one converter has to adjust the DC voltage while other converters operate in power control mode. However, single slack converter control is not acceptable for large DC distribution systems because it is difficult to assure power balance by only one converter station. Moreover, as mentioned above, it is not a reliable method to integrate all required controllers in a specific bus. On the other hand, the slack converter needs to be oversized and connected to a strong AC grid to deal with severe disturbances in DC distribution networks. Beside the technical issues, the geographic location of such a DC slack converter might be controversial, as one system operator would have to cope with all issues on the DC distribution system. In practice, distributing the task of voltage control among several converter stations (which results in multiple slack converters) can cover the abovementioned issues. Therefore, DC voltage control and power sharing techniques in DC systems can be divided into two categories: single slack converter and multiple slack converters. These two categories are explained in the following subsections.

11.5.2.3 Single Slack Converter Control

The single slack converter control or master/slave control is an extension of the conventional point-to-point control where one terminal controls the DC link voltage and the other terminal adjusts the power flow. If this method is used in a DC distribution network, it is necessary to employ a converter station to regulate the DC link voltage and ensure the power balance inside the grid. In this case, the rest of the converters should regulate power in the DC distribution network.

The main advantages and disadvantages of this method are:
- It is easy to implement and it is not necessary to coordinate the controller with other voltage regulating converters.
- External communication systems are not necessary, as only the local voltage measurement is enough for the slack bus to adjust its active power properly.
- The voltage control of a DC distribution network is lost after the fault occurrence in the slack converter of its corresponding bus.

Therefore, it is reasonable to distribute DC voltage control among several converter stations. This leads to the idea of multiple slack converters, presented in the next section.

11.5.2.4 Multiple Slack Converters Control

This control method employs two or more converter stations for DC voltage control. It means that for each voltage-regulating converter a voltage power or voltage current characteristic curve is assigned. The contribution of these power converters to the voltage control is determined according to their characteristics curve.

In DC distribution systems it is possible to control the DC voltage based on current or power. The current-based control uses an *I-V* characteristic curve to control the voltage, whereas the power-based control uses a *P-V* characteristic curve.

For each terminal that works in voltage droop mode, a linear relationship between its voltage and the power (or current) is assigned. Fig. 11.23 shows a typical linear characteristic curve. Using the assigned characteristic curve, the VSC of each terminal contributes in voltage/power balance control. The contribution of each terminal depends on the droop constant, which is the slope of the droop line.

If the droop constant is zero or infinite, the bus acts as a fixed-voltage or fixed-power bus, respectively. The DC voltage droop control is similar to the frequency droop control in an AC power network. In this regard, the DC voltage acts as an indicator for power balance as the frequency acts in AC grids. In all the terminals of an AC network the frequency is the same; as opposed to the bus voltages in DC distribution systems, the frequency is the same throughout the entire AC system. For a cable between two points in a DC distribution system, due to resistance there will be a certain voltage drop. This means that in all cases except zero power transmission, the voltage between two connected terminals will never be the same.

According to Fig. 11.23, in order to change the power-flow in each terminal, it is necessary to change the voltage of that terminal. For example, to increase the injected power of a specific VSC, it is necessary to increase the DC voltage at that the related bus.

The general structure of the droop control is illustrated in Fig. 11.24. The advantages and disadvantages of this method are summarized below.

- Easy and straightforward extension to DC networks.

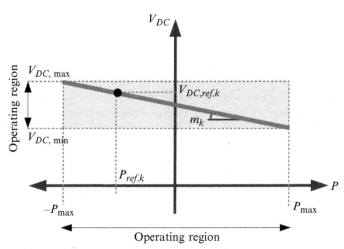

Fig. 11.23 Voltage droop characteristic [6].

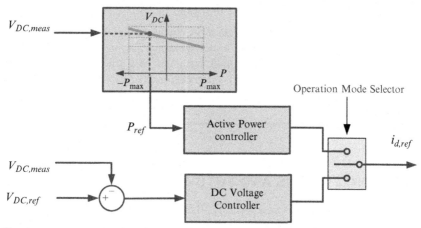

Fig. 11.24 Droop control structure [36].

- Uses simple structure controllers.
- Limited power flow control capabilities.
- Difficulties in coordination of distributed controllers, especially in large DC networks.

Example 11.7

A VSC station, connected to bus i, injects the DC power to a DC distribution system. The voltage droop characteristics curve of the corresponding bus can be described as follows:

$$P_{dc,i} = P_{dc,0,i} - \frac{1}{K_i}\left(V_{dc,i} - V_{dc,o,i}\right)$$

In this bus assume that the power and voltage reference are

$$P_{dc,0,i} = 0.35 \text{ PU}$$
$$V_{dc,o,i} = 1 \text{ PU}$$

To increase the power of this VSC to 0.75 PU, the nominal voltage of the corresponding bus decreases to 0.97 PU. If the droop characteristics of this bus are similar to Fig. 11.23, calculate the droop constant of this bus.

Solution

Using the related the voltage droop characteristics curve, the droop constant is expressed by

$$K_i = \frac{V_{dc,i} - V_{dc,o,i}}{P_{dc,i} - P_{dc,0,i}}$$

Therefore the droop constant is

$$K_i = 0.075 \text{ PU}$$

From the implementation point of view, the power and voltage reference values are obtained under normal operation while droop control is not activated. It should be noted that the smaller droop constant, K, results in a bigger contribution in power sharing. If K approaches zero, the converter behaves as a voltage controller (i.e., a slack bus).

11.6 CONCLUSIONS AND FUTURE TRENDS

In this chapter, we have explained the advantages of distribution systems and have introduced the reasons that prove that DC distribution system can be considered as an alternative to AC systems in future active and smart distribution systems. This chapter also provided an analysis of the fault current characteristics in DC distribution systems. The issues associated with the protection and control of these systems especially in the presence of DGs were explained as well. Moreover, the challenges of conventional fault detection methods and fault interruptions strategies have been explained. On the other hand, the different methods for control of these grids has been described.

It was found that, considering the requirements of future DC distribution systems, the realization of future DC systems requires further investigation about protection and control issues/solutions, especially in multiple terminal DC systems with DGs. The following are specific conclusions and recommendations of this study for future efforts:

- An important requirement for the future DC distribution system is providing suitable fault interruption strategies for these systems. It is mainly due to the inefficiency of conventional ACCBs for fault interruption in DC systems. Economic concerns and technical challenges related to the implementation and production of DCCBs is one of the main concerns about the protection of future DC systems. Therefore further studies are required to address the issues associated with fault breaking devices in DC distribution systems; the problem particularly manifests itself in medium voltage level systems and above.
- Most of the fault detection and fault location methods in AC systems are notpplicable in DC systems. The high rising rate of DC fault currents makes it difficult to coordinate current-based protective devices in these systems. Although the use of communication technologies may enhance

the performance of DC protection systems and improve their selectivity, there are other issues such as communication delays and the speed of fault detection methodology that should be taken into consideration. Moreover, because DC systems need to be protected by relatively faster protection, it is difficult to provide effective fault detection methods based on signal processing algorithms, such as ANN or wavelet-based methods. Furthermore, conventional protection methods, such as directional OC relays and distance protection, must be modified based on the characteristics of DC fault current. Therefore, worldwide application of DC distribution systems requires modification of existing protection methods and development of new fast and effective fault detection algorithms.

- It is important to investigate the effects of fault resistance on the specifications of the fault current. The impacts of fault resistance in an AC fault current is a well-known concept; however, the effect of fault resistance on AC and DC systems are not exactly the same. The fault resistance not only changes the maximum value of the fault current but also changes the response form of these currents. Accordingly, most of the high-impedance fault detection methods are not applicable in DC systems. Therefore it is necessary to analyze the impacts of fault resistance with more details and provide new methods for high-impedance fault detection in DC systems. On the other hand, the effects of grounding methods on fault behavior is another important subject that still needs more investigations.
- There are no existing standards specifically for MVDC distribution networks of commercial buildings. Therefore, before worldwide implementation of these grids, it is necessary to provide new suitable standards for these systems.

REFERENCES

[1] S.D.A. Fletcher, P.J. Norman, S.J. Galloway, P. Crolla, G.M. Burt, Optimizing the roles of unit and non-unit protection methods within DC microgrids, IEEE Trans. Smart Grid 3 (2012) 2079–2087.
[2] P. Karlsson, DC distributed power systems-Analysis, design and control for a renewable energy system, Lund University, 2002.
[3] W.L. Li, A. Monti, F. Ponci, Fault detection and classification in medium voltage dc shipboard power systems with wavelets and artificial neural networks, IEEE Trans. Instrum. Meas. 63 (Nov 2014) 2651–2665.
[4] F. Mura, R.W. De Doncker, Design aspects of a medium-voltage direct current (MVDC) grid for a university campus, in: 2011 IEEE 8th International Conference on Power Electronics and ECCE Asia (ICPE & ECCE), 2011, pp. 2359–2366.

[5] M. Monadi, M. Amin Zamani, J. Ignacio Candela, A. Luna, P. Rodriguez, Protection of AC and DC distribution systems embedding distributed energy resources: A comparative review and analysis, Renew. Sustain. Energy Rev. 51 (2015) 1578–1593.

[6] K. Rouzbehi, A. Miranian, A. Luna, P. Rodriguez, Optimized control of multiterminal DC grids using particle swarm optimization, Eur. Power Electron. Drive J. 24 (2014) 38–49.

[7] C.L.T. Borges, An overview of reliability models and methods for distribution systems with renewable energy distributed generation, Renew. Sustain. Energy Rev. 16 (2012) 4008–4015.

[8] A. Karabiber, C. Keles, A. Kaygusuz, B.B. Alago, An approach for the integration of renewable distributed generation in hybrid DC/AC microgrids, Renew. Energy 52 (2013) 251–259.

[9] J.D. Park, J. Candelaria, L.Y. Ma, K. Dunn, DC Ring-Bus Microgrid Fault Protection and Identification of Fault Location, IEEE Trans. Power Delivery 28 (Oct 2013) 2574–2584.

[10] D.M. Larruskain, I. Zamora, O. Abarrategui, A. Iturregi, VSC-HVDC configurations for converting AC distribution lines into DC lines, Int. J. Electr. Power Energy Syst. 54 (2014) 589–597.

[11] D. Nilsson, A. Sannino, Efficiency analysis of low-and medium-voltage dc distribution systems, in: Power Engineering Society General Meeting, IEEE, 2004, 2004, pp. 2315–2321.

[12] R. Majumder, Aggregation of microgrids with DC system, Electr. Pow. Syst. Res. 108 (2014) 134–143.

[13] IEEE Recommended Practice for 1 kV to 35 kV Medium-Voltage DC Power Systems on Ships, IEEE Std 1709-2010, 2010, pp. 1–54.

[14] R.W. De Doncker, C. Meyer, R.U. Lenke, F. Mura, Power electronics for future utility applications, in: 7th International Conference on Power Electronics and Drive Systems, 2007. pp. K-1–K-8.

[15] M. Monadi, Protection and fault management in active distribution systems, PhD Electrical Eng., Barcelona Tech-UPC University, 2016.

[16] J. Yang, J.E. Fletcher, J. O'Reilly, Multiterminal DC wind farm collection grid internal fault analysis and protection design, IEEE Trans. Power Delivery 25 (2010) 2308–2318.

[17] H. Li, W. Li, M. Luo, A. Monti, F. Ponci, Design of smart MVDC power grid protection, IEEE Trans. Instrum. Measure. 60 (2011) 3035–3046.

[18] S. Falcones, X. Mao, R. Ayyanar, Topology comparison for solid state transformer implementation, in: Power and Energy Society General Meeting, 2010 IEEE, 2010. pp. 1–8.

[19] F. Deng, Z. Chen, Design of protective inductors for HVDC transmission line within DC grid offshore wind farms, IEEE Trans. Power Delivery 28 (2013) 75–83.

[20] H. Wang, F. Blaabjerg, Reliability of capacitors for DC-link applications in power electronic converters—an overview, IEEE Trans. Ind. Appl. 50 (2014) 3569–3578.

[21] J.-D. Park, J. Candelaria, Fault detection and isolation in low-voltage DC-bus microgrid system, IEEE Trans. Power Delivery 28 (Apr 2013) 779–787.

[22] X. Li, Q. Song, W. Liu, H. Rao, S. Xu, L. Li, Protection of nonpermanent faults on DC overhead lines in MMC-based HVDC systems, IEEE Trans. Power Delivery 28 (2013) 483–490.

[23] T.N. Jimmy Ehnberg, Protection system design for MVDC collection grids for offshore wind farms, 2011. elforsk.se/Rapporter/?download=report&rid=12_02_2011.

[24] S. Fletcher, P. Norman, S. Galloway, G. Burt, Determination of protection system requirements for DC unmanned aerial vehicle electrical power networks for enhanced capability and survivability, IET Electr. Syst. Trans. 1 (2011) 137–147.

[25] J. Yang, J.E. Fletcher, J. O'Reilly, Short-circuit and ground fault analyses and location in VSC-based DC network cables, IEEE Trans. Ind. Electron. 59 (2012) 3827–3837.

[26] J. Candelaria, J.-D. Park, VSC-HVDC system protection: A review of current methods, in: Power Systems Conference and Exposition (PSCE), 2011 IEEE/PES, 2011. pp. 1–7.

[27] M.A.Z. Mehdi Monadi, Cosmin Koch-Ciobotaru, Jose Ignacio Candela, Pedro Rodriguez, A communication assisted protection scheme for directcurrent distribution networks, Energy 109 (2016) 578–591.

[28] W. Leterme, J. Beerten, D. Van Hertem, Non-unit protection of HVDC grids with inductive dc cable termination, IEEE Transactions on Power Delivery, vol. PP, 2015, pp. 1–1.

[29] C. Jin, R.A. Dougal, S. Liu, Solid-state over-current protection for industrial dc distribution systems, in: 4th International Energy Conversion Engineering Conference and Exhibit (IECEC), 2006, pp. 26–29.

[30] R. Cuzner, D. MacFarlin, D. Clinger, M. Rumney, G. Castles, Circuit breaker protection considerations in power converter-fed DC Systems, in: Electric Ship Technologies Symposium. ESTS 2009. IEEE, 2009, 2009. pp. 360–367.

[31] M. Monadi, C. Gavriluta, J.I. Candela, P. Rodriguez, A communication-assisted protection for mvdc distribution systems with distributed generation, in: Presented at the PES General Meeting, 2015.

[32] K. Rouzbehi, A. Miranian, A. Luna, P. Rodriguez, DC voltage control and power-sharing in multi-terminal DC grids based on optimal DC power flow and voltage droop strategy, IEEE J. Emerg. Selected Top. Power Electron. 2 (2014) 1171–1180.

[33] K. Rouzbehi, A. Miranian, J. Candela, A. Luna, P. Rodriguez, A generalized voltage droop strategy for control of multi-terminal DC grids, IEEE Trans. Ind. Appl. 51 (2015) 607–618.

[34] W. Zhang, K. Rouzbehi, A. Luna, G.B. Gharehpetian, P. Rodriguez, Multi-terminal HVDC grids with inertia mimicry capability, IET Renew. Power Gen. 10 (2016) 752–760.

[35] A. Yazdani, R. Iravani, Voltage-Sourced Converters in Power Systems: Modeling, Control, and Applications, Wiley-IEEE, New York City, United States, 2010.

[36] K. Rouzbehi, W. Zhang, J.I. Candela, A. Luna, P. Rodriguez, Unified Reference controller for flexible primary control and inertia sharing in multi-terminal VSC-HVDC grids, in: IET Generation Transmission & Distribution, 2016.

INDEX

Note: Page numbers followed by *f* indicate figures, *t* indicate tables, and *b* indicate boxes.

Printed in the United States
By Bookmasters